Science and Cultural Crisis

An Intellectual Biography
of Percy Williams Bridgman
(1882–1961)

STANFORD UNIVERSITY PRESS

Science and Cultural Crisis

• • •

An Intellectual Biography

of Percy Williams Bridgman

(1882–1961)

• • •

Maila L. Walter

STANFORD CALIFORNIA 1990

Stanford University Press
Stanford, California
© 1990 by the Board of Trustees
of the Leland Stanford Junior University
Printed in the United States of America

CIP data are at the end of the book

Preface

THE ADVENT OF relativity theory and quantum mechanics created a crisis of meaning for the classically educated physicist, one that was especially acute for the American experimentalist. The story of P. W. Bridgman's intellectual life is an account of how a physicist who believed not only in the truth of the Newtonian worldview, but also in the exclusively empirical origin of scientific knowledge, was forced to cope with revolutionary change in the foundations of his belief system. It is the story of a classical physicist who felt abandoned in an age that had lost its faith, and who, ironically, felt compelled to become a scientific skeptic against his will.

This book has been written for the general reader, as well as for historians, philosophers, and sociologists of science. Although many of the issues discussed necessarily involve highly technical scientific and philosophical problems, I believe they can be made understandable in principle to everyone. It has always been a delight to me to discover that even the most esoteric problems have a very simple structure, and once that structure is made clear, technical language need no longer obscure the issues.

To be sure, sometimes the reader may begin to wonder why academics concern themselves with such abstract problems, but a little patience will be rewarded when he or she realizes that many of them are nothing more than specialized cases of situations common to everyday life. For example, anyone who has been frustrated by the inflexibility of legal procedures, the insensitivity of medical practice, or the lack of common sense so characteristic of bureaucratic order will surely appreciate Bridgman's claim that "systems" and formal categories distort the truth—that they deliberately ignore the reality experienced by the individual. Similarly, feminists will learn that Bridgman had discovered the social oppression inherent in language well before it became a political issue for the feminist movement. The fact that Bridgman's critique originated in a scientific problem only demonstrates the generality of the issues and the relevance of apparently abstract academic exercises to life situations.

Scholars are in for some surprises, too. Literary critics, for example, may come away with a greater appreciation of science when they realize that thermodynamics involves some of the same principles they invoke in the analysis of literature. Social scientists will learn the history of the familiar methodological injunction they know as "operationalism." Intellectual historians and students of American history will have the opportunity not only to become acquainted with the ideas of an influential and controversial American physicist whose biography has not been written before, but also to be introduced to some subjects that conventional histories usually ignore. And physicists and engineers will come to know the human side of a physicist they probably associate only with his scientific achievements.

But most important, the individual who is interested in understanding the relationship between science and human values will gain insight into how intimately one's view of the origin and meaning of scientific knowledge is bound to moral and ethical judgments.

I would like to alert the reader to a few details of style and language. First, although it is general editorial practice to correct errors in quotations to avoid distraction and make them more readable (and in general this has been done), I have chosen to quote Bridgman's letters verbatim, errors and all. Bridgman's generation did not use the telephone as casually as we do today. Communications were more often written, and for Bridgman, his typewriter was what the telephone is to most of us today. He typed his own letters as offhandedly as we pick up the telephone. Bridgman was a good typist, and disregarding trivial errors such as "hte" for "the," I am guessing that when a flood of errors occurs, it indicates that Bridgman was becoming excited. Therefore the errors tell us something about the tone of the communication and are an important part of its meaning.

Second, since I believe that a historical account should convey, as much as possible, the sensibilities of the time, and furthermore, that the only way one begins to intuit the nuances of linguistic usage is by being immersed in it, I have made extensive use of quotation and close paraphrase. In addition, whenever it made rhetorical sense, I laced the narrative with the language of the time. My goal in using these devices was to present to the reader as much of an uninterpreted first impression as this kind of historical narrative comfortably permits. Consequently, I have made no attempt to adjust the language to accommodate present-day ideological trends. In my judgment, to make this kind of compromise is to pay too great a price in historical integrity and rhetorical freedom for an outcome of minimal political value.

The book is organized by topic. The topics are arranged as far as possible chronologically, according to the order they either affected Bridgman or were considered by him. The first two chapters introduce an individual and a setting. They present a New England personality and depict his scientific style and expectations as they were determined by personal inclination, American intellectual tradition, the practice of physics at Harvard University, and the state of industrial-materials development in the United States in the late nineteenth and early twentieth centuries. Chapter 3 describes how Bridgman's vision of science and early theoretical aspirations were made obsolete by the radical advances in quantum mechanics in the mid-1920's. Chapters 4 and 5 deal with the problem of scientific meaning—both metaphysical and practical—as it appeared in physics as a consequence of relativity theory. They describe how Bridgman devised a strategy, "operational analysis," to cope with what he understood to be the revolutionary implications of relativity. Chapters 6 and 7 discuss the wider cultural perception of a crisis of meaning as the occasion for the popular and the academic response to operational analysis. It is here we learn how Bridgman discovered to what extent language itself can be an instrument of social oppression and how he came to value his own subjective experience as being cognitively and morally more authentic than the categories of either science or common language. Chapters 8 and 9 probe more deeply into the specific epistemological rationale behind Bridgman's operational interpretation of scientific meaning. An examination of Bridgman's views on the nature of quantum mechanics and thermodynamics uncovers the conflict between the classical and the modern scientific epistemologies and shows in what respect the new physics was truly revolutionary. The last two chapters consider the political and social ramifications of the operational strategy as they were articulated by Bridgman, and finally, how operational thinking led Bridgman to an evaluation of the meaning of life itself.

I would like to express my appreciation for the help and moral support given to me by Professor Erwin Hiebert, Professor Donald Fleming, and Professor Silvan Schweber. In addition, I must add that without assistance from the library staff of the Harvard archives (Harley Holden, director) and especially from Clark Elliott, I would not have been able to complete the project. Special thanks are reserved for Mrs. Jane Koopman, P. W. Bridgman's daughter, who graciously offered her personal reminiscences, as well as access to important documents held by the family. I would also like to mention the encouragement I have received from Dr. Pietro Corsi, Bessie Zaban Jones, and Dr. Shirley Roe. There are a good many other friends, colleagues, and correspondents who have sustained my spirits, stimulated my inquiry in discussion, and provided me with information from their own research. To them I also convey my appreciation. Last, but decidedly not least, is my family—my children, who never lost their faith, and my husband, who made the whole effort possible.

M. L. W.

Contents

   Eight pages of photographs follow p. 26

Science and Cultural Crisis

An Intellectual Biography
of Percy Williams Bridgman
(1882–1961)

Bridgman, Science, and Cultural Crisis

*Ever since Copernicus man has been rolling down an incline, faster and
faster, away from the center—whither? Into the void? Into the "piercing
sense of his emptiness"? . . . all science, natural as well as* unnatural
*(by which I mean the self-scrutiny of the "knower"), is now determined to
talk man out of his former respect for himself, as though that respect had
been nothing but a bizarre presumption.*
      —Friedrich Nietzsche, *The Genealogy of Morals* (1887)

*When subjectivity, inwardness, is truth, the truth objectively defined
becomes a paradox; and the fact that the truth is objectively a paradox
shows in turn that subjectivity is the truth. . . . The paradoxical
character of the truth is its objective uncertainty; this uncertainty is an
expression for the passionate inwardness, and this passion is precisely
the truth.*
      —Søren Kierkegaard, *Concluding Unscientific
      Postscript to the "Philosophical Fragments"* (1846)

■ ■ ■

THIS IS A STORY about the meaning of science—its meaning for an individual in a particular culture in a particular era. And while it is an intellectual biography of an individual, it is at the same time an intellectual
account of an age. The time is roughly the first half of the twentieth
century, the place is New England, the individual—Percy W. Bridgman—is a Harvard physicist who won the Nobel Prize in 1946 for his
work in high-pressure physics.

The life of P. W. Bridgman (1882–1961) spanned an era of extraordinary change. The outstanding feature of this change was the challenge
to the classical, Newtonian worldview. The assault was two-pronged,

metaphysical and ideological, and its effect was to create a crisis of meaning in American, indeed all of Western, thought. The new physics of the twentieth century insinuated a radical anthropocentrism or secularism, more radical than the humanism of either the Renaissance or the Enlightenment. It appeared to uphold the sentence already pronounced by Darwin, condemning humanity to a godless world—a world without absolutes.

Similarly, as the twentieth century advanced, political and economic events joined to discredit the utopian promise of Enlightenment ideology, an ideology philosophically linked to Newtonianism and predicated upon the ideal of the perfectibility of man and his knowledge. World War I and the Depression severely undermined the faith that science and free enterprise could stand at the basis of the construction of the Good Society. The rise of Nazism and Marxism threatened American freedom precisely at this moment of metaphysical vulnerability. They seemed to be aimed directly at this cultural Achilles' heel.

Bridgman watched as the classicism of the nineteenth century gave way to a self-conscious and doubting modernism. With his entire being he felt the glow of nineteenth-century optimism and confidence fade into twentieth-century disillusionment. And with all of his might he tried to keep this from happening. Nevertheless, it overwhelmed him with the irresistible force of scientific logic. When it did, the best he could do was face up to it and bear its indignity with the pride of intellectual integrity. However, he was not left entirely to his own resources to find the strength required to carry this lonely burden. The ascetic Protestantism indigenous to New England provided a ready-made template of rationale and value to justify the stoicism that the situation demanded.

The intellectual life of P. W. Bridgman provides an example of what happens when the precarious equilibrium among individual consciousness, knowledge, and cultural forms is disrupted by value-destroying change, change that provokes in individual consciousness what psychologists call cognitive dissonance—a severe conflict between competing realities—and change that concomitantly threatens the stability of culture at large. The result is a crisis of meaning which cannot be left unresolved. The energies of thoughtful individuals become engaged in rediscovering the foundations of meaning. However, the available alternatives are not unlimited. There is no absolute freedom to create new meaning. No matter how revolutionary the disturbing event, the inertia of culture and of individual psychology constrain the resolution.

This is precisely the situation in which Bridgman found himself. Moreover, because science was at the center of the disruption of equilibrium, and because Bridgman had invested all his intellectual, professional, and spiritual capital in science, the problem created by the demise of classical Newtonianism was especially acute for him. In his search for a revised intellectual and cultural identity, a new foothold of psychological and social stability, he was constrained not only by American tradition but also by his personal refusal to admit the cogency of religious truth. Thus, within the boundaries of cultural possibilities his alternatives were even further limited.

The experience of cognitive dissonance thrust Bridgman into a state of intellectual skepticism, a disillusioned and overcautious condition of self-consciousness which withheld trust, balking at the chance of a second betrayal. Nevertheless, his skepticism was in fact disingenuous, not because he was being dishonest, but rather because such a state of suspended belief is psychologically impossible to maintain without inducing paralysis. In the end, science was for him still the arbiter of truth.

However, while the old physics was no longer true, the new physics painted a picture of the world that to Bridgman's eyes was devoid of stable meaning. Thus, he was stranded in a no-man's land where the only oasis of integrity was the self, and it was to the self, a self that defied the validity of universal and cultural norms, to which he turned in his search for meaning. Into this self he tried to absorb science. Only by inscribing science with the value of the self could he begin to undo its meaninglessness.

Bridgman struggled to extricate himself from the cultural and intellectual deceits which obscure truth. Yet bound by the paradoxes that limit rational clarity, his attempts carried him ever further into the depths of irony. Nevertheless, though Bridgman's experience was unique, it was not merely idiosyncratic. When viewed in the context of modern Western consciousness, it takes on a significance that reverberates throughout the entire culture. From a personal standpoint, it was unfortunate for him not to have known that outside the relatively cloistered scientific and philosophical community of his immediate social environment were thinkers who might have been sensitive to his perceptions and thoughts (had he been willing or able to open a channel of communication). However, they were among the literary and theological avant-garde who, by the positivistic standards prevailing in his intellectual world, had little to contribute to the understanding of truth.

As a study in the history of American ideas, Bridgman's intellectual life, in its progressive awakening to a crisis in twentieth-century values, is a modern allegory. It is also a tale of the loss of American innocence as the New World and the Old became culturally and politically reacquainted. In a narrower philosophical sense, it is an ironic Jamesian parable, underscoring the self-limiting horizon of pragmatic or positivistic philosophies. In the still narrower field of scientific philosophy, Bridgman's quest for cognitive purity forced two ideals of knowledge, the empiricist and rationalist, into critical contrast. And finally, in the context of American ideals, Bridgman's experience is a confrontation between two fundamental American traditions—the bright, egalitarian optimism of the Enlightenment and the dark, deterministic individualism of the Puritan ethic, each of which counted science in its camp.

Nevertheless, it was change within science, change for which Bridgman personally was intellectually and emotionally unprepared, that prompted his examination of meaning and value. The new physics shattered the security he had taken for granted when he placed his trust in the authority of classical Newtonianism. He was not alone. The timeliness of his concern was confirmed by the receptivity of the reading public to his ideas. But as he reached deeper toward the source of knowledge and value, he discovered that others could not follow him into the solitude of his own thoughts. His ability to communicate broke down, and he was left with a feeling of frustration and an acute sense of alienation.

Bridgman's story provides a unique window upon the eruption of cognitive and moral crisis, first in science and then in the wider American culture. Because he felt the impact so keenly, and because its consequences were so poignant, his life teaches us with special force that such events are not mere abstractions invented by historians, but tangible realities, realities that touch individuals, yet also realities that individuals help to create.

One of the most problematic aspects of writing this biography was the need to reconcile Bridgman's innocent scientific outlook with the darker content of his introspective discoveries. As I came to realize what was at stake, I understood that I was also being confronted with a challenge to the form and meaning of historical writing itself. Hidden behind what appeared outwardly to be the prosaic, orderly life of an experimental scientist who spent his time carefully adjusting gauges and tabulating the hard, dry, facts of measurement, was an intensely experienced subjectivity, an exquisitely felt self-consciousness. It was this inner world of human subjectivity that Bridgman dared the scientific community to recognize and value and that, by extension, also presented the challenge to historical writing. How could I effectively convey the stark existential reality of his encounter with subjectivity? How could I avoid the factitious categorizations he so vehemently resisted? For history, like science, is a public genre. History is the record of a public memory, an account of public events. The existential subject is not the material of science or of history. Thus, again I found myself directly facing the limitations of discursive writing, not to say the boundaries of my own discipline—the history of science.

Bridgman's scientific and philosophical contemporaries did not acknowledge the radical nature of his dare. Many of his interests and judgments appeared to be "irrational" or incomprehensible, sometimes not even entirely "respectable," especially for a scientist of his high professional and social stature. However, Bridgman was not a frivolous or impulsive person. He was not prone to making superficial or thoughtless statements. He was rational and deliberate in his way of thinking and doing. He was earnest and intelligent. And he did not offer his challenge lightheartedly. It was, in his mind, essential not only in his *search* for truth, but also for making the truth *known*. Therefore, in order to do justice to Bridgman's situation I knew that I needed to write something more than a scientific biography—something more than intellectual history. I had to write a *story*, the story of a human being. But what kind of story?

As Bridgman's life unfolded before me, I found that the answer to this question was not to be my choice. The story of Bridgman's life, as he himself made plainly evident time and again, in both thought and deed, was the story about the relationship between consciousness and freedom, knowledge and responsibility—a relationship which in the narrative of human action is the universal material of tragic drama. Moreover, it was a story that was characterized by a unity of thought and action usually found only in fiction, and that possessed all of the elements of a complex ironic tragedy. This was so not only because the problem of consciousness and freedom was something over which Bridgman privately and publicly anguished—and in most explicit terms—but also because in his life he was self-consciously acting out his own understanding of this relationship.

Yet tragedy is a literary, and not an historical category. I could try to argue that in this case, the distinction is one of name only. Or, perhaps it would be more germane to suggest that the story I have told is historical evidence for the legitimacy or authenticity of the tragic genre, that it shows that tragedy is true mimesis. But that would be to miss the point that the full impact of the story clearly depends in the first place not on the details of historical fact but, instead, on the way it conforms to the literary structure known as tragedy. There is a striking irony in the circumstance that it is an implicit literary form, rather than the sum of evidential particulars, which is powerful enough to communicate the existential urgency of Bridgman's consciousness of his subjectivity, and transparent enough to translate his private dilemma into a public language.

However, while the story is clearly tragic in a literary sense, it conforms to neither the high classical form—where, in the end, the hero, who is an individual of nobility, is at least privileged to understand the cosmic meaning of his fall—nor the modern ironic form, where the hero, who is now merely an ordinary person, "unheroically" suffers a meaningless or an ignominious death with no saving insight, for the simple reason that neither nobility nor cosmic meaning exists any longer. Bridgman is neither an Oedipus nor a Willie Loman. In fact, one of the more subtle ironies of Bridgman's story is that while the tides of history were inexorably edging his life into an era where only the ironic form of tragedy could speak, he did everything in his power to hold back this impending meaninglessness, and in so doing, effected a subtle inversion in the structure of the tragic plot.

A formal closure to this story is brought about with the realization of the tragic plot. But actual closure, while temporally coincident with the experience of "the end of the story," occurs not in the universe of narrative form, but in the universe of real historical action. Because the events narrated in the story happen not only formally, or "onstage" as spectacle, but also actually, or "offstage" as real historical action, and because the protagonist is none other than the historical individual himself, there is the possibility of another resolution to the sequence of events, a resolution which takes place offstage, rather than onstage. This resolution, though ironic, is not a tragic one. In fact, it is reconciliatory.

To appreciate this we must recognize that in Bridgman's mind there can be no tragedy. The facts are what they are. There is no underlying meaning, cosmic or social. There is only the individual, alone in an indifferent universe. Bridgman believed that the individual is the sole agent of action and the one upon whom all responsibility ultimately devolves. There are no forces which are beyond the control of the individual. And there can be no cause for guilt—personal, genetic, historical, scientific, or otherwise. One simply does the best he can and there is nothing more to be said.

But if this is all there is to it, if the actions of the individual mean no more than what they accomplish for him, we are compelled to ask, for whom does the tragic experience exist? In what sense can it be understood that Bridgman's life story is tragedy? The answer to this question constitutes the final irony, for the tragic experience is conceived, if not in terms of a relationship to the divine, at least in terms of a relationship to society. It does not have any meaning without this larger context. In that sense, Bridgman's perception was correct. If the individual is truly alone, there can be no tragedy. Granting this premise, then, it is I as storyteller, together with you as audience, who are responsible for the final irony. Telling the story is what makes it a tragedy. But telling the story does something more. It accepts Bridgman's life as being meaningful. By drawing Bridgman into the social order which allows the judgment that his life was tragic, telling the story overcomes the isolation which was the original condition for tragedy. It affirms his place in society and sanctions his actions as being humanly understandable.

PART I

A Career in Physics

CHAPTER 1

Percy Williams Bridgman:
Harvard Physicist

*An inner devotion to the task, and that alone, should lift the scientist to
the height and dignity of the subject he pretends to serve.*
                    —Max Weber, "Science as a Vocation" (1918)

■ ■ ■

PERCY W. BRIDGMAN, physicist and philosopher, was born April 21,
1882, in Cambridge, Massachusetts.[1] He was the older of two children.
His sister, Florence, was born June 20, 1883. His parents, Raymond
Landon Bridgman and Mary Ann Maria Williams, were both descended
from New England families. Percy received his early education in the
public schools of Newton, Massachusetts, the town where he grew up.
He entered Harvard College in 1900, graduated *summa cum laude* in 1904,
and remained at Harvard as a graduate student in physics. Bridgman received his M.A. in 1905 and his Ph.D. in 1908. That same year he joined
the staff of the Harvard physics department as a research fellow. He
went on to become instructor in 1910, assistant professor in 1913, professor in 1919, Hollis Professor of Mathematics and Natural Philosophy
in 1926, and Higgins University Professor in 1950. He retired in 1954
and became professor emeritus.

Bridgman's long and immensely productive career spanned an era
during which American physics experienced vast changes in both style
and content as well as in its relationship to society and government.[2] In
1904, the year Bridgman graduated, physics was still only a budding
profession. Academic research was not yet regarded as vital to national
interests, nor had the university yet become established as a center for
research. The bulk of scientific research was motivated by practical
concerns and was carried out under the auspices of federal agencies

such as the Weather Bureau or the Geological Survey whose responsibility was to oversee the development and exploitation of natural resources. University research, practiced for the more dignified goal of discovering truth, depended on private support. The merger of academic with broader democratic social interests was only just beginning at the dawn of the century. Science had not yet saturated American culture. Modern America was still around the corner.

Prior to the twentieth century, mechanical skill and ingenuity had been regarded as the characteristic requisites for the practice of physics. Theoretical sophistication was not emphasized in physics education, and it was customary to travel to Europe for training in mathematical physics. In America, science was generally thought of as an empirical activity seeking to produce useful results. Thomas Edison, who was not an academic physicist, was commonly viewed as the exemplary scientist. Indeed, in the minds of many, there was no difference between an inventor and a physicist. To be sure, change had already taken root, but it would take war and the creation of a technologically dependent society for academic physics to demonstrate its social value.

The impulse to modernism, however, had been activated and was growing in strength, with the university functioning as a vital relay. Science was both instrument and beneficiary in this movement. The educational reform of the 1870's had enhanced the importance of science in its own right. Harvard University, under the presidency of Charles W. Eliot, was among the pacesetters. During Eliot's administration, physics at Harvard grew into a modern experimental science. His enthusiasm for the merits of laboratory experience as a way to combat dogmatic thinking was the force that led to the improvement of science instruction at Harvard. Eliot was responsible for hiring John Trowbridge, who had established the first instructional laboratory at MIT. Under Trowbridge's energetic and aggressive guidance, the Harvard physics department acquired new facilities, expanded its offerings, and made laboratory instruction an integral part of the physics curriculum.

The changed perspective expressed in the call for educational reform reflected America's impatience with the constraints of nineteenth-century formalism, both intellectual and social, as the pressures of increasing urbanization, industrialization, and the expansion of the middle class created new social perceptions and vocational expectations. An agrarian nation built around village mores, America needed new guidelines to cope with its changed social realities. Science could answer to this need. The rational principles of science and its empirical methodology found a natural place within the framework of readjustment. Congenial to emerging bureaucratic tendencies, science was nevertheless still linked to a conservative tradition.

Indeed, science, enveloped in the virtues of the experimental method, was a flexible ideology. It was broad enough to encompass egalitarian democratic principles and the traditional Puritan values of hard work, honesty, and self-sufficiency. At the same time, it was iconoclastic enough to be invoked on behalf of reform and yet dignified enough to be adaptable to the classical pedagogical and cultural ideals of character building and the development of right thinking. Science was moral because it was practical; it was also moral because it was not practical. Thus Harvard's president Eliot could proclaim the moral worth of pure research and hail the ethical benefits of cultivating a scientific attitude. He encouraged scientific study for the development of "candid, fearless, truth-seeking" habits of thought.[3] Science was to be practiced as much for the development of character as for the sake of knowledge.

The reference point of the experimental spirit was "fact." Science dealt with fact, useful fact or true fact, and who could dispute either the practical or moral priority of the truth? The dedication of science to fact was a powerful social and intellectual liberating force that freed the man of ability from crusty social conventions based on wealth and privilege. It was an antidote to political corruption and economic opportunism. It could restore honesty, industry, and integrity to American life. In short, it nourished the ideals of the Progressivist era, an era of action, reform, and social progress that would discard the romanticism and idealism of the nineteenth century and adjust social institutions to the facts of reality. It would confront life as it "really" is, correcting, purifying, and modernizing. Progressivism, inspired by the moral conscience of Protestantism and buoyed by a confidence in science, would fulfill the American promise.

It was within this academic and social atmosphere that Bridgman was initiated into science. The youthful Bridgman imbibed the spirit of Eliot's Harvard and, as a young adult, was surrounded by the enthusiasms of the Progressivist movement. The scientific ideology held great appeal for the young Bridgman, whose upbringing had instilled in him the virtues of honesty, self-discipline, and industry, but who had rebelled against the strict orthodox Congregationalism that was the source of these values. Shy and unpretentious, Bridgman was nevertheless fiercely independent and resentful of arbitrary authority. His daughter remembers him as a man who was shy and given to solitary activities:

My father was full of energy and high spirits, but at the same time was shy—at least as a boy he was shy, and he never was gregarious as a man. He grew up in a family which had a sense of mission . . . his aunt wrote hymns, his father wrote books on world union and poems some of which were collected as "World Brotherhood Verse." His mother, on the other hand, was a more down to earth character. . . . A lot was expected of him, and he demanded a lot of himself. He did well in school and college, also played chess, liked to whittle, was on the track team, climbed mountains—long over-night hikes rather than expert rock-climbing—played the piano. His activities demanded much of mind and body, but were seldom group activities.[4]

Bridgman embraced the scientific spirit. By dedicating himself to the pursuit of scientific truth, he believed that he had found a rational and realistic alternative to traditional authority. Bridgman's defiance of authority was not a blind or thoughtless emotional reaction, but a rationally considered response. His rejection of religion (and with it all "metaphysics"), as he wrote, was based on

the method of superficial probability. I made the discovery that this method is necessary quite early in life, when in Sunday School I was confronted with various expert "proofs" of the historical validity of many of the things in the bible. I saw that if I did a thorough job at deciding for myself the validity of these claims my time would be spent on nothing else. I was not willing to give my life to this, so I merely rejected with sufficiently logical argument, many of the claims of the religiously minded [of?] most of my elders, on the basis of inherent improbability, because of their failure to jibe with the external world as I found it.[5]

Bridgman's father, Raymond, a writer and journalist, was a man whose faith was absolute and concrete. His diary contains references to a spiritual world whose immediate presence was palpable and with which communication was possible. When asked on a Eugenics Society questionnaire to state his father's most outstanding trait, Percy responded, "was a most religious man."[6] Raymond Bridgman was extremely disappointed with his son's rejection of his religious views. Near the end of his life, however, he offered a conciliatory interpretation that allowed him to accept Percy's commitment to honesty and integrity as a moral equivalent to religion. He expressed his feelings in a letter written in 1924.

The human rebellion against religious dictation may lead a person to say that he is an atheist or agnostic and he may refuse to follow observances which are regarded as essential by the professedly religious.

But, if there is a sincere determination to follow the inner leading and to be honest and faithful in all things, to my mind that is the repentance which is meant, it is the real conversion, it is the yielding of the personal will to the divine, it is the harmony of the personal life with the pull of the love of God, notwithstanding there is intellectual failure to understand or accept the common teachings or to order the life according to current orthodox standards.[7]

In one respect, the elder Bridgman's hopes were vindicated, for despite his son's overt renunciation of the religion of his youth, the Puritan values he absorbed maintained a strong hold on his conscience for the remainder of his life. In another sense, however, there was no possibility of reconciliation, for the younger Bridgman saw nothing to be achieved by yielding to divine will but the abdication of human responsibility.

P. W. Bridgman's rebellion against authority went beyond the father-son relationship. It was a determining element in his intellectual, social, and political judgments. A striking example of this was his refusal to be swept away by the mania of patriotism that overcame the nation in World War I. Bridgman was not a man to fall victim to the pressures of mass psychology, no matter under what authority acquiescence was sought. He made this clear shortly after World War I, in an indignant letter to a Captain Defrees of the U.S. Naval Experimental Station in New London, Connecticut. During the war, Bridgman had conducted research in New London for the Navy on submarine detection devices. Now, with the war over, he was being asked to sign an oath of secrecy. This, he said, he was unwilling to do.

In the first place, I dislike on principle to take any oath. You know and I know perfectly well what is required of a man who has been in the confidential relation to the Navy that I have. I do not like the imputation that my sense of the requirements of the situation is not sufficient to ensure the correctness of my behavior, and that any such mere formality as going before a notary and holding up my right hand is any more likely to make me act in the way which I should.

In the second place, I signed for Commander MacDowell last summer a pledge of secrecy. My position then was that the Navy had a right to ask such a pledge while I was in the service of the Navy, although I objected to such a proceeding on principle. This signed statement from me, which you now have, should be sufficient. A gentleman is not in the habit of being asked to give his word of honor twice.

Finally, I do not understand the attitude of the Navy in asking a man who has left its empy [sic] to give a guaranteewhih [sic] should have been asked of him, if at all, before entering its employ. . . . At the conference of the submarine situation with the foreign scientific mission in June 1917, important military secrets were disclosed in this spirit to a large number of civilians, with no attempt to bind anyone by a formal oath. Neither was any such formal assurance required for nearly a year after the founding of the Experimental Station, although it was vastly more important that secrets should be kept then than now.[8]

Bridgman was impeccably honest and sufficiently idealistic to expect that others would respect his code of conduct. It was natural that he would have been hurt and insulted by any suggestion that he might not be trustworthy. To him it was probably little short of an accusation. Just as important, however, was the logic of the situation. The whole idea, as he argued, was irrational. It simply did not make sense. Bridgman possessed a deep hostility to deception of any kind, and it was unthinkable that he should ever be guilty of a betrayal. He did not concern himself with subtleties of meaning or intent. He expected his word to be taken literally and at face value. In the words of the theoretical physicist John C. Slater, "There is nothing devious or fuzzy about his thinking. If something is true and makes sense, that is the end of it."[9]

This attitude was typical. Thus, for example, in 1914, when a paper he had submitted to the *Journal of the Franklin Institute* was edited, he demanded that the authorship be changed from "by P. W. Bridgman" to "amplified and edited from the manuscript of P. W. Bridgman."[10] He had said exactly what he had meant to say. It was not to be modified in any way.

It might be easy to underestimate the courage behind such statements if we forget Bridgman's shyness. He was shy, yet highly principled. He was not a flamboyant individual who enjoyed calling attention to himself by making outrageous statements; rather, despite a strong competitive drive, he was a modest and private person. Those who have never experienced shyness rarely appreciate the psychic energy required to overcome the barrier it imposes. Try to imagine with what difficulty Bridgman confessed to G. B. Pegram, professor of physics at Columbia, the reason for his failure to contribute a promised item to the exhibit of the American Association for the Advancement of Science in 1916: "On seeing the exhibit I was overcome with shyness at the unpretentiousness of my offering and decided it was better not to show it."[11]

Bridgman's shyness also caused him extreme discomfort in public speaking situations; in fact, not only did he dislike lecturing, he was not very good at it. Although he put a great deal of work into his lectures, his audiences often failed to respond, and consequently he felt no gratification for his conscientious efforts. In later years, he would often make remarks about his failure to "get his point across." He became especially disillusioned by his inability to communicate with undergraduates, a problem exacerbated by what he perceived as their immaturity and general lack of self-discipline. "Callow pups," he called them and wondered how they were educable at all. In a parlor game in which the task was to identify a person from a list of five likes and five dislikes, "undergraduates" in the dislike column was enough to single out Bridgman. However, Bridgman admitted (in what seems a clumsy attempt at humor), the hazards of university lecturing included something even more odious—first on his list had been "Bare Legs in the Front Row, referring to the nether limbs of [E.R.—a female student], who is taking the course that I am giving for Kemble this half-year. Sometimes the sun shines on them and there are long bristly hairs."[12] Evidently Bridgman's attention was not always on physics.

Physically, Bridgman was of slight stature, but he was lean and wiry, strong and energetic—compact, according to one description. He was rarely sick. Neither a smoker nor a drinker, he made a point of keeping himself in good physical condition, attending gym classes regularly during the school year. He took pride in his stamina and his ability to outdistance his summertime mountain-climbing companions and was not above letting them know it. This was a situation where Bridgman must have judged that humility was out of place.

Bridgman's dedication to high principles went well beyond matters of abstract judgment. He also felt a strong sense of duty to his family. As devoted and ambitious as he was with respect to his work, and as firm as he was in his commitment to intellectual honesty, he made every attempt to be attentive to the needs of family life, even when they competed with professional obligations. He was not tied to his vocation to the extent that it might destroy the unity of the family, and he was disciplined and energetic enough to live in both worlds at once.

In 1912 Percy Bridgman married Olive Ware (the occasion was duly recorded in his lab notebook).* The Bridgmans had two children, Jane and Robert, born January 15, 1914, and December 9, 1915, respectively. Except for a brief stay in New London, Connecticut, during the war, the family lived in Cambridge, Massachusetts. They spent their summers in Randolph, New Hampshire, a small town in the White Mountains, not too far from the scenic Franconia Notch, where Peter, as Bridgman was known to his close friends, refashioned a barn into a comfortable vacation home.

Olive Ware, born July 10, 1881, was the daughter of Sarah Jane Twitchell and Edmund Asa Ware. Edmund Ware was the first president of Atlanta University, a school founded for the purpose of educating the recently freed blacks. Olive's family lived in a black section of town, and Olive spent her first year of school as the only white child in a black kindergarten. Her father died when she was just four years old. Her mother had tuberculosis and moved the family—herself and four children—to Connecticut, where she had relatives. In Connecticut, Olive lived with two spinsters in a small boarding school. However, she attended a local public school. After graduating from Smith College, she spent a year as a governess in the family of a Scotch Presbyterian missionary in Spain. On returning to the United States, she attended Simmons College and, after graduation, became a secretary to the Harvard physicist Wallace Sabine, who was a member of Bridgman's doctoral committee. It was while she was working for Sabine that she and Percy met. Olive maintained her loyalty to Atlanta University and to Smith College for many years, participating in fund-raising benefits for both institutions.

In contrast to her husband, Olive was gregarious and considerably less self-disciplined. She enjoyed socializing, and it is recorded that she even smoked a cigarette and read a naughty book or two. However, it was her habit of overspending that gave her husband cause for concern. During the early years of marriage when the children were young and just beginning their schooling, the family chronicles contain frequent lamentations that Olive's spending was undermining her husband's budgetary regime. However, the simple fact was that the salary of a young Harvard physics professor was hardly sufficient to maintain a lifestyle commensurate with his social company.

*Notebook IV, p. 137, May 16, 1912: "getting married and writing paper."

With respect to the children, too, Olive appears to have been less of a disciplinarian. While Bridgman expected his children to conform to strict standards of behavior, Olive had more patience, tending more to believe that they would, in time, outgrow their careless habits or asocial behavior.

Moreover, Olive did not have the physical stamina to keep up the demanding pace Bridgman imposed on himself, nor did he expect it of her. He did not consider it demeaning to help with the chores or baby-sitting when his wife was overwhelmed, or even if she had engagements that took her away from home. (Olive, Bridgman remarked, commenting on his wife's charm, had a knack for getting even [college] presidents to wipe dishes for her.) On the other hand, he did not attempt to hide the feeling of superiority he derived from being the one upon whom final responsibility always fell. Nevertheless, staying home with the children was not always conducive to thinking about physics problems—how could he ever become a "theoretiker" if he had no peace and quiet? Still, he was not willing to sacrifice family unity to achieve that goal.

Indeed, the writing of the family chronicle was a conscious device he instituted for cementing family bonds. Another was the weekly "toot," a standing engagement with Olive (rarely canceled) that was usually fulfilled by taking leisurely, extended walks or trolley rides to outlying towns—Arlington Heights or Belmont, perhaps, or Franklin Park—sometimes to see the sights, other times, to sit or nap under their favorite tree. In winter, if the Charles River were frozen, they might go skating.

Olive was also a churchgoer and a sometimes Sunday School teacher, and despite his own antipathy toward religion, Bridgman always respected her choice, without intimations of disparagement, and did not object, either, to having the children attend Sunday School. He added one stipulation, however, and that was that although they were free to choose whether or not to attend, once the decision was made, they must follow through. He would tolerate no half-hearted commitment. While his wife and children went to church, Bridgman cleaned the yard or chopped wood.

For Bridgman, life in Cambridge was dominated by the demands of the laboratory. The hours were long, and the work tedious and exacting. Teaching was an added burden. Then there were meetings—the Shop Club, the American Academy of Arts and Sciences, talks in the laboratory—and conferences that took him out of town (not all of which he could afford). For Olive, there were endless benefits, rummage sales, courtesy calls to be made on friends and neighbors, social teas, meetings of the Mothers' Study Club (Bridgman called it the Mothers' Snob Club), and the children, whose schooling and health were always in the forefront of her mind.

However, life was not all that dreary or routine. The Bridgmans' Cambridge scintillated with social and cultural riches—dinner parties with prominent people, theater, lectures, and concerts. Never a week went by without some stellar event involving both or one of the Bridgmans. Not all such events, however, were necessarily to Bridgman's liking, for he sometimes ran out of patience with too much socializing. When, for example, Olive "went to a lecture at the Longfellow house, and met all the elite of Cambridge in evening gowns, . . . [he] went to bed in a grouch, not feeling any desire to get himself mixed up in any more damned things, having all he can attend to in the Laboratory."[13]

The end of each academic year was met with great relief and a sense of anticipation. The pressure was off and the mountains beckoned. The Bridgmans packed their trunks, gathered up the children, the maid, and the pets, and moved to Randolph, New Hampshire. Randolph was no ordinary vacation community. The annual summer residents were a closely knit group (highly class conscious, it might be added). They were mostly academics—college presidents, professors, and ministers who brought along their families and, occasionally, a paramour. A scandalous moment might be created by a college president's revelation that "he didn't mind talking to his former High School classmates who might now be clerks in jewelry or grocery stores"[14] or by the realization that the musicians hired to play for a wedding were black. Bridgman himself was disturbed by the uncomfortable thought that the Koopmans, who were also Randolph regulars, might have Jewish blood in the family. (Years later, Bernard Koopman became his son-in-law.)

Nevertheless, the atmosphere was filled with stimulating conversation and the spirit of good clean fun—hiking, swimming, picking berries, sharing meals, and of course, for everyone but Bridgman, churchgoing. Bridgman enjoyed the relaxed mode of socializing, shunning only the baseball games and the afternoon teas, and took particular delight in charades, a game that must have been challenging enough for him to have some "serious" fun. In fact, he was a local celebrity at charades, well known for his ability to excite waves of hilarity with his antics. Picture, for example, our reserved and taciturn professor Percy W. Bridgman, standing before his audience "with a wreath of ferns around his middle and a chaplet on his head, suggesting a Hawaiian costume, and a smile on his face," acting out "persecute." [15] Yet despite all of this, Bridgman followed his own star.

Indeed, for Bridgman, the vacation lifestyle, though more rural and rugged, was every bit as orderly and vigorous as the more formal academic regimen of university life. Work first, play later, was the guiding rule. "There was a regular daily summer routine," his daughter, Jane, remembers. The morning began with stories for the children before breakfast. When breakfast was finished, there would be two hours in the garden, followed by a cold bath. Bridgman would then retire to his study to write up the results of the previous year's experimentation. He stayed there until four in the afternoon, "with just an interruption for lunch with family at one o'clock. . . . After four Mother often had callers, but Father was excused from the formality of tea, working some more on his garden, or on some improvement to the house. In the evening his callers might come—Leigh Page, J. Q. Stewart, Harvey Davis, or Dr. [G. N.] Lewis possibly—and the conversation would be interesting but over our heads." [16]

Outside the family, those still alive in Randolph who knew Bridgman remember him as a man who was held in great respect, who was cordial if not sociable, but who was also somewhat intimidating. The younger people found his acerbic sense of humor incomprehensible and sometimes even frightening. They were made uncomfortable by his abrupt manner, a habit that reflected his impatience with indecision or with speculations that contradicted his principles. He was too quick to judge, it was said. He would pronounce before considering the qualifications.

With undisciplined children—and not just his own—Bridgman had even less patience and did not hesitate to act when he regarded their behavior unacceptable. Jack Stewart, the son of the Princeton astrophysicist John Q. Stewart, remembers this from firsthand experience. The occasion was the annual picnic of the Randolph Mountain Club. Jack, six or seven years old at the time, was playing at the river, throwing stones, oblivious to the fact he was splashing water on the guests. Bridgman asked him to stop. When Jack did not obey (Jack claims that he did not hear the request), Bridgman marched over to the river and promptly sat him down in the water, saying, "If you wet others, you should be wet yourself." Not surprisingly, Jack's parents were offended and did not speak to Bridgman for several weeks afterward.[17]

Indeed, Bridgman's high expectations created a problem for both himself and others. He expected of others what he demanded of himself, and when they failed to meet his standards, he was disappointed. He felt let down, and not just because he couldn't have it his own way, but because he was doing his best and thought everyone else should as well. He was, in a way, hurt when his high values were disregarded or spurned. The undergraduates at Harvard were too remote as individuals, and the general public too abstract an entity, to feel or know this personal response, but his children experienced it all too intensely. It was not easy for Jane and Robert to live with the seriousness of the goals their father had set for them. In their view, his expectations were unreasonable, and his enforcement of them too strict. They wanted to have some fun, too.

On the other hand, as a parent, especially one for whom the family was so clearly an extension of himself, it must have been equally difficult for Bridgman to know just when to relinquish control. Well into their college years, he persisted in chastising the children for putting a good time ahead of intellectual achievement and for not living up to his own "work first, play second" ethic. He wanted to be able to admire them, and the kind of person he admired, he said, was one possessing "unusual ability, or fanatical devotion to a high vision, or the resolution to force through against man and devil an admirable course of action, or even any decently intelligent singleness of purpose"—a weighty expectation, indeed.[18] Bridgman did not seem to appreciate that such extreme goal-directedness cannot be imposed from the outside, nor did he seem to consider the ordinary enjoyment of life a value worthy of his admiration.

In late September, just as the autumn colors were coming into full display, summer vacation ended. For the Bridgman family, leaving Randolph was a sad time. The return to Cambridge signaled the imminent onset of winter and its confinements. Once again, the time had come to rejoin the hectic pace of formal obligations and to face the almost certain reappearance of fevers, coughs, and grippe. "The shadows of the prison house are closing around us," wrote Olive in September 1923, "and we are beginning to feel decidedly uncomfortable as our date of departure for Cambridge approaches near."[19] But a vacation cannot be forever, so after the good-by calls were made, the belongings packed, and the house closed up tightly, the Bridgmans would head back to Cambridge—the children to school and music lessons, Olive to her social activities, and Bridgman to Harvard and physics.

Besides being a family man, Bridgman was also a physicist, one whose scientific style was a model of the traditional American concept of science. Classically "Baconian" in its goal of collecting data from from concrete measurements, it was an economical one-man program of pioneering exploration. Its results came from the exploitation of a mechanical discovery that he realized provided him with a unique opportunity. It was, in fact, an unintended consequence of "a problem proposed for a doctoral thesis, [during which investigation he] hit upon a packing which allowed higher pressure than ever before," he explained to a correspondent, P. J. Risdon, in 1922.[20] Thus, as he wrote in his textbook on high-pressure physics in 1931, "The whole high-pressure field opened almost at once before me, like a vision of a promised land, with the discovery of the unsupported area principle of packing, by which the only limit to the pressures attainable was set by the strength of the metal parts of the apparatus."[21]

The principle of unsupported area was a way to incorporate into the design of the pressure-transmitting piston the physical law (actually, a definition) that with a given force, the pressure increases as the area upon which it acts is decreased. By constructing the piston in two pieces, with the "unsupported area" (an open space) in between, Bridgman was able to take advantage of this principle and to create a situation where the packing, or washer, was always under higher pressure than the substance being compressed. The seal automatically became tighter as the pressure was increased. In this manner, leaking, the nemesis of high-pressure experimentation, was prevented (see Fig. 1).[22]

Endowed with the capacity for long hours of highly disciplined concentration, Bridgman refined his measuring techniques and began his exploration of the newly accessible physical terrain. He was soon able to map out previously unsuspected states of matter. While professional interest in his work was concentrated among industrial metallurgists and physical chemists, general interest was aroused by some of its more exotic prospects.

One such discovery was the existence of allotropic forms (different molecular arrangements) of water ice, the most spectacular being "hot ice"—solid water at 80° C (176° F).[23] Bridgman received many inquiries

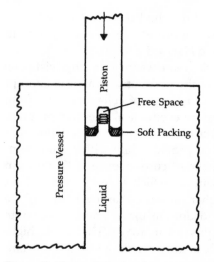

Application to a piston of the principle by which the pressure in the packing is automatically maintained at a pressure greater by a fixed percentage than the pressure in the liquid. Leaks therefore cannot occur.

Fig. 1. Bridgman's design for a piston using the principle of the unsupported area. Because the packing is under higher pressure than the substance being compressed, the seal becomes tighter as the pressure is increased. This allowed Bridgman to study the characteristics of substances under higher pressures than previously possible. Source: Diagram from P. W. Bridgman, "Some Results in the Field of High Pressure Physics," *Endeavour* 10, no. 38 (Apr. 1951). See also the discussion in P. W. Bridgman, *The Physics of High Pressure* (London: G. Bell and Sons, 1931), pp. 32–33, 35.

from trade journals such as *Ice and Refrigeration Illustrated* and *Cold Storage and Ice Trade Journal* asking for articles. E. F. McPike, manager of Central Fruit Despatch, a division of the Illinois Central Railroad Company, even thought it might be useful in the transport of perishables. In 1912 he wrote to Bridgman:

I have read a brief note in the newspaper regarding your production of a solidified form of hot water.

Might we inquire if in your opinion this process has commercial possibilities for use in transit in a manner similar to ice to protect fruits and vegetables and other perishable products against cold instead of heat. Portable heaters of various kinds have been used to some extent by carriers and if there is any prospect of a better process we would be glad to know about it. How long does the new solid retain its form and heat? How rapidly does it cool off?[24]

Unfortunately for Mr. McPike, hot ice was only of academic, not practical, significance.

And, of course, there was the tantalizing possibility that Bridgman might have discovered the long-sought method of making diamonds (a possibility, in fact, never too far from Bridgman's mind). Besides those he received from various inventors, one inquiry came from an English schoolgirl:

I have read with great interest the account of your High Pressure ex-periments, which has been published in "Conquest," but could you refer me to a more detailed account? . . .
. . . If intensely hot carbon—heated with an electric arc let us say—were subjected to an enormous pressure might not diamonds be formed? I know that minute diamonds have been made.
If by a happy chance you have not already made this experiment, and you should think it worth your while to try it; and if by a happier chance it should succeed, . . . would you send me a diamond so made? I who am a girl still at school would esteem it an immense favour on your part.[25]

To this request, Bridgman graciously replied:

I was much pleased to get your nice letter. I have thought a good deal about the possibility of making diamonds by subjecting carbon to high pressures, and whenever I have had a new kind of apparatus made that presented features that I had not tried before, I have always serreptitiously [sic] put into it a little piece of carbon before applying the pressure, but a diamond has never greeted my expectant gaze on unscrewing the apparatus at the end. No doubt the dia-mond problem can be solved in some way as you suggest, but there are great difficulties in the way of a practical character. It is difficult to attain any great degree of heat in the midst of a medium which is transmitting great pressure. No doubt the difficulties could be surmounted if one tried hard enough and were willing to spend enough money in the preliminary work, and in any event I will certainly send you a diamond when I find how to make them—the second one—the first has been promised for a long time to Mrs. Bridgman.[26]

This young lady never received her diamond (neither did Mrs. Bridgman), but P. W. Bridgman's achievements in high-pressure phys-ics would win him a Nobel Prize in 1946. When diamonds were finally synthesized in 1955 by General Electric (for whom Bridgman had been a consultant for many years), it was in no small measure due to the efforts of Bridgman, whose work dominated the field of high-pressure physics during his lifetime. However, though he had begun his re-search in 1908 and was nominated twice without success for the Nobel Prize, it was not until the development of solid-state physics in the 1940's that the importance of his research was formally recognized. In the meantime, he carried out his work with the diligence of a man who respected his calling.

Bridgman was a Harvard man in every respect and did not express any desire to leave. Nevertheless, the fact that he had so quickly distinguished himself as a brilliant and original experimentalist made him an attractive candidate for solicitations from competing institutions. At the same time, his uncompromising independence and single-mindedness were becoming evident. His goal was to do pure research, and he would settle for nothing less. To Arthur Day, director of the Geophysical Laboratory of the Carnegie Institute, whose offer he declined in 1909, and again in 1916, he declared, "My interest is almost entirely in experimental research and much less in teaching. Neither is the prospect of theoretical study abroad inviting." He went on to express his doubt "as to whether in a laboratory dedicated to a special purpose the same freedom of choice of subjects would be possible . . . as in a University." [27]

The University of Michigan was interested, too, and in 1915 asked the Nobel Prize–winning Harvard chemist T. W. Richards to write a letter of recommendation. What Richards had to say confirms Day's experience.

Bridgman . . . is only 33 years old, and has had, comparatively speaking, almost no experience with the world and its ways. By a curious paradox, although a slight and insignificant looking person, he has dealt with the greatest pressures ever studied carefully by physicists. His mind seems to me rather of the one-track type. His work with these high pressures is epoch making in my opinion, but his interests, in so far as I have talked with him, seem to lie almost entirely in this direction alone. I should say also, however, that he seems to me a very expert mathematician, with solid and deep knowledge of thermodynamics. [28]

In later years, in fact, thermodynamics was the only subject that Bridgman was called upon to teach, and his course, offered every other year to graduate students, was well known for its difficulty and challenging content. What Richards could not know at the time, however, was that Bridgman, despite his unworldliness, was soon to step into the world of philosophy. Here, his straight-talking independent manner, if refreshing, was often clumsy and insufficiently refined to communicate his insights.

This inability to communicate—in this case, not simply with mere undergraduates—was to become an ever-increasing source of frustration to Bridgman. He came to feel that attempts to enlighten others

Bridgman's laboratory in Lyman Hall at Harvard, ca. late 1930's. Photo courtesy Harvard University: Cruft Photo Lab—Paul Donaldson

P. W. Bridgman, June 18, 1886, at age 4

Bridgman as a Harvard student, ca. 1905

Bridgman, ca. 1912

Bridgman, ca. 1920. Photo courtesy Harvard University: Cruft Photo Lab—Paul Donaldson

Participants in the 1923 Solvay Conference. (*front row*) E. Rutherford, M. Curie, E. H. Hall, H. A. Lorentz, W. H. Bragg, M. Brillouin, W. H. Reesom, E. van Augel, (*middle row*) P. Debye, A. Joffe, O. W. Richardson, W. Broniewski, W. Rosenhain, P. Langevin, G. de Hevesy, (*back row*) L. Brillouin, E. Henriot, T. Dedonder, H. E. G. Bauer, E. Herzen, A. Picard, E. Schrödinger, P. W. Bridgman, J. Verschaffelt

Participants in the Conference of Science and the Modern World View—Toward a Common Understanding of Science and the Humanities, May 5–6, 1956. (*l. to r.*) Gerald Holton, Charles Morris, Nathan M. Pusey, I. I. Rabi, J. Robert Oppenheimer, Detlev Bronk, John E. Burchard, P. W. Bridgman, Perry Miller, Philipp Frank, Howard Mumford Jones, W. V. Quine, Harcourt Brown, Giorgio De Santillana

Olive Ware Bridgman

Bridgman addressing the Cornell Section of the American Chemical Society on "Recent Experimental Work at High Pressures," February 19, 1948. Photo courtesy Harvard University: Cruft Photo Lab—Paul Donaldson

Bridgman receiving the Nobel Prize from King Gustav V of Sweden, December 11, 1946

might be useless. He expressed his disillusionment, for example, in a letter to the editors of the *Bulletin of the Atomic Scientists* in 1950. The *Bulletin* was in some financial difficulty and Bridgman was being asked to make a contribution. In response, Bridgman stated his opinion that the journal's publication should be discontinued if there was no public recognition of its value and, consequently, no public financial support for it. He noted that the Ford Foundation had turned down a request for a grant and then continued:

I have never been much of a missionary but have always thought that a man should be allowed to go to the devil if he is that kind of man and has had his attention called to all the facts in the situation. It seems to me that the public has had its chance, and that the scientists of this country have fully discharged what may reasonably be claimed to be their responsibility. . . . Any further attempt to cram down their throats something which they plainly do not want would, in my mind, smatters [sic] of failure of decent self-respect and I am in favor of discontinuing the Bulletin unless there be some radical change in the situation.[29]

This statement should be taken not as a sign of cynicism but rather as an expression of Bridgman's pride in science. He certainly did not, in general, eschew attempts to bring science to the attention of the layman.

The first decades of the twentieth century during which Bridgman was establishing his career were a time when American physics was struggling to increase the rigor and quality of instruction. In particular, America had a wide gap to close in theoretical physics in order to reach the level of sophistication already achieved in European physics. Except for J. Willard Gibbs, the notable American thermodynamicist, the prominent American physicists were experimentalists. The new physics of Einstein and Bohr could be intelligible only to the mathematically literate. The American physics community was unprepared to grasp the new mode of physical thinking. The physics curriculum was not up to the task.

With the establishment of laboratory instruction, Harvard physics had reached a new level of modernity but it had not yet become internationally competitive in the area of theory. Harvard was trying to remedy that shortcoming. And despite his claim that he was not interested in teaching, Bridgman did not isolate himself from these efforts of Harvard's physics department to strengthen its theoretical capability. This is evident from his correspondence with Edwin Kemble, a promising young theoretical physicist (Kemble had done his doctoral thesis under

Bridgman) whom Harvard was courting. The following letter, written on behalf of the department, describes the situation as it was in 1919.

My Dear Kemble,

As Professor Hall has written you, the Physics Department has picked on me to tell you in detail what work we had planned for you in case you come to Cambridge next year, as we all very much hope. I suppose I have been selected to do this because I have had more hand in inventing the courses which we hope you can give.

The list of courses has been pretty thoroughly revised; partly in consequence of Professor Sabine's death, but the probability is that the revision would have been made anyway, because he as well as the rest of us felt the need of it. Our aim in the revision has been to make more courses accessible to the undergraduate, and to bring the graduate courses up to date. [Then follows a list of courses and who will teach each one. Kemble was assigned special topics in theoretical physics for graduate students. This included relativity and gravitational and quantum theory.]

. . . . .

I am really enthusiastic about this scheme of courses. It comes pretty close to what I have been wanting for a long time. If we can get the courses well given, it ought to put Harvard pretty near the top in this country. What is more, it is a good beginning to putting this country on the map in theoretical physics. . . . But you see that you are an essential part of this program. Don't you want to be a member of a Department that is trying to do this, and don't you feel the challenge in this?[30]

Kemble accepted the challenge, joined the Harvard faculty, and went on to distinguish himself in the field of quantum mechanics. The Harvard physics department soon produced two brilliant young theoreticians, John H. Van Vleck (a student of Kemble) and John C. Slater (a student of Bridgman); later they were joined by J. R. Oppenheimer (also a student of Bridgman).

Bridgman's own theoretical ambitions were short-lived. Early on, his measurements of the effect of pressure on electrical resistance in metals suggested to him a possible explanatory mechanism for metallic conduction, a subject on which he first published in 1917. He continued to pursue this topic throughout the 1920's, but his interest faded away in the 1930's. By then it had become unequivocally clear, to his disappointment, that the phenomenon of metallic conduction required a quantum mechanical treatment, and this was beyond the range of his expertise. In America, it was the province of the next generation of physicists, those who had benefited from the efforts of men like Bridgman to upgrade the theoretical competence of the American physicist. He himself was constrained by the education he had received; he was, as well, awed by the rapid pace of innovation with which he had to contend. (See Chapter 3 for Bridgman's encounter with the difficulties of theorizing.)

Bridgman had felt his limitations already in 1922 when he admitted to fellow physicist Ludwik Silberstein, after having read some of his papers, "I think I got considerable profit from them, particularly from the one on the aspherical nucleus of hydrogen, but the chief feeling aroused by them was one of envy for the theoretical physicist. I find it almost impossible to devote most of my time to experimental work and still keep my theoretical tools sharp, and this limitation is a source of keen regret."[31] Specialization was beginning to appear in American physics. Theoretical and experimental physics were turning into separate activities, and as the change took hold, the secure expectation that mastery of physics was within the reach of any one individual was vanishing.

But as theory worked its way into American physics, something else happened, something more subtle and profound. The concrete and absolute reality of traditional science was brought into question. Along with theory came metaphysical questions. As a consequence, the meaning of scientific truth could no longer be taken for granted. The relationship between empirical findings and theoretical formulations had to be clarified. What exactly was it that science talked about?

This unsettling situation was the starting point for Bridgman's excursion into philosophy. Here, too, he would explore new territory— new, at least, to the thinking of the American physicist. For in contrast to Europe, in America there was no well-established tradition to which the physicist-philosopher belonged. Here, the physicist was a practical man. He did not waste his time on idle speculation.

Bridgman's answer to the philosophical problem he faced became known as "operationism," or "operationalism." Operationism attracted much attention in the intellectual community and among the reading public. But for Bridgman, it was the cause of much personal anguish. Twenty-five years after its enunciation, he felt the need to protest:

I feel that I have created a Frankenstein, which has certainly got away from me. I abhor the word operationalism or operationism, which seems to imply a dogma, or at least a thesis of some kind.[32]

Professionally, physics was good to Bridgman. The beginning years held great promise; they were filled with optimism and confidence.

Success was quick to appear. Bridgman's work received international acclaim early in his career, and it continued to carry him to ever-greater distinction. On the other hand, philosophically, science was not at all kind to Bridgman. No sooner did he step onto philosophical territory than he found himself entangled in a web of logical dictates, verbal nuance, and metaphysical prejudice. For this, he was unprepared.

CHAPTER 2

The Experimental Method: Pioneering a Research Program

There is so much we do not know about this subject that we would rather try everything in sight and trust to learning the explanation later.
—W. R. Whitney, Physicist, G.E. Research Laboratory (1914)

• • •

A FIRST STEP IN the modernization of physics at Harvard University had been its transformation, under the leadership of President Eliot, from a liberal arts subject to a laboratory science; that is, from a canonical set of established truths to an investigative activity. However, the university was still supposed to be a gathering place for gentlemen, not a vocational institution. Therefore, although Eliot could invoke the ideology of empiricism to underwrite and sanction the reform, the classical ideal of a humanistic education continued to prevail. The university was a place of culture, not a practical training ground. Thus laboratory education had to be rationalized in terms compatible with traditional pedagogical goals. The key idea was "pure (empirical) research," a concept that allowed the academic scientist to become actively engaged in inquiry directed toward matters of mundane existence without professing utilitarian goals. Accordingly, the objective of pure research was not practical gain but, rather, the pursuit of truth and the cultivation of virtuous habits of thought.[1]

Contemporary debates indicate that Eliot's gentlemanly and elite image of science was by no means unanimously shared by the community of scientists at large. However well suited to the objectives of university presidents, Eliot's academic and moral appropriation of science had the effect of institutionalizing the polarization of science at the expense of both philosophical and social reality. Besides resting precariously on the tenuous distinction between pure science and useful science, his representation of science lacked sensitivity to the interconnections between individual activities and societal dynamics. Furthermore, by emphasizing the merits of the activity itself and casually referring to the outcome as "truth," he was bound to encourage the confusion of method with metaphysics, of practice with outcome.

Indeed, this is precisely the confusion that would characterize Bridgman's operational interpretation of physical reality; it would also account for much of its popularity. Nevertheless, when method was elevated to the metaphysical status of knowledge, there could be no truth, only method. For the moment, however, moral and ideological motives prevailed, postponing consideration of the metaphysical questions. If it was supposed that science is true because it is factual, and ethical because it is true, the metaphysical and practical ramifications of scientific knowledge could be disregarded by treating science as a method or attitude, a process of right thinking, rather than a body of universal or useful knowledge.

Bridgman accepted this model of science, and in so doing, he also accepted, without criticism, the ambiguous distinction between pure and applied science. At the same time, this model inspired the idealistic young Bridgman, who had rejected tradition and dogma, and it fostered his belief that the method of science could replace the inherited authority that had been the traditional arbiter of social and philosophical truth. A scientist, Bridgman believed, begins with a neutrally skeptical mind and accepts as truth only those facts grounded in experience. Strictly speaking, a scientist does not *believe* in anything.* To "believe" would be to surrender to unsubstantiated assertions. This is something a scientist does not do. In Bridgman's words, "The first business of a man of science is to proclaim the truth as he finds it, and let the world adjust itself as best it can to the new knowledge."[2] The role of the man of science, according to Bridgman, was to uncover truth in its purest form—empirical fact. The academic physicist is committed to doing pure research, without any intention of aiding utilitarian enterprises. He is engaged in the project of adding to the storehouse of pure knowledge, knowledge for the sake of knowledge.

Nevertheless, as we follow the young Bridgman step-by-step

*Bridgman to Paul Sabine, June 16, 1953: "I. N. Richards, in a lecture at New York University this spring, is reported to have said 'I do not *believe* in anything at all.' He expressed my attitude exactly. Literally I do not *believe* anything."

through the process of establishing his research program, and later, as we note his attempts, as a struggling householder, to supplement his university income by doing consulting, it will become clear just how arbitrary the distinction between pure and applied science was and how difficult it was to faithfully observe whatever one imagined to be the line of demarcation. This was especially the case for Bridgman's research, which depended on commercial and industrial developments in metallurgy. In most cases, in Bridgman's experience, the distinction was enforced by means of a monetary convention. What was done for the sake of pure research did not entail a fee, either from the supplier of materials or from the researcher providing information. A gentleman, of course, does not take money for his contribution to pure knowledge.

At the same time, if Bridgman took for granted the moral and ideological values associated with empiricism, he had no use for its academic or philosophical form. It was as irrelevant to his understanding of the experimental method as it was to the moral and ideological rationale supporting the idea of "pure research." His was a methodologically raw, practitioner's conception of empiricism, lacking the artifice of philosophical sophistication, tuned not to logical problems but to active, pragmatic ones. Only when he came to consider the absolute meaning of primary measurement did he open himself to the philosophical dimension, and even here, practical experimental problems were of overriding concern. It was obvious that no amount of philosophizing could solve the immediate practical problems that faced him as an experimentalist. There was no logical formula that could tell him when his results were "good" and when they were not. He was left to rely on technique and a certain intuitive sense of physical discrimination to make his decisions, and these decisions did not leave room for philosophical subtleties. This is the case for all experimentalists, but it was particularly true in Bridgman's situation, where there were no theoretical guidelines to direct or corroborate his results. As Bridgman described it, once he understood the potential of his new device—the leak-proof seal—he "endeavored to use this new technique to the limit, attacking with it any problems in which the information to be expected from the behavior under high pressures seemed likely to be of significance. This is not the usual procedure in scientific work, in which the problem usually presents itself, and the suitable technique discovered."[3]

Here technique preceded theory. This was empiricism in one of its more rudimentary forms. Bridgman did not set out to test theory, for there was no theory to test. There were, of course, suggestive analogies that could be drawn from the theoretical behavior of ideal gases, for example, but as every physical chemist well knows, these were woefully inadequate as guides for research into the behavior of liquids and solids. What theory Bridgman had consisted of the phenomenological categories of thermodynamics. In his field he was both pioneer and mapmaker. Bridgman's insistence that experiment is prior to theory was not an arbitrary claim. Neither was his assertion that all physical phenomena are unique.

Bridgman was of necessity deeply immersed in what have been called the "tacit" or "craft" dimensions of scientific practice. These include the practical details not accommodated in the logical version of empiricism and frequently omitted in journal reports. He was exquisitely aware of how much rehearsal is necessary and upon what banal considerations the staging of reproducible physical phenomena depends. He never tired of warning against the folly of oversimplifying the world of particulars, especially for the sake of elegance. His habit of personally attending to the details of even the most elementary aspects of experimentation reflected his appreciation of the precious individuality of physical events. "There are," he declared, "too many pitfalls of unanticipated sources of error, which often require ingenuity for their elimination, and which may take much time to discover if one is only watching from the sidelines."[4]

And if any principles of streamlining were to be invoked, they were "considerations of instrumental ease and simplicity,"[5] the experimentalist's standard of simplicity, rather than the formal conceptual simplicity valued by the theoretician. According to Bridgman, simplicity is quite a different idea for the experimentalist than for the theoretician. In fact, Bridgman believed, the reasoning of the experimentalist does not depend in the first place on language or formal symbolic representations. For himself he claimed that the thought patterns guiding his laboratory operations were "motor-visual," not verbal.* It would not be

---

*Bridgman to Bentley, Nov. 28, 1932: "A good deal of this cerebration [of an experimentalist like myself], I find by analysis of my own activity, is apparently divorced from any verbal element, but is almost entirely motor and visual in its character, with reactions when confronting any specific situation almost as definite and clean cut as the words of a conscious language. . . . these motor reactions . . . often rise to full consciousness, and . . . seem to involve recognizable and repeatable elements."

reasonable to assume that Bridgman was prepared to renounce the validity of all formal conceptual categories—at the very minimum he recognized that they enter by means of the measurement process. However, he wished to emphasize his understanding that staging a physical event was not a problem of language, but a physical procedure dependent on physical intuition.

These were all perceptions that he would bring to bear on his philosophical interpretation of scientific knowledge, and that would involve him in a very one-sided controversy with his philosophical adversaries. His conception of empiricism was not the formalistic one held by philosophers. It was based on his experience with laboratory practice, an experience that seemed to him to be richer and more complex than that capturable through the categories and assertions of rational exposition. It is to this we now turn.

Bridgman's original scientific project, begun as a graduate student, was to measure the theoretically predicted dependence of refractive index on volume. This was an example of what he characterized as the "usual procedure" in science. However, as he explained, that goal was soon abandoned for richer prospects.

When my work was started in 1905 it was my intention to study certain optical effects. I had no expectation of reaching pressures anywhere near the limits set by [French physicist Emile-Hilaire] Amagat, since, for one thing, it was necessary to use glass for visibility. After my apparatus was constructed and some preliminary manipulations were made, there was an explosion—something very likely to happen with glass, which is most capricious. This destroyed an essential part of the apparatus, which had to be reordered from Europe; the United States had not at that time acquired its present degree of instrumental independence. In the interval of waiting for the replacement I tried to make other use of my apparatus for generating pressure. While designing a closure for a pressure vessel, so that it could be rapidly assembled or taken apart, I saw that the design hit upon did more than originally intended; the vessel automatically became tighter when pressure was increased, so that there was no reason why it should ever leak. . . . This at once opened an entirely new pressure field, limited only by the strength of the containing vessels and not by leak. My intended optical experiment was therefore dropped; the laboratory wrote off the expense of the replacement part and of the apparatus already constructed, and development of the new field was begun.[6]

Nevertheless, while Bridgman was standing at the threshold of unexplored territory, the rewards of significant new discovery lay yet far ahead. Before him appeared only a promise, obscured by a thicket of practical details. Again, there was no logical formula to tell him how to proceed, to direct the design of his equipment, or to prescribe suitable materials and help in their procurement. And certainly nothing in the world of philosophy could assist in relieving the burden of the uncountable and tedious details he faced, or even guarantee that his efforts would be fruitful. These were all Bridgman's own personal responsibility. For Bridgman, gathering new scientific facts was much more than a collector's expedition.

First, the equipment needed to be redesigned and a more efficient way to produce the pressure had to be created. The original compressor was clumsy and inefficient:

In my early work up to 6000 kgm, the piston (1/2 inch in diameter) was driven by a screw. This screw had a pitch of 8 threads to the inch, and needed a six foot wrench to turn it. Its efficiency was less than 5%.[7]

The inefficient screw compressor was soon replaced by a hydraulic press. But by far the most challenging problems facing him were finding a steel strong enough to withstand the high pressure that he could now produce, and building a reliable device for measuring these pressures.

The search for a suitable steel was tedious and prolonged. The piston transmitting the pressure had to be hard enough so that it would not collapse or crack, and the vessel containing the pressure had to be tough enough to resist rupture. The steel companies were anxious to help and usually supplied samples free of charge or at a nominal cost. The vice-president of Bethlehem Steel, for example, wrote in 1906, "Understanding this is for your personal use in making experiments and is not furnished on order from the University, we will make a nominal charge."[8]

Although Bridgman had the good luck to be starting at a time when advances in industrial metallurgy were producing stronger steels, there were many disappointments. The Baldwin Steel Company wrote, "We do not think, however, that the steel we send you will show any deflection before rupture nor will it be as tough as you would like."[9] At another time Bethlehem Steel sent its regrets "that as we have nothing on stock at the present time suitable for your experiments, it will be impossible for us to accommodate you with the samples requested."[10] And Braeburn Steel Company stated, "We do not make any alloy steel, therefore, samples of our steel would not prove satisfactory for your requirements."[11]

Furthermore, precise characterization of the properties of the steels was not available. In a progress report written in 1909 to Charles Cross, chairman of the Rumford Committee and administrator of the Rumford Fund of the American Academy of Arts and Sciences, Bridgman explained, "Upwards of 60 tests of all the leading brands of commercial steels have been made in the search for pistons producing the pressure. These somewhat wearisome tests were made necessary by the fact that at these high pressures all theories of elasticity fail, and there are no empirical results of value."[12]

The search was not over yet, either, for as Bridgman informed Professor Cole of Ohio State University that same year, the apparatus broke down at 7,000 kg/cm^2.[13] This was a pressure only about two times higher than other investigators had achieved. More work would be necessary before Bridgman could produce the brilliant and novel results that attracted the interest of a surprisingly diverse audience.

It was not, however, just the number of tests that made the procedure so wearisome. Constructing the apparatus was a slow and painstaking job. It required not only the hands of a skilled machinist but also an almost superhuman ingenuity and persistence. For example, when Bridgman learned that the metal tubing he needed was not commercially available, he made it himself, drilling out solid rod. He later described the technique:

The inside diameter of the tubing is 1/16 of an inch, and it is quite possible with a little practice to drill pieces at least 17 inches long. . . . Two essentials in successfully drilling a long piece of tube are to start with the drill accurately central, and to use as homogeneous stock as possible. . . . Great care must be observed that the drill does not become clogged with chips. I have found that it pays to carefully clean out the hole with a swab after drilling no more than 1/8 of an inch. It is easy, if all precautions are observed, to drill a hole 1/16 of an inch in diameter 17 inches long in from seven to eight hours.[14]

It is difficult to imagine how anyone could sustain such a level of concentrated effort on so tedious a task.

Nor was Bridgman's experimental program free from serious danger. The threat of explosion was always present, due, for example, to the instability of glass, to the chemical or mechanical properties of materials reacting under the stress of high pressure, or the mishandling of highly reactive chemicals. This was tragically underscored in 1922, when two researchers were killed in an unfortunate accident, which also severely damaged a laboratory.* Moreover, the leakproof seal itself

*Laboratory notebook, May 19, 1922: "explosion in Dunbar's room—Dunbar and Mr. Connell killed. Good part of laboratory wrecked." The Family Chronicles (pp. 148–

created a unique hazard. In becoming tighter as the pressure was increased, the seal could squeeze off the outside end of a tube or piston, which could then be expelled as a projectile, sometimes reaching a speed comparable to that of a rifle bullet. Bridgman called this the "pinching off effect."[15] According to Charles Chase, Bridgman's machinist, it was Bridgman's habit "to duck when walking by the apparatus when the pressure was on." Many anecdotes circulating among Bridgman's friends and acquaintances are variations on this reminiscence.[16]

Nevertheless, Bridgman's ability to produce pressures higher than those achieved by his predecessors, while a remarkable accomplishment, was not in itself sufficient grounds upon which to build a scientific research program. The scientific value of his technique depended upon being able to *measure* the pressure accurately and reliably. Therefore, Bridgman had to turn his attention to designing a pressure gauge. The problems of measurement in a pressure range never before experienced were the subject of Bridgman's Ph.D. dissertation, "Mercury Resistance as a Pressure Gauge," submitted in 1908 and signed by John Trowbridge, Wallace Sabine, G. W. Pierce, B. O. Peirce, H. W. Morse, H. N. Davis, and Theodore Lyman. This material was subsequently refined and published in the *Proceedings of the American Academy of Arts and Sciences* in 1909 as three successive papers.[17] The research presented in these three papers constituted the essential groundwork for his later accomplishments, but not until 1914 did Bridgman feel confident enough of his position to publish a summary article describing his technique in detail.

Indeed, one searches in vain through the earlier papers for more than just a brief mention of the packing that was of such central importance. Bridgman appears to have been reluctant to disclose its design. As early as 1908, the president of Watson-Stillman, a firm dealing in

---

49) give a fuller description: "Half the windows on the north side were fractured, the middle section of the basement was completely wrecked. Dunbar was blown to pieces, and the carpenter Mr. Connell was instantly killed. Eight students in the room above were injured, the floor being lifted bodily, and a heavy dynamo turned over onto some of them. The accident was in Dunbar's room, where he was pumping oxygen into a heavy tank. There may have been oil in the tank, which is the official explanation published in the papers, but P. W. thinks more likely a suggestion of Baxter's, got from van der Pyle, that Dunbar had not completely removed the methane on which he was previously experimenting. There seems no other way of accounting for the explosion of the gas balloon which was outside the building, and which must have exploded or else the windows in the law school [the law school is situated across the yard from the physics building] could not have been shattered as they were."

hydraulic equipment, inquired, "What kind of a packing do you expect to use and retain 15,000 pounds to the inch?" There was no reply. Two weeks later he repeated his inquiry: "We would be very much pleased to know what form of packing you are using that you are having such satisfaction with up to pressures 100,000 pounds and over."[18] Again Bridgman made no reply.

In February 1912, John Johnston, a physicist at the Geophysical Laboratory, commented on this lack of information. Referring to articles published in the *Proceedings*, he wrote, "In your measurements of the volume change on freezing you speak of the packing of the moving piston being free from leak throughout the pressure range, but so far as I can discover, you do not give a description of this packing." Incidentally, he suggested that Bridgman should publish his results in "some place more generally accessible (especially to Europeans) for example, Ann Physik or perhaps better, Z. anorg. Chem."[19] Bridgman must have agreed to disclose his technique to Johnston provided that he would keep the information confidential, since this letter is followed by another in which Johnston says, "I am much obliged to you for letting me know the details of your packing, which I can see is very simple and efficient; and shall be careful to respect your wishes as to keeping it quiet and preserving your letter."[20]

Yet Bridgman denied that he was purposely withholding a complete description. Only three months later when Jerome Greene, general manager at the Rockefeller Institute for Medical Research, wrote asking if Bridgman would set up an apparatus for experimentation at the Rockefeller Institute, he politely broached the issue. "I should not want to press the latter alternative as against any disinclination of yours to let the work pass out of your control at this stage. I suppose, however, that sooner or later you will publish your technique. Have you discussed with Professor Sabine the idea of patenting your apparatus?"[21] Bridgman's response was to say, "With regard to setting up an apparatus for you for the high pressure work, I have no objection at all to imparting the technique to others. The reason that I have not published it as yet is because of lack of time as much as anything else."[22] Or, possibly, as Greene had suggested, Bridgman had not yet decided whether he wanted to patent his device.

Nevertheless, the work described in the three 1909 papers on pressure gauges is, in all other respects, characteristically thorough and detailed. It is marked by the meticulous attention to methodological particulars so typically Bridgman.

The task of constructing a suitable measuring device was much more subtle than simply selecting a primary standard and marking off intervals along an extended or extrapolated scale. Contrary to popular assumptions about establishing a measurement procedure, Bridgman was free neither to choose his standard arbitrarily nor to calibrate his reference scale in simple multiples of a primary quantity in advance of empirical determinations. He was constrained by both formal and practical contingencies.

For the construction of a primary gauge, the subject he discussed in the first 1909 paper, Bridgman had to select a physical mode of measurement that would conform to a formal, theoretical requirement, namely, the mechanical definition of pressure. In other words, he had to arrange his experimental apparatus in such a way that the measurable empirical parameters could be translated unambiguously into the theoretical categories relating mechanical motion to the definition of pressure. He could satisfy this requirement with the free-piston gauge, whose response to pressure changes was simple mechanical motion in one dimension. Once he had met this requirement, he could then construct a secondary gauge that correlated the values of a conveniently (not necessarily linearly) varying physical property with the corresponding primary or directly determined values of pressure. The electrical resistance of mercury, Bridgman discovered, was just such a property, and the construction of a pressure scale based on the variation of resistance with pressure was the subject of the second paper.

The procedure, however, in either case was not simply a matter of recording and tabulating the direct values of measurements. The empirical readings had to be "corrected"; that is, a compensation had to be made for the imperfections of actual experience, imperfections whose effects could not, in general, be known *a priori*, but that still had to be eliminated for the sake of accuracy. Thus, any interfering influences that might produce deviations from "ideal" behavior had to be either calculated away or minimized procedurally. In the case of the primary free-piston gauge, for example, corrections had to be made for the distortion of the piston, the stretching of the cylinder, and even for the liquid that might leak into the minute space between the cylinder and the piston, which could be as small as 1/16 inch in diameter. Bridgman often conceptualized experimental corrections such as these in terms of convenient physical equivalencies imagined in such a way as to maintain dimensional consistency. For example, he pretended that a leak past the end of the piston was the same as an addition to the effective area of the piston.[23]

Upon a little reflection, however, one might begin to wonder what could be meant by a "correction" to a primary, or direct, measurement. Corrections determined against what standard? What, indeed, is a "mistake," given that every empirical outcome is true, if only by virtue of having happened? Bridgman's answer to this problem provides additional insights into his views on the foundations of science. Corrections, Bridgman noted in his 1931 work, *The Physics of High Pressure*, are made against another independent measurement (or set of measurements, he might have said). "Indeed," Bridgman pointed out, "there is no primary gauge in the strict sense, for corrections, even though slight, which demand an approximate knowledge of the pressure, must be applied to the readings of all such gauges."[24] On the one hand, Bridgman was saying, the validity of a given set of measurements depends on being able to reach the same results by another method. We trust our measurements if independent methods produce convergence. On the other hand, all results still need to be similarly "corrected." (Bridgman seems not to have noticed that his statement about corrections implied yet another standard—a theoretical one—to which the experimenter can refer to know what corrections must be made.)

With the secondary gauge, then, corrections also had to be made, and precautions taken to filter out effects that were not due exclusively to the influence of pressure on the property being observed. In the case of the mercury resistance gauge, Bridgman found that impurities in the mercury might alter the resistance and that too great a current in the resistance circuit might produce unwanted heat effects. Determining how much variation was acceptable without compromising accuracy was a practical and personal decision, not a formal one, and it relied very much on a refined physical intuition cultivated through the accumulation of laboratory experience.

Furthermore, since Bridgman had to consider that the mercury resistance gauge itself was not immune to the compressive action of the pressure, he had to determine the compressibility of the mercury and of the glass from which the capillary tube containing the mercury was constructed. He found that the characteristics of the glass were a significant source of error. Measurement was further complicated by the ability of mercury to form an amalgam with the steel containing vessel. The compressibility problems he treated separately in the third paper. The investigation of the action of mercury on steel was not published until two years later.[25]

The mercury resistance gauge was not a device which was calibrated as, say, a thermometer, where the pressure readings, in the same manner as temperature readings, could be taken immediately from the device itself. The actual calibration was done on paper. The relationship between pressure and electrical resistance was first represented as a set of points on a graph, and a smooth curve was drawn through these points. Then an empirical formula—a custom-made, specially constructed mathematical expression—was fitted to the curve. In subsequent trials, either the graph or the empirical formula could be used to determine the values of pressure corresponding to the measured values of resistance.

The use of mathematics as an empirical tool differs considerably from its theoretical use. Empirical formulas such as the one Bridgman employed are generally derived by curve-fitting techniques; the set of experimental results is matched with the expression that can most closely reproduce them. This method has a strong element of trial and error. It involves introducing arbitrary numerical constants and discarding data judged to be erratic; both acts depend on the form of the curve the investigator chooses to represent the relationship.

In fact, "derivation" in this context does not connote the deterministic logical thrust of a formal theoretical derivation. It looks more like a practical adjustment to the facts, something like the way a tailor cuts a pattern to fit an individual body. Furthermore, there is always inherent in the results an unavoidable "fuzziness" or margin of error. Not only is each experimentally measured value surrounded by a statistical "halo" of error, but also the mathematical formula or "smoothed out" curve seldom ever touches all recorded experimental values exactly, even within the boundaries of error. Indeed, it may touch none of them, and speculation upon the philosophical question of which might be approximate, the curve as a whole or the individual experimental values, would be futile. In a purely empirical situation, there is no alternative but to accept the results as thus determined, error and all. The influence of this practice on Bridgman's scientific philosophy as well as upon his attitude toward mathematics will become clear in subsequent chapters.

Moreover, the empirical technique is not limited only to measurement procedures, and it is common practice to extend it to what might be called empirical calculation. Therefore we should not be surprised at Bridgman's use of graphical techniques to "derive," or construct, higher-order thermodynamic quantities. Roughly speaking, he interpolated, or "read," data from his primary empirically fitted curves and used these data to graph other thermodynamic relationships of interest to him. This process could be repeated as many times as he was confident that his interpolations would not introduce an unacceptable amount of error into the results.[26] The procedure is not formal in the logical sense, but it is regular or systematic in the empirical sense and is accepted laboratory practice.

In sum, as the foregoing has illustrated, practical matters consumed Bridgman's time and energy. His attention was dominated by the demands of immediate laboratory exigencies, where philosophical considerations were of negligible interest or relevance. The questions of importance were those related to getting his research program under way, to getting things to work. What liquid was best for transmitting the pressure? One might freeze too soon; another might be chemically reactive. What was the effect of temperature? Some impurities made a difference; some did not. Crucial parts of the apparatus, Bridgman found, had to be "seasoned," or submitted to preparatory pressure, before they would behave in a reproducible manner. And a flaw in the steel was ruinous.

Adding to the difficulty of his task was the failure of Bridgman's university education to have provided him with the technical background his research program required. For this, he was left to his own initiative, and he learned the necessary metallurgical techniques as he went, often from the industrial engineers with whom he established contact. As we have seen, a symbiotic relationship with industry characterized Bridgman's scientific project from the very beginning. (Indeed, it was not uncommon to find university-educated physicists working in industry and government.) Bridgman communicated extensively with commercial and industrial firms as well as with government agencies, taking advantage of whatever materials and knowledge he was able to obtain, and they, in turn, solicited from him what new information he could provide. Early on, however, Bridgman learned more from industry than industry learned from him. For example, from Mr. Fielder Clarke of the Firth-Sterling Steel Company, himself a former Harvard man, Bridgman learned about the properties of alloys and the technique for hardening steel.[27] Marcus T. Lothrup, metal-

lurgical engineer for the Halcomb Steel Company, was especially aggressive in suggesting improvements to Bridgman's methods for testing steel samples. He criticized Bridgman for the lack of uniformity of
his samples and for not considering the processes by which they were
hardened and drawn. He also questioned the results Bridgman had obtained and recommended a recalibration of the gauge used. He reminded Bridgman that "all results in which test pieces bend under
compression represent not a maximum strength, but a combination of
several stresses." [28] Bridgman, in Lothrup's opinion, should have been
more careful to avoid oversimplification.

Soon, however, the professionals began to learn from Bridgman.
This is well illustrated in his correspondence with Dr. Tuttle, melter
and refiner of the United States Mint in Philadelphia. The occasion was
an attempt to procure samples for the study of the electrical properties
of metals under pressure. Although it would be some time yet before
Bridgman would turn his full attention to this problem, he was already,
in 1909, starting to inquire about the availability of samples for testing.
It was imperative that these samples be metals of the highest purity
drawn into the form of fine wire. Suppliers, however, often did not
stock metals of the required purity. Moreover, wire made by being
drawn through steel dies was likely to be contaminated by the process.
For this reason Bridgman asked that the wires be drawn through jeweled dies instead.

When in 1909 Bridgman wrote to Tuttle requesting samples of pure
gold and silver wire, he indicated his concern about the problem of
contamination. Tuttle replied, modestly, but nevertheless declaring his
credentials—B.S. Harvard, 1855; Ph.D. Göttingen, 1857—by suggesting that perhaps Bridgman was mistaken. From his experience, Tuttle
claimed, "steel picks up gold more readily than gold picks up iron." [29]
He must have tested this assertion, however, because a month later
Bridgman received a letter from Tuttle admitting that the "objections to
steel dies are well founded." [30]

In fact, it had been only a short time earlier that Bridgman had
learned wire could be drawn through jeweled dies. He made no attempt to hide his pleasure when he acknowledged receipt of this information from the New England Electrical Works. "It is a very great satisfaction to me that this work can be done as easily as your remarks lead
me to suppose; I had expected that it would be almost impossible of
accomplishment." [31]

As it happened, however, the work would be harder and more expensive than Bridgman was led to believe. And when in 1915, the drawer for the New England Electrical Works complained about the difficulty of meeting Bridgman's specifications—.003 inches in diameter for silver, and .004 for gold—the president of the company felt it necessary to explain the situation to Bridgman.

He says that he will do the best he can with them but will not guarantee to get the special stuff down to the sizes specified; the sizes are so small that there may not be anything but scrap before the required size is reached.

We have written him to save even the smallest pieces and return all the scrap. . . . He expects to charge us 60 cents per hour for the time he puts on the job. This is very reasonable for this kind of work.[32]

In the meantime, Bridgman had set his mind to improving his apparatus. Already in 1909 he had reported to Charles Cross that the original hydraulic ram had been replaced by one with six times the capacity.[33] The problem was that the steel used at the time would not support pressures higher than 7,000 kg/cm². However, with use of Krupp nickel steel (later replaced by electric furnace chrome-vanadium steel), he was able to extend the measurable pressure range up to 20,000 kg/cm². But since his measuring gauges had been constructed for much lower pressures, they were of no use in the higher pressure range. This meant that new ones had to be devised.

Again the work resulted in a trilogy of papers, published in 1911 in the *Proceedings of the American Academy of Arts and Sciences*, and again the fundamental problem was measuring technique. Bridgman had designed a new free-piston gauge that was operable up to 13,000 kg/cm². But more significantly, he had developed the manganin resistance gauge, which permitted accurate measurement of pressures above 13,000 kg. (Manganin is an alloy composed of copper, manganese, and nickel.) The manganin resistance gauge was compact and easy to manipulate, and the manganin did not need to be enclosed in glass as did the mercury. There were, however, a few disadvantages. The gauge required a long process of seasoning; sometimes as much as a month of pressure applications was necessary before its behavior remained consistent. Moreover, the manganin gauge was not readily reproducible. Each piece of wire had to be individually calibrated.

But the ultimate property that recommended its use at higher pressures was the linearity of the relationship between pressure and electrical resistance. This allowed Bridgman to be confident that not too much accuracy would be sacrificed if the gauge were used to measure pressure by extrapolation in the range beyond the reach of the absolute gauge; that is, in the range where no corroborative information was available to validate the calibration. If the relationship between resistance and pressure continued to be a linear one, the 20,000 kg limit was imposed by the steel, not the gauge.

"Of course," admitted Bridgman, "the use of any standard by extrapolation is undesirable, but at present any means of measuring these very high pressures with probable accuracy is welcome." Extrapolation, Bridgman knew, meant going out on a limb. It meant trusting that the behavior of the system would continue to be the same under extended conditions. Obviously this could not be guaranteed, but Bridgman reasoned that "in any event, the extrapolation from 12,000 to 20,000 is very much less than the extrapolation from the previous maximum of 4000 to 12,000, which is here shown by actual experiment to be justified." [34]

At this point, Bridgman was ready to proceed to the next phase of his research program. The preparations had finally been completed. He was now able to explore the properties of matter under very high pressure and trust that his new facts would be reliable and accurate. Thus, the second paper of the trilogy was a report of his study of the transition of mercury from the liquid to the solid phase under pressure, and the third reported a similar study of the liquid-solid phase transitions for water. It was this work on water that led Bridgman to discover the previously unknown forms of ice, a discovery that brought him popular, as well as scientific, attention.

More than a hundred pages long, "Water, in the Liquid and Five Solid Forms Under Pressure" was impressive in its wealth of detail, all of which Bridgman included so that "any one could, if he wished, check the computations for himself." "Every one of the original observations," he claimed, "except those marred by obvious accidents has been given." [35]

The content of this work underscored the inadequacy of existing theory to provide either predictive or explanatory support for his results. "All present theories of liquids are incompetent to explain known facts, and as for the theory of the equilibrium liquid-solid, even the fundamental facts are unknown." [36] To be sure, speculation based on an analogy with liquid–gas equilibria could be used to suggest a very general direction for investigation, but even the qualitative nature of the actual phenomena repudiated the possibility of much similarity. For example, although conditions can be found under which the transformation from liquid to gas is continuous and insensible (the critical point), Bridgman found no such stage in the solid–liquid transition. This change was always phenomenologically abrupt, marked by a discontinuity. Similarly, thermodynamics was of no help in locating the regions of pressure and temperature where allotropic forms might exist. Therefore, just as in a geographical exploration, the maps had to be made as the expedition advanced.

However, although Bridgman was now able to specify his procedure clearly and demonstrate its potential, the labor involved was exacting and too time-consuming for one person alone. He needed some assistance. "The nature of the work," he wrote in a grant application, "is such that the readings must be made continuously over intervals usually a day long, and for such work as this, particularly when it is considered that I have teaching duties in addition to those of pure research it is evident that at times an assistant is almost absolutely necessary." [37] With grants from the Bache Fund of the National Academy of Sciences and the Rumford Fund of the American Academy of Arts and Sciences he was able to secure enough money, about $1,000 per year, to hire a mechanic and purchase supplies.

Bridgman was especially fortunate that the National Academy and the American Academy were of a mind to provide financial assistance, for in spite of the efforts on his behalf by John Trowbridge and E. C. Pickering, the director of the Harvard Observatory, the Carnegie Institute, recently established with the professed goal of searching out and supporting "the exceptional man," was not as accommodating. Twice, Trowbridge and Pickering were rebuffed. Pickering submitted the first application on the recommendation of Trowbridge when Bridgman was a graduate student. It was refused on the basis of a policy decision not to "give aid to students in colleges and universities who are pursuing work for higher degrees." [38] Besides, the institute complained, it was swamped with too many appeals. The second application, also submitted by Pickering upon the request of Trowbridge, was made after the value of Bridgman's research had been clearly established. It received a response warm in praise of his work, but nevertheless a negative one. This time the reason given was that the institute was trying to avoid entanglement with the business of education:

Our experience forces us to draw the line sharply between men who are primarily devoted to research and those who are primarily devoted to the work of instruction. Any attempt on our part to play the role of paternalism for the benefit of educational institutions would clearly result unfavorably for all concerned. . . . We are not disposed, therefore, to appoint any man connected with an educational institution to a Research Associateship unless he is able to give all or nearly all of his time directly to research. Moreover, we cannot afford to have our annual budgets and their construction conditioned by or in any wise tangled up with those of educational institutions.[39]

The circumstances surrounding Bridgman's research, it seems, were not pure enough to qualify him for support from the institute.

If we are looking for factual evidence of the social and scientific deceits inherent in the idea of "pure research," we hardly need look further. The institute's reasoning was based on a deliberately narrow conception of what activities contribute to scientific development, and its elitist assumptions were made even more obvious when it later confessed its annoyance at the tendency of so many to "confound research with education, with invention, with patent rights, copyrights, and the necromancies of 'geniuses' and 'male witches.'" All this could be put aright if there were "a considerable number of institutions and hence individuals committed to the business of research, and consequently responsible to themselves and to society for the application of funds designed not to give poor boys and inventors special advantages but to attack obvious problems in research with continuity of effort and intelligent appreciation of well known effective methods."[40] To be fair, it must be recognized that the Carnegie Institute was attempting to set and enforce some standards for scientific excellence above and beyond the common notion that any sort of amateur tinkering was equivalent to doing science, but, even so, its reasoning seems unduly restrictive.

On the other hand, Bridgman evidently felt ill at ease with the thought of asking for what to him must have seemed a gratuity. His discomfort is indicated by the apologetic manner in which he requested a renewal of his grant from the Bache Fund in 1915. "If it is not too grasping," he said, "I should like to make application for another such grant for this year."[41] He must have made a similar remark to the Rumford Committee, since Charles Cross felt it necessary to assure him, "You need not fear wearing out your welcome so far as the Rumford Committee is concerned although we may not always be able completely to meet your wishes."[42] The specific incident that may have prompted these expressions of unworthiness is unknown. They do, however, serve to call attention to Bridgman's personal aversion to taking more than his due. To be sure, a gentleman was not in the habit of expecting gratuities, but Bridgman's reluctance, given his shyness and his strict Puritan upbringing, was more likely to have originated in a feeling deeper than mere gentlemanly pride, more akin to a personal embarrassment in asking for favors, a feeling engendered by a belief that one is due only what he has earned.

By 1914 Bridgman had established a distinctive career pattern, and the work from here on would build on the foundation he had so carefully laid. Bridgman, now an assistant professor of physics, had mastered his technique and achieved recognition as an experimentalist of international stature. Judging by the papers he published that year, 1914 was the culmination of a period of high productivity. He reviewed his technique, describing systematically the procedure for high-pressure experimentation. He presented more studies of melting curves and on the thermodynamic properties of liquids. He had also discovered that phosphorus was capable of existing in previously unsuspected allotropic forms, one of which was stable under ordinary conditions; that is, it did not revert back to its previous form after the pressure was released. The results of the studies on water had been translated into German and published in both *Zeitschrift für anorganische und allgemeine Chemie* and *Zeitschrift für physikalische Chemie*. So confident was Bridgman of the accuracy of his measurements that he challenged the correctness of work done by Gustav Tammann, a prominent German physical chemist and an established authority in the field.

Bridgman had passed through the initiation rites and was now an accepted member of the international scientific community. He had won a place in the elite and gentlemanly world of pure academic science, a place his subsequent research would continue to strengthen. Even more significantly, on the foundations of his work alone the new field of high-pressure physics was established.

Nevertheless, we would be guilty of a serious oversight if, in celebrating Bridgman's achievement in the field of pure research, we failed to acknowledge the contributions made by commerce and industry. Bridgman's research program was not, after all, a Socratic inquiry or a Victorian fossil-hunting expedition; it drew heavily on the resources made available to him through a strong industrial presence and a vigorous commercial interest in scientific production. Although he, perhaps, had no immediate utilitarian goals directing his experimentation, he certainly benefited from the results of application-oriented research—techniques, and especially materials—without which his own research could not have succeeded. Indeed, in addition to the extensive contacts with steel companies, Bridgman maintained an ongoing communication with manufacturing firms such as General Electric, Westinghouse, Fansteel Products Company, and New England Electrical Works, soliciting both technical advice and materials.[43]

Nor was Bridgman lacking in entrepreneurial spirit. He did not disdain selling either his skills or their concrete products, although he occasionally complained about the time taken away from more personally rewarding pursuits. But as a young husband and father supporting the social and cultural activities of his family, and a Harvard professor who was expected to conform to a lifestyle beyond his basic financial means, the simple reality was that he needed the money. The following examples, which are not exhaustive, illustrate the scope and nature of his undertakings.

Early on in his scientific career, for instance, Bridgman conducted a study of the potential problem of submarine signaling for the Submarine Signal Company of Boston; for this he was paid a fee of $800. Later, during World War I, while employed by the Special Board on Anti-Submarine Devices of the United States Naval Experimental Station at New London, Connecticut, he invented a vibration-absorbing mounting, which he subsequently patented and offered for commercial use to the Stevens Sound-Proofing Company in Chicago and also to the Korfund Company in New York, a firm specializing in vibration-damping problems. It appears that neither company went beyond expressing some initial interest in the device. The Korfund Company, in particular, thought its own devices superior insofar as they required less maintenance and were thus more practical in situations where maintenance was not given much attention.[44]

From November 1919 to February 1921, Bridgman was under contract to the Watertown Arsenal (Massachusetts) to determine the cubic compressibility of steel samples. He thought this job a bit dreary and would have preferred working in his own laboratory, but his expenses made the sacrifice necessary. (During World War II, the Watertown Arsenal would again call on him, this time, to study the properties of armor plate penetrated by projectiles. He felt scientifically stifled by bureaucratic practices and did not hesitate to say so in his reports to the

National Defense Research Committee.) In addition, Bridgman's early experience with the behavior of thick cylinders stretched beyond the elastic limit suggested to him a method for the one-piece construction of gun barrels, a technique successfully tested by the Navy during World War I but not adopted. However, he made it clear in an article published after the war in the journal *Mining and Metallurgy* that he saw no reason "to think that the technique should not be successfully applied on a commercial scale."[45]

When the Russ Manufacturing Company of Cleveland wrote to him about a problem with an evaporator in a refrigeration system, Bridgman's reply was accompanied by a bill for $50. Standard Development Company (Standard Oil) paid him $100 a day plus expenses for consultation about testing high-pressure bombs to destruction. To Rheinische Metallwaaren und Maschinenfabrik in Dusseldorf and to Schneider and Cie in New York, he prepared and sold calibrated manganin pressure gauges, charging $50 for the first and $30 for each additional one.[46] From Mr. C. H. Greenewalt of du Pont de Nemours, who wanted to get together with him to "talk over high pressure problems," Bridgman requested clarification what Greenewalt meant by "talking over" his problems, that is, whether his goals were commercial or purely scientific.

I am not sure from your letter just what should be the basis for our conference. If the interest in your proposed experiments is commercial, then I think we ought to proceed on a consultation basis, for which I should make a charge of $100 per day. If, however, the interest in your experiments is purely scientific, as for example some of the work done at the research laboratory of the General Electric Co. in Schenectady, then of course I would be glad [to] give what help I can.[47]

In his reply, Greenewalt noted,

The boundary between fundamental and commercial research is a very broad one and I would hesitate to make the distinction in this or in any of the problems we are undertaking. We will certainly use the data we obtain in the design of commercial equipment and yet we expect to make our determination much more accurately than is necessary for such purposes. In any case, however, I would be very reluctant to ask you [to] make us a present of your time, and I have therefore obtained authorization to place our consultation on the basis suggested in your letter i.e. $100 per day.[48]

Bridgman insisted on drawing a distinction, however indeterminate, between pure and applied scientific knowledge, as if instances of each category could be tallied in separate columns on a ledger: pure knowledge was free; useful knowledge cost money. Moreover, laboratory time was not to be taken up with projects of commercial interest. It was to be given only to subjects of purely scientific interest. This was the reason, Bridgman explained to George H. Lee of Omaha, Nebraska, that he had not undertaken the attempt to make diamonds on University time. However, he had no objection to pursuing the matter independently. Thus, in 1916, when Lee, an inventor and dealer in poultry and stock supplies, suggested that he and Bridgman collaborate on a diamond-making project, Bridgman responded positively. His letter is revealing. "Dear Sir," he wrote:

The ruptured steel cylinder which you sent me arrived safely, and I have examined it with interest. It certainly is evidence that you have reached rather high pressures. I would be much interested to learn in what way you have overcome the difficulty of leak, as this was a matter which caused me much difficulty in my own preliminary work. I have found at least one way of overcoming the difficulty; it would be of interest if you have adopted the same method, or if there are other methods. . . .

With regard to the matter of crystallizing carbon, I have been much interested in this matter for some time, but have not been able to make any experiments on it myself. The reason for this is that I have not thought it fair to use the resources of the Laboratory in investigating a question whose solution is a matter of commercial rather than purely scientific interest. As far as I can see at present, the solution of this problem would not materially advance our knowledge of the properties of matter in general, but would be a contribution only to our knowledge of the behavior of one particular element.

I have nevertheless felt that the problem was one which can certainly be solved, and doubtless will be solved by the application of high pressures. It has been a matter of considerable regret to me that my present position did not justify me in attacking a problem whose solution would bring such great rewards.

Since this is a purely commercial matter, the problem should be undertaken by some concern as a commercial proposition, with the clear understanding that it involves rather more of the speculative element than most purely commercial propositions. I have even gone so far as to get certain parties interested in this proposition and a scheme was drawn up by which the backers of the scheme were to contribute the money for experiments and I was to contribute my experience with high pressure manipulation, and any profits were to be divided in an equitable manner. This scheme fell through however, because certain other inventions in which these same men were interested turned out to be more profitable and demanded much attention.

So far as I am concerned, I am entirely free to take up such a proposition again, and it has occurred to me that possibly you might be willing to consider such an arrangement. It seemed to me that your own great interest in the scientific aspects of the problem, apart from the commercial aspects, would make you the more interested.

May I repeat again my absolute conviction that the problem is not only one capable of solution, but that some one is going to solve it, and solve it soon. Why should not we be the ones to do it? If the problem is one that can be solved by the aid of high pressures, as seems most probable, I think I may fairly claim, and my reprints will bear out this claim, that I have a much better chance than any one else anywhere of reaching the solution.[49]

For reasons unascertainable, this proposed collaboration never materialized. The Family Chronicles also mention one other such scheme —a project camouflaged, why, we are not told, under the name "isotopic tungsten."[50] The details and outcome are not given. It would not be until 1941, when three industrial corporations, General Electric, Carborundum, and Norton, joined hands in the effort to create artificial diamonds, that Bridgman finally had the opportunity to become seriously involved in the diamond-making project.[51] All three companies were motivated by a desire to have an independent source of industrial-grade diamond abrasives and to free themselves financially from the arbitrary pricing of their South African supplier. Through the intervention of Zay Jeffries, a metallurgical consultant to General Electric, it was arranged for Bridgman to work with General Electric on the hunt for the artificial diamond.

However, the diamond proved more elusive than expected, and carbon would not respond to Bridgman's straightforward method of applying heat and pressure. A decade later, the research team at General Electric learned that the key was chemical catalysis. In the meantime, the effort was judged a failure. After five years of experimentation, an executive decision was made to halt the project. The sponsors had become weary of watching "the professor" turn diamonds into graphite, but not graphite into diamonds. Their money was being spent on making a cheap product out of expensive material. Of course, the professor did not see it that way. There was much valuable scientific knowledge to be had, even if there were no diamonds, and the technique was at hand. But the sponsors wanted diamonds, not scientific data. The consortium split up, and the project was ended. The professor went home.*

---

*Eventually, through the concerted efforts of a team of physicists and chemists dedicated to producing diamonds, General Electric achieved success. Bridgman's method had been the starting point. High pressure was indeed an essential factor. But the process for

In Bridgman's collaboration with industry, one would be hard-pressed to find anything in the science itself that distinguishes "applied" research from "pure" research. Regardless of the locus of the research or its generality, the interaction of the researchers with natural phenomena remains the same, as do its supposed moral benefits. The production of the artificial diamond is of no less scientific value for having occurred in an industrial setting or for having been the result of research on a specific reaction. After all, Bridgman considered his work on water and phosphorus to be pure research but hoped that General Electric might have a commercial use for his newly discovered allotropic form of phosphorus.[52] Because "the professor" was willing to be satisfied with something less than reaching the target, because he wanted to explore other possibilities within the reach of a technique that had been developed with a single goal in mind, does this really distinguish his approach as being more dignified or virtuous?[53]

Perhaps Bridgman's concept of "pure research" owes something to the traditional view that a morally uplifting activity is supposed to have a divine object—for science, nature as God's creation. However, in the twentieth century all that remained of this discredited natural theology was a meaningless residue. Truncated and rationalized through the medium of the Puritan work ethic, it became an ideologically redeemed methodology deprived of its original morally justifying objective—a means with no end to justify it or, rather, a means that justified the means.

It is difficult to see how Bridgman's criteria might be used to differentiate pure and applied science. In either case, someone has to pay for it. If the goal makes the difference, how is having no particular goal or a generic one better than having a specific, utilitarian one? In any case, specific and general are probably matters of context or perspective. If it

---

making diamonds was chemically more subtle than his experience with single-element pressure-temperature phase transitions could have taught him. The brute heat and squeeze approach, guided by mechanical intuitions, ignored the need to create a favorable chemical environment for crystallization. The General Electric researchers found that this environment required, besides high pressures, the presence of a metal such as iron or nickel to act as a solvent and catalyst, and that the metal-carbon mixture had to be heated above the melting point of the metal solvent and then cooled at a rate favoring the formation of diamond rather than the metallic carbide. Even after initial empirical success, it was many years before the process was understood in principle. The development of the diamond-making process into a commercially profitable venture was the next challenge for General Electric. That is another story. George Wise, "Research and Results: A History of the GE Research and Development Center, 1945–1978" (Unpublished ms., 1982, on file at GE Research and Development Center, Schenectady, N.Y.).

is the scientific activity itself—the work—from which the moral benefits accrue, why should it matter if the goal is utilitarian or not? By this criterion all scientific research is "pure." Perhaps the only difference, other than a certain remove from the final application, has to do simply with profits, and money (later, add national defense) is the distinguishing criterion. To state the same point more generously, if the facts discovered in research are intended to earn money, the research is "applied," and if these facts, no matter how they are eventually used (and surely they are meant to be used), are given away free, it is "pure."

Nevertheless, however ambiguous the moral status of pure research may be, the efficacy of the ideal cannot be disputed. Bridgman's scientific research program stands as proof and exemplar. And at the heart of his science was his methodology. Having no theoretical framework to function as a supporting structure, he had to rely on the strength of his method to ensure the trustworthiness of the results. This meant that he had to control all facets of the problem under consideration. Nothing could be casually left to chance. Bridgman's close attention to detail and concern with even minute corrections were directed precisely toward this goal. And irrespective of any qualifications his faith was not misplaced, for the merit of his scientific achievements rested directly on the enduring value of the fruits of this methodology.

For the exceptionally high caliber of his experimental work, Bridgman was nominated in 1917 and again in 1933 for the Nobel Prize, an award he would not receive until 1946.[54] This well-earned honor came his way for a sustained program of experimentation: "for the invention of an apparatus to produce extremely high pressures, and for the discoveries he made therewith in the field of high-pressure physics."[55] The presentation underscored the empirical nature of his accomplishment:

Professor Bridgman. In awarding you this year's Nobel Prize for Physics, The Royal Swedish Academy of Sciences desires to express its unreserved acknowledgment of your outstanding pioneer work in the field of high-pressure physics. By means of your ingenious apparatus, combined with a brilliant experimental technique, you have, by your intense research work and the resulting manifold and remarkable discoveries, very greatly enriched our knowledge of the properties of matter at high pressures.

On behalf of The Royal Swedish Academy of Sciences, I congratulate you on your important and successful work in the service of science.[56]

In his Nobel lecture, Bridgman reviewed his forty years of achievements in high-pressure research. From successfully surpassing the

3,000 kg/cm^2 limit established in the nineteenth century by French investigators, he had gradually extended the range of pressures to 100,000 kg/cm^2 and even to 500,000 kg in special cases. Each stage of advancement had required innovations in the design of the equipment, which in turn involved the exploitation of both general physical principles and the particular properties of the materials being used. Of especial importance in creating new possibilities for research was the development by General Electric of Carboloy, a sintered aggregate of tungsten carbide held together with cobalt. By far the most scientifically productive extension of pressure range was made when Bridgman used Carboloy for the pressure-transmitting piston together with various methods of providing external support for the pressure-containing vessel.[57] Overall, Bridgman carried out countless measurements of the physical properties of matter under varying pressure conditions. His results, many yet to be theoretically assimilated, are of interest to investigators in fields ranging from solid-state physics to geophysics and cosmology.

Neither elegance of concept nor theoretical innovation was responsible for Bridgman's unique scientific accomplishment. Rather, his achievement was the result of a progressive, cumulative program of pure research, built upon tens of thousands of hours of dedicated hard work and attention to detail. That he was aided by advances in industrial technology, or even that this highest recognition awaited the development of a new discipline—solid-state physics—does not detract from the indispensable importance of the meticulous, time-consuming labor that Bridgman invested in his scientific pursuit of truth or from the distinctive value of his contribution. In the end, his concept of science was vindicated. Bridgman's scientific career, as publicly acknowledged and honored, testified to the fruitfulness of the scientific ideal of pure empirical research.

CHAPTER 3

Theory and the Generation Gap: Explaining Metallic Conduction

Those who have handled science have been either men of experiment or men of dogmas. The men of experiment are like the ant, they only collect and use; the reasoners resemble spiders, who make cobwebs out of their own substance. But the bee takes a middle course: it gathers its material from the flowers of the garden and of the field, but transforms and digests it by a power of its own. Not unlike this is the true business of philosophy; for it neither relies solely or chiefly on the powers of the mind, nor does it take the matter which it gathers from natural history and mechanical experiments and lay it up in the memory whole, as it finds it, but lays it up in the understanding altered and digested.

— Francis Bacon, *The New Organon* (1620)

■ ■ ■

THE SECOND STEP IN the modernization of American physics was the maturation of its theoretical capability. This was a process of many dimensions. Not only was it necessary to put more emphasis on educating young physicists to become theorists, but at the same time, because of the radical innovations introduced into physics by Einstein and Bohr, the older generation had to be re-educated. Re-education meant more than just learning the language of the new physics. It required the development of new attitudes toward the meaning of theory and the process by which theory comes into being. Physicists with an inductivist conception of science found their ideas of physical reality and scientific progress severely strained. Their problems reached a crisis in the mid-1920's with the quantum revolution. Bridgman's experience is a vivid example of the kind of difficulties his generation faced.

Bridgman was a man who knew how to make things work, whether in the laboratory, garden, or household. His practical ingenuity never ceased to elicit the unqualified admiration of his colleagues. He did not need fancy or expensive equipment; he kept to the simple and basic. It was in the environment of the practical where he felt most at home, and where, typically, he would take refuge when his philosophical adventures drew criticism. Bridgman's attitude is conveyed in the following excerpt from a letter written shortly after the publication of *The Logic of Modern Physics* in 1927.

You must realize that I do not spend much of my time doing things like my book [*The Logic of Modern Physics*], but I am one of those dirty physicists, all of whose time is occupied with the highly unabstract work of discovering whiskers on the suspensions of galvanometers or rubbing dirt from electrical contacts.[1]

It would be a mistake, however, to think that Bridgman possessed no theoretical aspirations. His early interest in measuring the electrical properties of metals pointed toward something more than just mechanical inventions and laboratory discoveries. Indeed, Bridgman was trenching upon one of the outstanding contemporary theoretical problems in physics: the explanation of metallic conduction. However, he did not take up the problem as basically a theoretical investigation. His approach was conditioned by his Baconian expectations. He would proceed in an orderly fashion, first collecting the necessary data and only then advancing to theoretical judgments. Nonetheless, he was already constrained by theory.

J. J. Thomson's discovery of the electron in 1897, together with the conceptual apparatus developed in the kinetic theory of gases, had provided the raw material for the classical free-electron theory of metals, first enunciated by Paul Drude in 1900 and refined in subsequent years by H. A. Lorentz. In the Drude-Lorentz model it was assumed that in a metal there were free electrons that were responsible for its characteristic properties and that these electrons could be treated theoretically as particles of a gas moving randomly through the lattice, entering into collisions with essentially immovable atoms.

Using this model, it was possible to calculate a reasonably good value for the Wiedemann-Franz ratio, a relationship, discovered in 1853, that expressed the empirical law that the ratio of electrical to thermal conductivity at a given temperature is the same for all metals. The fact that the Wiedemann-Franz ratio could be derived on the basis of the Drude-Lorentz theory was probably its greatest strength. But in many other respects the theory failed.

Among the most serious flaws of the classical free-electron theory was its prediction of an electronic contribution to specific heat (energy taken up in temperature changes) that was much too high. Involved in the prediction of the electrical resistance of a metal were two crucial parameters whose values had to be determined—the number of conduction electrons and the distance the electron travels between collisions (the "mean free path"). Neither value could be established independently theoretically, and each had to be assigned on the basis of assumptions that could not otherwise be validated. If it was assumed that each metal atom contributed one free electron, the computed specific heat (given that each electron would contribute 3/2 kT to the internal energy of the metal) was much higher than the experimental value.* Roughly stated, the classical theory postulated too many free electrons. If, alternatively, the value selected to represent the number of free electrons was small enough to be compatible with the accepted experimentally measured specific heats, the mean free path of an electron (related, in turn, to the electrical resistance) had to be assigned a much greater value than could theoretically be justified. It was not possible to understand how an electron could travel so far without colliding with an atom.

On the other hand, beginning in 1912, the approach taken by Einstein, Debye, Born, von Karman, and Nernst accounted satisfactorily for the specific heat in terms of the quantized vibrations of the ions (atom minus the electron) in the lattice, but ignored the presence of free electrons. In their version, it appeared that electrons had essentially no role to play. This, briefly, was the dilemma that Bridgman faced when he began his investigations of metallic conduction.[2]

Working within the classical Drude-Lorentz framework, Bridgman chose to attack the problem from the side of the electron path and focused on measuring resistance under varying conditions of pressure. These measurements, he expected, would provide information about the free path of electronic motion and consequently contribute to the resolution of the classical conundrum. Bridgman presented his theoretical speculations in 1917 in an article published in the *Physical Review*

---

*k = Boltzmann constant; T = absolute temperature. kT expresses the energy a classical particle has when in surroundings at temperature T.

entitled "Theoretical Considerations on the Nature of Metallic Resistance, with Especial Regard to Pressure Effects." Referring to data published earlier in the *Proceedings of the American Academy*, he presented a list of experimental relationships requiring explanation. His intention, he modestly declared, was "to present a view of the nature of metallic conduction which had its origin in an attempt to bring into line the facts at high pressures, but which also accounts for other facts not intimately connected with pressure effects."[3]

As a starting assumption, he postulated that all metals are "naturally perfect conductors in the sense that the electrons may pass without resistance from atom to atom when the atoms are in contact at rest."[4] He stated further that it was not necessary to know the precise mechanism of this transfer. According to Bridgman's theory, at absolute zero conduction is unimpeded.* But when the temperature is raised and atomic vibration begins, the atoms are periodically separated and electrons encounter "gaps" that make passage more difficult. When a certain critical amplitude of vibration is reached, electrical resistance occurs. Electrons, he said, do not collide with atoms, they collide with "gaps." This resistance, he hypothesized, is proportional to the number of "gaps," which, in turn, is proportional to the amplitude of vibration. However, at high temperatures, he suggested that collisions of unusual violence may actually facilitate conduction. Some combination of these two effects is required to explain the behavior of both normal and abnormal metals.

Furthermore, he argued, it is unnecessary to assume, as in the "old theory," that just because good electrical conductors are also good thermal conductors, the electrons themselves play a large part in ther-

*It is likely that the idea derived from the concept of an ideal gas. In an article entitled "Thermodynamic Properties of Twelve Liquids Between 20° and 80° and up to 12,000 kgm. per cm.," *Proceedings of the American Academy of Arts and Sciences* 49, pp. 3–114 (1913), he had already speculated about the utility of introducing the concept of a perfect liquid: "The results have exhibited one striking feature which has been frequently emphasized, namely that at high pressures all twelve liquids become more nearly like each other. This suggests that it might be useful in developing a theory of liquids to arbitrarily construct a 'perfect liquid' and to discuss its properties. Certainly the conception of a 'perfect gas' has been of great service in the kinetic theory of gases; and the reason is that all actual gases approximate closely to the 'perfect gas'. In the same way, at high pressures all liquids approximate to one and the same thing, which may be called by analogy the 'perfect liquid.' It seems to offer at least a promising line of attack to discuss the properties of this 'perfect liquid', and then to invent the simplest possible mechanism to explain them."

mal conductivity. Whatever mechanism allows easy passage of electrons may at the same time permit easy heat transfer. He did not at this time offer an explanation of the Wiedemann-Franz ratio, but only hinted that this was a possibility. He also noted that under certain conditions of overlapping atomic fields, the electrons may be regarded as effectively losing their individuality.[5] However, the most important feature of his model, he emphasized, was the straightforward relationship between the amplitude of atomic vibration and the variation in resistance brought about either by pressure or by temperature changes.

In his summary, he concluded:

The most important result of this paper is the observation that the variations of resistance of a normal solid metal are preeminently concerned with one factor only, the average amplitude of vibration of the atoms, irrespective of whether the change of amplitude is brought about by a change of pressure or of temperature. The proportional change of resistance is approximately twice the proportional change of amplitude. This suggests that a successful theory of metallic conduction must discard the old viewpoint, which explained resistance in terms of the properties of an assemblage of electrons little affected by the inert framework of atoms, and substitute an explanation in terms of the properties of the atomic framework.[6]

Without worrying about internal consistency or problematic behavior at extreme conditions, especially at low temperatures, Bridgman based his mathematical exposition on initial assumptions that could be trusted to hold only in the temperature range where "the classical statistical mechanics and quantum theory are not essentially in conflict."[7] He saw no difficulty in combining features from theories based on very different assumptions. It seemed quite reasonable to attempt a compromise that exploited the partial successes of other investigators. And a clever compromise it was, for qualitatively at least, he had devised a way to relate the motion of electrons and the vibrational activity of the atomic particles that made up the metallic lattice. In his model, the electrons interacted with the atomic vibrations to the extent that the electronic motion was impeded by the "holes" or "gaps" created by these vibrations.

In the light of the contemporary theoretical situation, Bridgman's synthesis seemed promising indeed. As things stood, it was certainly far from obvious what approach would ultimately prove successful. Despite the wealth of scientific intelligence being applied to the problem, no one appeared to be any closer to a solution.

Bridgman continued his measurements of the effect of pressure on electrical resistance in metals, and in 1921 he published an updated version of his theory, supporting it with his experimental findings.[8] He argued that he was able to incorporate the successful features of the Drude-Lorentz theory into his own and to overcome the difficulty of the long free path that was necessary to explain specific heats but found no natural place within the constraints of the classical assumptions. This was accomplished by postulating the concept of a perfect conductor, whose plausibility, he pointed out, was additionally supported by quantum theory, which permitted resistanceless paths or orbits for the movement of electrons.

He went on to indicate that not only was Ohm's law understandable in the framework of his theory, but that the possibility of superconductivity at low temperatures remained open. Furthermore, he was able to explain the "extra resistance" developed in mixed crystals by pointing to the irregularities created in the crystal lattice by impurities. But of particular importance, he believed, was that his mechanism allowed him to account for the measured difference in conductivity between the liquid and solid phases of a metal without requiring a change in the number of free electrons. He regarded this as confirmation of the mechanism for conduction he was proposing.

In this paper, he also sharpened this theoretical distinction between normal (pressure coefficient negative) and abnormal (pressure coefficient positive) metals. In "normal" conduction, he explained, electrons pass through the substance of the atom, whereas in abnormal conduction, electrons travel through channels that are "spaces left between atomic centers after the impenetrable nucleus has been subtracted."[9]

Nevertheless, when faced with the necessity of deriving an expression for the Wiedemann-Franz ratio, Bridgman did not reason from his first principles, as might be expected, but adopted the classical formula for electrical conductivity completely unmodified. The rationale for this decision was essentially that an encounter between an electron and a "gap" is analogous to a collision between an election and an atom, and thus it can be treated formally as the same thing. He did the same with the expression for thermal conductivity, but qualified his action by stating that the classical expression is not complete, since it "neglects the part of conduction done by the atoms, and hence will be expected to fail particularly at low temperatures." To develop the theory further, he asserted, it was necessary to possess "more intimate knowledge of atomic structure than we have at present." Still he thought that his theory was an advancement insofar as it brought "a very large number of facts qualitatively into line."[10]

Yet Bridgman was troubled. He was not getting the expected results from his experiments. He described his feelings in a letter written to a friend in May 1921.

I am now in one of my short time periods of depression over the experiment. Every few days I think that I am all through and ready to publish, and then some little thing turns up that I don't like the look of. This time it is the fact that the departure from Ohm's law seems to be very much the same in silver and gold. I would be less suspicious if they were considerably different. However, I cannot find anything the matter anywhere, and of course there may be here a law of nature that the departure from Ohm's law of all metals is the same, just as their temperature coefficient of resistance is the same. In which case, all the more glory to the discoverer.[11]

More revelatory of his theoretical goals, however, is the letter he wrote in July that same year to his colleague E. H. Hall, with whom he maintained an ongoing correspondence regarding the subject, and who was also working on the problem of metallic conduction. After informing Hall that the task he had assigned himself for the summer was "to coordinate the evidence that I got last winter from three lines of experiment, namely the pressure coefficient of thermal conduction, the change of resistance under tension, and the departures from Ohm's law at high current densities," he went on to declare:

I still regard the Wiedemann-Franz ratio as the most striking fact to be explained by any theory of thermal conduction. Repeated attempts on my part have failed to explain it on any except the classical basis, namely that the carrier of the current is also the carrier of thermal energy, and that the thermal energy of a single electron must be the same as that of a gas molecule at the same temperature. But at the same time it must be recognized that some of the thermal energy is handed on from atom to atom, as in an insulator. My present feeling is that my pressure experiments mean that a larger part of the thermal energy is carried by the atoms than I had previously supposed. As far as the details of conduction by the electrons go, I believe that I would still stand by my last Physical Review paper.[12]

Bridgman accomplished his goal, and over the next year in the *Proceedings of the American Academy of Arts and Sciences* he published a series of papers dealing with each of the experimental categories he had outlined to Hall.[13] In fact, the misgivings he had experienced gave way to the fancy that the papers contained something so significant that he felt compelled to press for immediate publication. On August 7, 1921, he wrote E. B. Wilson, then editor of the *Proceedings of the National Academy of Science*:

I am enclosing a paper for the National Acad Proceeding. Are you getting caught up at all with your publication, and is there any chance of getting this out before Einstein gets home and tells his friends all about it, and some miserable German publishes it?[14]

The reply from Wilson included remarks far from admiring of German science, probably a reflection of lingering postwar anti-German sentiment:

I should be glad to run this into the Proceedings as promptly as possible. . . . I am inclined to think I should not worry about Einstein and the rest. I have been working up my course on Constitution of Matter during the summer and I must say that I consider a great deal of the German work which has been put out in the last seven years as exceedingly low grade in respect to such a thing as critical self-examination. I have lost my confidence in some of these fellows and in others I have never had any confidence.[15]

Bridgman's article, "Measurements of the Deviation from Ohm's Law in Metals at High Current Densities," appeared in the *Proceedings of the National Academy of Science* later in 1921. It appears not to have created a stir.[16]

As a programmatic guide, Bridgman maintained a firm conviction that the Wiedemann-Franz ratio was the single most important fact requiring explanation; having dealt first with the problem with electrical conductivity, he saw the next item on his program as explaining thermal conductivity. He did this in an article published in the *Physical Review* in 1922 entitled "The Electron Theory of Metals in the Light of New Experimental Data."[17] His discovery that thermal conductivity increases less rapidly than electrical conductivity as the pressure increases prompted Bridgman to conclude that thermal energy is transferred by a shift in the position of a "gap," in much the same way that a separation in a row of billiard balls might move after the row is hit from one end or the other. Postulating this mechanism allowed Bridgman to estimate that the rate of propagation would be on the order of the speed of sound.

In retrospect, one is struck by how close Bridgman's many intuitive conclusions came to their mark,[18] especially given the limited scope of his theoretical reach. However, his quantitative arguments notwithstanding, much of his treatment was qualitative and rested on plausibility claims. He virtually ignored the possible relevance of optical phenomena (which would prove to be of central importance), relegating them to the province of quantum theory, in which he also showed little interest. "We are probably justified in disregarding the line of attack from the optical side until quantum theory has become more developed."[19] In the order of things, optical phenomena could wait. As a theoretical guide, Bridgman had set his sights on explaining one phenomenon—the Wiedemann-Franz ratio—and he pressed single-mindedly toward that goal.

Yet for all the prescient physical intuition one might wish to attribute to Bridgman, in retrospect his theoretical insights must be judged unproductive. His "free path" theory of metallic conduction had nowhere to go. Mathematically he had accomplished no more than what was contained in the classical free-electron theory. For what was potentially creative—the hypothesis of the perfect conductor—he had no means for translating it into a quantitative assertion, and therefore he had no way to develop it to definitive, testable conclusions. It functioned merely as a qualitative principle to justify his assertions about the length of the free path of the electron. As for the amplitude of atomic vibration, the idea to which he attached so much importance, this was a cul-de-sac, a concept with no independent predictive power.

No wonder his old friend, physicist Edgar Buckingham, expressed dismay. To his mind, Bridgman had lost sight of what an empirical science was all about and was taking the invisible microscopic world all too seriously, making it the goal of scientific research rather than its instrument. Buckingham did not hesitate to scold Bridgman for his aberrance.

Thank you very much for the batch of papers I received yesterday. I have read the two general ones and found them very interesting. . . .
The strange thing about it all is that you should be just as crazy as all the rest of the world except me and a few other fossils. For you seem to regard your experimental results as interesting only so far as they bear on the structure of matter, whereas I regard the question of the structure of matter as interesting only insofar as it may help to explain and interpret the properties and behavior of bodies—matter being only an abstract name for what bodies are made of,—just as heat is an abstraction invented to account for changes of temperature which are the things we really observe. The structure of matter seems to me about as interesting as the nature of God or the origin of the universe.[20]

Be that as it may, Bridgman's theoretical aspirations were being encouraged. His confidence was high, and with good reason. In the summer of 1922, he received an invitation from the prominent theoretical physicist H. A. Lorentz to participate in the upcoming Solvay Conference, which was to be held in Brussels in 1924. This was an enviable mark of prestige, especially for an American physicist. His correspondence became embroidered with references to this honor. He was proud, and rightly so. He had achieved a status that placed him among the leading European physicists of the time.

Indeed, he may have been a little too proud, for he experienced a slight dampening of spirit, perhaps a twinge of jealousy, when he learned that he was not the only American to have been invited. On July 24, 1923, E. H. Hall wrote to Bridgman:

> I suppose that you have by this time received the provisional program of the Fourth Solvay Council, to be held in Brussels about next Easter, and discovered that you and I are the U.S. representatives invited to attend and participate.
>
> I have written Mr. Lowell [Abbott Lawrence Lowell, president of Harvard, 1909–34] expressing the hope that matters will be so arranged that you will be able to attend, for this invitation is an honor to Harvard as well as to you and me. . . .
>
> I observe that in the list of those invited to take part your name comes next to that of Bohr and mine next to that of Einstein. We seem to be in good company.[21]

Bridgman confessed the diminution of elation that he felt upon learning he would be sharing the honor with Hall to a close friend. The letter not only is a good account of the events but also gives insight into Bridgman's anticipations.

> Dear Howard [Trueblood],
>
> . . . Perhaps you have heard of the Solvay Conferences in Physics. If you haven't, you ought to. They are held in Brussels every few years with a fund left by [Belgian chemist Ernest] Solvay, and they are planning to hold the fourth next spring at about Easter time. The subject is to be Electrical Conduction in Metals, and soon after your letter came last summer, I received an informal invitation from Lorentz to attend this Conference. If I accepted, Lorentz said that he would send me an official invitation later. I could not at once accept without consulting Lyman as to the possibility of my getting away. Lyman [Theodore Lyman, director of the Jefferson Physical Laboratory] was in Europe, where I wrote him, and from there he immediately communicated with Lorentz his willingness to have some one else perform my duties while I was away. All this was by the middle of last August. Since then I have been waiting for the official invitation, which has not come, nor any other communication from Lorentz. But last Sunday a distinguished Belgian physiologist, M. Seger, who I since have found is expected to be the honorary Chairman at the Confer-

ence, bobbed up in Cambridge with the personal word from Lorentz that he hoped I was coming, with the statement also that political circumstances were so unsettled that it was not yet perfectly certain that the Conference could be held, and that I would receive an official invitation later. Professor E. H. Hall is also invited to attend, I suppose because of his connection with the Hall effect and the fact that he entertained Lorentz for five days at his own house a couple of years ago. *For some strange reason his participation somewhat assuages the intensity of my satisfaction in my own invitation* [emphasis added], although I believe that Millikan [Robert Millikan, experimental physicist, Nobel Prize 1923] is the only American who has gone previously as a regularly invited delegate, with [Albert] Michelson [the first American to win a Nobel Prize in a science, 1907] and R. W. Wood [Robert Williams Wood, Johns Hopkins physicist] invited by special dispensation to watch from the side-lines.[22]

The fourth Solvay Conference did take place as planned, but its achievements were disappointing. It has been remarked that confusion reigned at the 1924 Solvay Conference,[23] and understandably so, for at the time no one could have foreseen that ultimately a new language, the language of wave mechanics, would be the key to creating a self-consistent theory of metals.

Indeed, the Solvay Conference met on the eve of an explosion in scientific creativity, a quantum mechanical revolution that would overturn some of the most cherished concepts of physical truth. In the words of Banesh Hoffman, "Professional physicists, swept off their feet by the swift currents, were carried they knew not where, and it was years before the survivors recovered sufficiently to see, with the beginnings of perspective, that what had so overwhelmed their science had been the convulsive birth pangs of a new and greater era."[24]

Bridgman was among these physicists who would feel overtaken by a great tidal wave of change. In 1930 he complained that new theories "come crowding on each other's heels with ever-increasing unmannerliness, until the average physicist, for whom I venture to speak, flounders in bewilderment."[25] At the same time, he could not help but feel some resentment, for he had accepted the challenge in the terms it was offered and the rules of the game had changed. To make matters worse, they had changed at breakneck speed.

During the years following the Solvay Conference, Bridgman's theorizing languished. His experimentation continued to generate results that demanded explanation, but without fresh conceptual input, he could only repeat what he had already proposed and indicate in what manner his data challenged the current model. The fertility of his own theory had quickly been exhausted. His ideas, he later judged, had been "stillborn."[26] It is not surprising that he experienced a sense of bewilderment and disillusionment. His opportunity to contribute theoretically to the understanding of metallic conduction had been rendered obsolete by developments outside his familiar world.

In contrast, two of his pupils, J. R. Oppenheimer and John Slater, were participants in the ferment that left him so unsettled. Bridgman was keenly aware of this generation gap. It is, he noted in 1931, "a striking fact that modern physics is a young man's subject; the most brilliant successes are being attained by men young enough to have received all their scientific impressions since the general acceptance of the theory of relativity."[27] Indeed, as Bridgman recognized, the younger physicists viewed physical events from a changed theoretical vantage point. This did not fit into the Baconian picture of scientific knowledge as a result of the accumulation of observed fact. It appeared that the role of experiment was not quite what Bridgman had thought. He must have spent some time reconsidering its importance, for in an address as retiring vice-president and chairman of the physics section of the American Academy of Arts and Sciences, he declared, as if to defend a commitment to a program of experimentation, that the experimentalist "may feel a renewed sense of the importance of [his] contribution" when he realizes that an increase in the range of measurement may yield unexpected new facts.[28]

Yet Bridgman could not have failed to notice that the "new facts" responsible for the scientific upheaval were not experimentally generated novelties but highly abstract theoretical formulations drawing on data beyond the narrow scope of his experimental program. And what unexpectedly strange facts they were, these new quantum theoretical ideas.

First there were Louis de Broglie's new matter waves, proposed already in 1923, but largely ignored until Einstein and Erwin Schrödinger pointed out their importance. The concept of matter waves prepared the ground for wave mechanics. Then in 1925 came the Pauli exclusion principle, the crucial quantum discovery which provided rules for dealing with internal atomic organization, and also showed why matter would not collapse. This was a product of a field of investigation—complex spectra—that appeared to have no immediate bearing on the type of relationships Bridgman was studying. He had brushed aside consideration of optical phenomena, believing it was a subject that could wait. In any case, there was no place in Bridgman's reasoning for the quantum theory, with or without the Pauli principle. The closest he came to recognizing quantum states was a suggestion that electrons move in "tracks somehow connected with quantum conditions," but this, he said, was too complicated.[29] Just as far removed from Bridgman's immediate concerns was the development in 1926 of the Fermi-Dirac statistics, a modified quantum statistical mechanics based on the exclusion principle, which subsequently formed the basis for Arnold Sommerfeld's successful reformulation, in 1927–28, of the Drude-Lorentz free-electron gas theory of metals.

Such drastic theoretical revision of physical reality stood outside of Bridgman's scientific anticipations. His scientific standard was built upon empiricist and inductivist premises. With respect to progress in physics, he did not accept the idea of "revolution," nor did he possess any sense that the time had to be "ripe." Bridgman expected progress to be a steady, forward advance toward understanding, a cumulative approach to truth that could be carried out systematically, requiring only patience and hard work.

Therefore, he did not expect to reconstitute scientific truth when he considered the theoretical problem of metallic conduction. He took for granted the classical delineation of the situation, and his experiments were undertaken with the idea that he could fill in the blanks or, at most, make additions or modifications. When his resistance measurements failed to produce definitive results, the alternative he considered was still classically defined. In 1923 he wrote to Lorentz suggesting that perhaps Heike Kamerlingh Onnes might be persuaded to perform an experiment to count the electrons in conduction. The experiment was not made because of flaws in its conception.[30]

Thus, Bridgman's theoretical aspirations faded. Nevertheless, Oppenheimer, writing to Bridgman in February 1927 from Göttingen, tried to offer encouragement.

You may remember that when I was at Harvard two years ago I was very much interested in your theory of metallic conduction. Recently, in the course of some work in quantum mechanics, an idea has turned up which seems to offer a certain support to your theory. I think it will be some time before a complete quantum theory of conduction is possible, but perhaps I may tell you briefly of this one point.

On the classical quantum theory, an electron in one of two regions of low potential which was separated by a region of high potential, would not cross to the other without receiving enough energy to clear the "impediment." On the new theory that is no longer true: the electron will spend part of its time in one region, and part in the other. IF the impediment is not very high, the electron will jump back and forth between the two regions quite often; if the impediment is higher, it will do so more seldom; but if one waits long enough, one can be sure of finding it, at some time in each of the two regions. [Here follows an example of a hydrogen atom in a homogeneous electric field.]

· · · · ·

From the former of these considerations, it follows that a valence electron in a metal is not to be thought of as associated with any one ion; it wanders about from atom to atom, never spending much time in the interstices, where the potential is high. In this sense the electron is free. Further, if one puts on an electric field, the electrons will tend to be in the part of the metal where the potential is lower; eventually they will be sure to get there; and the rate at which they wander is great when the potential between two ions is not too high, and falls when this rises. *Is this not just your "gap theory"?* [Emphasis added][31]

Still, Bridgman was discouraged. He did not see any similarity. In April he replied to Oppenheimer's letter.

I am glad to get your letter and was much interested in your suggestion of a connection between the ideas of the new quantum mechanics and electrical conduction in metals. The action which you suggest of an electron in passing from one minimum of potential to another under the action of even the weakest fields was a new idea to me, but I find on inquiry from Schrödinger, who has just been here, and Slater that this is quite to be expected. The idea will doubtless prove to be a fruitful one, and perhaps one can make connection with my ideas of "gaps" as you suggest, or perhaps it may prove to strengthen Professor Hall's ideas of the important part played by the ions in metallic conduction.

My own ideas as to the nature of metallic conduction have not been progressing very fast lately. I find very significant the fact that I have recently discovered the existence of an internal Peltier heat in crystals, that is, of a reversible heating effect inside a crystal when the direction of current flow changes, and am finding it difficult to fit in with other phenomena. I am inclined to think that the simplest meaning is that the electrons move on some sort of guided path, perhaps the remains of the quantum orbits which must be responsible for supra-conductivity. The same sort of thing is indicated by some other phenomena, although not at all unambiguously.[32]

It was only shortly after this, in September 1927 at the prestigious Volta Congress in Como (which Bridgman had been invited to attend, but decided not to) that Sommerfeld announced his achievement. By treating the electrons as a gas obeying the Fermi-Dirac statistics, Sommerfeld was able to derive a theoretical expression for the Wiedemann-Franz ratio that yielded values in much closer agreement with experimental values. The new statistics also revealed that an electron gas must be in a state of high degeneracy—the electrons are almost frozen in their lowest energy state—at normal temperatures. This result removed the difficulty surrounding the specific heat, since it meant that at ordinary temperatures the electrons make no more than a negligible contribution to the specific heat of a metal. Their effects become significant only at very low temperatures when the vibrations of the atoms in the lattice have become attenuated and no longer swamp them out.[33]

However, although Sommerfeld had resolved some of the most egregious difficulties, his account was not yet satisfactory. Among the unexplained phenomena were the variations of the Hall coefficient and the temperature dependence of the electrical resistance, a relationship that involves the mean free path. Sommerfeld had assumed a relatively long free path for the electrons but had not inquired why this was justified. Nor had he given a reason for ignoring the interactions between the electrons and the metal ions. Thus, while his theory represented a marked advance over the classical theory, it was nonetheless a quantum mechanical modification of it rather than a fundamentally new theory.[34]

These shortcomings were not lost on Bridgman (nor on Sommerfeld, as the following letter indicates), and he did not hesitate to point them out. Shortly after the Volta Congress, he received the following note from Sommerfeld.

Dear Colleague,

Do you remember our last conversation in Munich when I was taking you to the trolley station? It was about electronic conduction in metals. At the time I was of the opinion that it was hopeless. In the meantime I have become convinced that through the new statistics of wave mechanics the problem can be handled without additional new assumptions. The note that I sent you— please be good enough to pass on the extra copies to interested persons—is only provisional. A detailed work will soon follow.

Naturally I do not mean that the new statistics can do everything. In order to master the full range of your experimental results, surely special hypotheses will be required about the interaction between electrons and metallic atoms in the direction that you have worked them out. But it seems noteworthy to me that certain dominant traits of the problem can already be retained schematically without special hypothesis.[35]

Bridgman's reply was polite and reserved.

Dear Professor Sommerfeld;

Your new electron theory of metals has aroused my greatest interest, and I am very grateful to you for sending me the reprints so promptly. My couriosity [*sic*] has been aroused ever since I saw the title announced for your paper at Como. In fact, after seeing this, I very nearly reconsidered my decision not to make the trip to Como, for the express purpose of hearing your paper.

Your theory does indeed remove many of the difficulties of the old gas theory, and justifies the optimistic expectation that light is beginning to break on this most perplexing subject. I shall await with interest your next paper giving a more detailed working out of some of the points. I suppose that in this paper you will give some special consideration to the method of interaction between the electrons and the metal atoms, as you suggest in your letter. It seems to me that considerations of this sort will be necessary to explain such phenomena as the different resistance of crystals in different directions.

Shall you offer in your next paper any explanation of the discrepancy of sign shown by the Volta effect? and if so, I believe that phenomena in crystals shed some new light on this question. I tried to express some of these ideas in a paper which I sent to be read at Como, but which Professor Duane tells me, was not read. . . . It is interesting that the phenomena in crystals had compelled me to give up the hypothesis of equipartition, which was one of the assumptions of my own theory of conduction, so that from this point of view I was prepared for the sacrifice of equipartition which the new statistics demands.[36]

Bridgman voiced similar criticisms of Sommerfeld's theory in a letter to William Hume-Rothery, lecturer of metallurgical chemistry at Oxford. (Hume-Rothery had, incidentally, just recently finally received the printed report of the discussions at the 1924 Solvay Conference, which were now obsolete.)

I have read the papers of Sommerfeld, and regard them as making a very important step forward. At the same time, I cannot be [as] enthusiastic about his theory as some people, or feel that he has practically solved all our old difficulties. I do not see that Sommerfeld's theory, at least in its present form, can explain the difference of resistance in different directions any more than could the old gas-theory. It seems to me that the wave mechanics must be made a more integral part of the theory before we have success, instead of being only a more or less superficial graft, as it is in Sommerfeld's theory.[37]

Oppenheimer confessed to Bridgman that he felt similarly, but instead of emphasizing the unsolved difficulties, he outlined what he thought the theoretical approach should be.

Your opinion of Sommerfeld's theory I more or less share. I think that the only valuable point is the proof that the Wiedemann-Franz law does not necessarily involve equipartition; this seems to me physically correct; and the rest would appear to be algebraic divagation [divagation]. I think that in a proper theory one should start by solving the wave equation for a single electron in the ion lattice, assumed perfect; one should then quantize these waves, taking into account binary interaction by the method of Jordan, Klein, and Wigner. One should then consider the excited lattice and compute the rate at which the lattice absorbs energy from the electron wave. This should give unambiguously the [word illegible] for a single crystal. But the program is not easy.[38]

Common to these considerations, over and above the problem that Sommerfeld's theory failed to account for important experimental data, was the judgment that it was not a "proper" theory. The long mean free path required by the electrons had to be treated as an *ad hoc* assumption, and this violated a major canon of theoretical adequacy— sufficient reason. There was no *explanation* for how the electrons could move about with as little obstruction as appeared to be the case.

The solution to this dilemma came in 1928 from Heisenberg's first graduate student, Felix Bloch, who later recalled:

Except for the replacement of classical statistics and the inclusion of the spin, however, Pauli and Sommerfeld both accepted the old ideas of Drude and Lorentz, who treated the conduction electrons as an ideal gas of free particles. The high conductivity and reflectivity of metals of course strongly supported the assumption of very mobile electrons but I had never understood how anything like free motion could be even approximately true. After all, a metal wire with all its densely packed ions is far from being a hollow tube and as I started to think about it, I felt that the first thing to be done in my thesis was to face this striking paradox.[39]

It is in Bloch's work that the break with the classical picture occurs unequivocally. With no predisposition to favor the free-electron model, Bloch preferred to explore the behavior of electrons in solids in terms of the recently developed wave mechanics. His physical cue, he claimed, was the Heitler-London model for the covalent bond, where "valency electrons in a molecule were not confined to stay on a single atom."[40] However, the significance of Bloch's analysis consists not merely in his application of wave mechanics to electric conduction, but in his discovery of a rational foundation—now known as the Bloch Theorem—for the assumption of the long free path of the electron. According to this principle, an electron moves freely and unimpeded under the influence of a periodic potential, a condition that obtains in a perfect crystal. Resistance, now interpreted as scattering of electron waves, appears as a consequence of irregularities, due either to thermal vibrations of atoms or to impurities present in the crystal. The new model provided the "explanation" of the efficacy of Sommerfeld's assumptions and pointed the way to a more complete understanding of the properties of the solid state.

Superficially, it might appear that Bridgman's fundamental hypothesis, along with many of his qualitative inferences, had been vindicated. The wave-theoretical treatment of Bloch did indeed, albeit more "naturally," assume an ideal or perfect state of conduction as its point of reference. But the "realities," that is, the objects of physical discourse and the space in which they are manifested, had become transformed into entities alien to Bridgman's earlier assumptions. The comfortable visual space of ordinary mechanics, populated by particles moving about and entering into collisions, had been replaced by an ordering in momentum space, a statistical order not readily accessible to the physical imagination, populated not by "things," but by "states."

By what conceivable chain of inferences, by what pathway of reason, could the experience of ordinary physical intuition have become transfigured into this strange setting, a universe where concrete events are dissolved into probabilities and no sharp boundaries exist, where material identity has lost its familiar grounding, or, what is equivalent, where one may not even take it for granted that one knows how to count? Surely nothing in "commonsense" experience could warrant such an extrapolation. What could be more contrary to Bridgman's expectation that scientific knowledge should be constructed systematically upon well-established experimental facts?

Bridgman must have spent some time pondering such questions, for in the introduction to a small book entitled *The Thermodynamics of Electrical Phenomena in Metals*, published in 1934, he addressed the subject of change in physics, change that for him was taking place at a pace much too fast to permit critical evaluation. "The progress of physics," he wrote,

is unsystematic. The activities of the moment are determined by the most compelling interests of physicists at that moment, and into this enter many complex and human elements. There is a little of the element of sheer fashion, for most physicists are gregarious and enjoy talking over common activities with their fellows; there is the strategical element, for it is only human prudence to cultivate the fields in which success is most probable, and this usually means a new field; and there is the economic element, which demands that an experiment shall not involve too costly an apparatus. *The development of physics is thus not always in that direction which would be taken by a competent dictator, charged with the task of getting intellectual mastery of the physical world as rapidly as possible*, nor, indeed, is it in the direction which would be chosen by the majority of physicists themselves, if they could be freed from ulterior considerations. *The result is that physics sometimes passes on to new territory before sufficiently consolidating territory already entered; it assumes sometimes too easily that results are secure and bases further advance on them, thereby laying itself open to future possible retreat.* This is easy to understand in a subject in which development of the great fundamental concepts is often slow; a new generation appears before the concept has been really salted down, and assumes in the uncritical enthusiasm of youth that everything taught it in school is gospel truth, and forgets the doubts and tentative gropings of the great founders in its eagerness to make applications of the concepts and pass on to the next triumph. [Emphases added][41]

In particular, Bridgman added, "all this has been true of the development of our theories of the electrical properties of matter."[42] He made his point even more explicit when he outlined what he saw as the train of events. This passage is especially revealing because it underscores Bridgman's expectation that ideally scientific development should be an orderly process.

Historically, the development of new points of view and the discovery of new experimental facts came too rapidly for complete assimilation into what was already known. A further difficulty, of course, is that *the experimental facts are often not discovered in the logical order.* The electron theory came crowding on the heels of the formulation of the field equations, with its thesis that the properties of all matter could be explained in terms of the motion of concealed discrete electrical particles, and that the motion of these particles was controlled by the field equations, extrapolated to microscopic dimensions hopelessly beyond direct experimental verification, *and before their validity had been checked even over the entire experimental domain.* And finally, crowding on the heels of electron theory, is wave mechanics, forced on us by the new experimental facts of atomic physics, in which we give up the idea of discrete electrical particles with individuality, but retain the concept of the electrodynamic field to control the motion of what replaces the electron, and determine the magnitude of the field (as for example, in the neighborhood of the nucleus of an atom) in terms of a fictitious discrete elementary charge acting after the fashion of the charges of large scale experience. [Emphases added][43]

These statements carry a double message. They serve both as an explanation and a warning, an explanation of the failure of the electron gas model to provide a satisfactory theory of metallic conduction, and a warning that too much haste in theorizing could cause the same problem to occur again.

There is a poignant irony in the reluctance with which Bridgman viewed the new theoretical advances. For only a few years earlier, he had been asked to comment on the relationship of his own work on compressibility to that of T. W. Richards (Nobel Prize, 1914), for whom a memorial lecture was being prepared by the British chemist Harold Hartley. After making it clear that Richards had not been an influence on him, Bridgman went on to observe that although Richards might have been much ahead of his time when he first advocated the compressible atom instead of the "hard indivisible thing of tradition," later on

with the discovery of the electronic structure of the atom, things moved in a direction favorable to the early view of Richards faster than even he realized, and I and many others have felt that in his last years much unnecessary stress was put on emphasizing points of view which practically all of his audience accepted. He was a long time in accepting the electron, and I believe never whole-heartedly accepted the electronic structure of the atom, or saw that the fact of its electronic structure made its compressibility inevitable.

. . . The important part of Richards' theory I have always felt to be in its qualitative picture of the situation, rather than in its formal mathematical expression. Richards was not at ease with mathematical manipulations, and I do not believe that the mathematical form which he gave to his theory appealed particularly to physicists. In particular, his division of the atomic forces into four groups, external pressure, cohesive pressure, repulsive pressure, and thermal expansive pressure, I have always felt to be artificial, and not to be justified by any picture of the mechanism which Richards himself had. Our discussions often turned on this point, and he was never able to satisfy me. But I do think that he had a very vivid qualitative grasp of the situation and that nearly everyone would give assent to the qualitative background of his equations, although perhaps not subscribing to the equations themselves.[44]

In an important way, Bridgman might have been talking about himself. Perhaps he even had himself in mind. For he, too, was left behind by rapid theoretical change, even though his intuition was sure and led him to qualitatively valid conclusions. This was not a failure of intellectual capacity, but rather a consequence of scientific culture. Bridgman, like Richards, was part of a generation of American scientists who were trained to be experimentalists.

Bridgman was a product of Eliot's Harvard, where the modernization of physics meant laboratory training, not cultivation of the theoretical imagination. He was steeped in the empiricist scientific tradition which holds that truth is to be found by observation, by counting and measuring. His scientific program was first and foremost directed toward extending the reach of measurement. His was the mind-set not of a theoretician, but of an experimentalist. Thus, it would very likely come to be said of Bridgman, just as it was of Richards, "There is no doubt he will be remembered mainly as a great maker of measurements in the same class as men like [the nineteenth-century physicist/chemists] Regnault, Thomsen, and Kohlrausch."[45]

PART II

The Meaning of Measurement

In classical mechanics, as well as in the special theory of relativity, the co-ordinates of space and time have a direct physical meaning. . . . In the general theory of relativity, space and time cannot be defined in such a way that differences of the spatial co-ordinates can be directly measured by the unit measuring-rod, or differences in the time co-ordinate by a standard clock. —Albert Einstein, *"Die Grundlage der allgemeinen Relativitätstheorie"* (1916)

▪ ▪ ▪

P. W. BRIDGMAN WAS INDEED a maker of measurements. Neither theoretician nor scientist-metaphysician, he believed that the meaning of physics should be straightforward and self-evident. But because the development of twentieth-century physics thrust the question of the meaning of measurement into prominence, he was forced to consider philosophical problems he probably would have preferred to dismiss as unworthy of his task as a physicist. These problems, he had to concede, had grown out of physics itself and were not imposed from without by the speculative imagination of philosophical fancy. Still, it was more than two decades after Einstein's first paper on relativity before he made the following admission.

One of the most noteworthy movements in recent physics is a change of attitude toward what may be called the interpretative aspect of physics. It is being increasingly recognized, both in the writings and conversation of physicists, that the world of experiment is not understandable without some examination of the purpose of physics and of the nature of its fundamental concepts. It is no new thing to attempt a more critical understanding of the nature of physics, but until recently all such attempts have been regarded with a certain suspicion or even sometimes contempt. The average physicist is likely to deprecate his own concern with such questions, and is inclined to dismiss the speculations of fellow physicists with the epithet "metaphysical." This attitude has no doubt had a certain justification in the utter unintelligibility to the physicist of many metaphysical speculations and the sterility of such speculations in yielding physical results. However, the growing reaction favoring a better understanding of the interpretative fundamentals of physics is not a pendulum swing of the fashion of thought toward metaphysics, originating in the upheaval of moral values produced by the great war, or anything of the sort, but is a reaction absolutely forced upon us by a rapidly increasing array of cold experimental facts.

This reaction, or rather new movement, was without doubt initiated by the restricted theory of relativity of Einstein.[1]

The restricted (or special) theory of relativity, as Bridgman rightly perceived, had called attention to an underlying discrepancy between common expectations about the objects of physical knowledge and the implications of the mathematization of natural knowledge. The problem centered on the meaning of measurement. In one respect, Einstein was merely carrying to its logical conclusion a process begun by Galileo. But he introduced an epistemological element far from Galileo's way of thinking.

The seventeenth-century Scientific Revolution had recast the qualitative Aristotelian world into a mathematical order, a quantitative ontological structure in which measurement was the physical link in an epistemological procedure culminating in exact knowledge. Space and time were the stage for causal action between material bodies, a drama to which mankind was a privileged spectator. Physical laws were truths that captured the first principles of this reality in mathematical language, the language the Creator allowed humanity alone to understand. The physics of Galileo and Newton stood as a proud monument to man's ability to combine his measuring techniques with mathematical reasoning and thereby come to know the wonders of creation. Yet little by little, man banished God to a place of decreasing importance, forgetting that the possibility of objective knowledge rested in the first place on a divine guarantee.

As knowledge became secularized, the traditional meaning of metaphysics was gradually degraded. Ontology became ever less relevant. Hume and Kant notwithstanding, probably the most effective agent in its downfall was the evolutionary naturalism of Darwin, which demoted knowledge to a function of human activity, a mere consequence of the interaction of the individual with the environment. In the evolutionary framework, truth was no longer eternal and transcendent, but a time-dependent state of human physiology or psychology.

The impact of this blow was also felt in physics. Ernst Mach's well-known critique of the foundations of physics, for example, was unmistakably infused with the spirit of evolution. Nevertheless, despite difficulties presented by the nature of light, heat, and electricity, the Newtonian edifice remained in place. For most physicists, the idea that Newton's universe was not the real objective universe was simply unthinkable. Even those who viewed the mechanistic ontology with skepticism did not abandon the basic Newtonian spatiotemporal substratum. All this changed with the advent of Einstein's special theory of relativity. In a certain philosophical sense, physics caught up with the life sciences. Special relativity robbed Newtonian space and time of their objective standing. Consequently, measurement could no longer be thought of as being carried out against an independently existing framework.

This change did not happen as abruptly or as completely as we are sometimes led to believe, certainly not in America. And, it will become evident, neither had the Scientific Revolution completely overturned the Aristotelian world of qualitative elements.

When Einstein completed the transformation of the qualitative world of substance into a universe of metrical relations, he brought to fruition what was already implicit in the idea of the mathematization of natural knowledge. The outcome of his theoretical accomplishment was to repudiate the lingering qualitative ontological associations which were still clinging to the categories of measured quantities. The debates over the meaning of dimensional analysis, which centered on the ontological status of measured quantities, illustrate this very clearly. These debates are the subject of the next chapter.

Furthermore, when Einstein abolished the universal distinction between rest and uniform motion, the absolute dichotomy between matter and motion was similarly destroyed, challenging common assumptions about the meaning of physical objectivity. The nonconstancy of the fundamental categories of mechanics—mass, length, and time—was interpreted as implying the subjectivity of physical knowledge. This, however, was a judgment based on the social aspect of knowledge, the suggestion that each person may experience a different reality.

On the other hand, at the heart of the relativistic challenge was a critique of measurement, one that did, in fact, predispose special relativity to a subjectivist interpretation in a more universal sense. But that was because its focus, being epistemological, did not ask what was real and how it might relate to appearances. Instead, Einstein began his paper with a general analysis of measurement from a physical point of view and concluded that its physical meaning depended ultimately on the judgment of simultaneity, which had no absolute metaphysical basis.[2] By doing so, he showed precisely in what way Newton's absolute space and time (the existence of which were supported by their presentation to God's sensorium) were ideas that could not be realized *experimentally*. At the same time, he inverted the metaphysical priority of measurement and its object. That is, he showed that measurement does not mark off intervals against an independently existing space and time which merely wait to be measured, but that space and time are relationships between sensible events.

However, this realization raised difficult problems about the physical meaning of measured quantities. What is the physicist doing when he measures something? What does measurement have to do with physical reality? What, indeed, are the objects of measurement? These are questions that Bridgman took up during the dimensional analysis controversy. The solutions he devised were subsequently applied in his highly original interpretation of relativity theory, the subject of Chapter 5.

CHAPTER 4

Dimensional Analysis:
The Building Blocks of Physical Reality

We are assured that all things are wisely adjusted in number, weight and measure, yet with such complex circumstances as require many data from experiments, whereon to found just calculations.
                    —Stephen Hales, *Statical Essays* (1733)

■ ■ ■

WHEN EINSTEIN'S FIRST PAPER on special relativity was published in 1905, American scientists made very little of it. Not until 1908, largely through the efforts of the MIT physical chemist G. N. Lewis and his student R. C. Tolman, was relativity finally properly introduced to the American physics community, only to be greeted with hostility and resistance, not to say considerable misunderstanding. American physicists tried to deny its cogency. They could only view as false a theory that imputed to physical knowledge an underlying subjectivity.[1]

Central to the repugnance felt by these physicists was the sense that the nonconstancy of the fundamental units of mechanics—mass, length, and time—would render the world that human beings shared nonexistent. It was incomprehensible that these sturdy building blocks of the concrete physical world should suddenly become phantoms, apparitions whose manifestations changed from one person to another. If mass, length, and time could not be depended upon, it was hard to see how physics was even possible.

O. M. Stewart, professor of physics at Cornell, reasoned that "if we retain the old idea that the velocity of light is independent of the velocity of the source, we apparently must accept the relativity principle with its rejection of a fixed medium filling all space *together with an entire revision of our concepts of the fundamental and derived units*" (emphasis added).[2]

W. F. Magie, professor of physics at Princeton, went into more detail.

A theory becomes intelligible when it is expressed in terms of the primary concepts of force, space, and time, as they are understood by the whole race of man. . . . I do not believe that there is any man now living who can assert with truth that he can conceive a time which is a function of velocity or is willing to go to the stake for the conviction that his "now" is another man's "future" or still another man's "past." . . . I believe that these ultimate perceptions are the same for all men now, have been for all men in the past, and will be the same for all men in the future. I believe, further, that this is true because the universe has a real existence apart from our perceptions of it, and that through its relations to our minds it imposes upon us certain common elementary notions which are true and shared by everybody.[3]

Lewis and Tolman met the objection by offering a psychological interpretation. To them, the distortion of a moving body could not be an actual physical change, but must be only apparent. Lorentz, they asserted, might have considered the distortion real, but Einstein did not.

Let us emphasize once more, that these changes in the units of time and length, as well as the changes in the units of mass, force, and energy which we are about to discuss, possess in a certain sense a purely factitious significance; although, as we shall show, this is equally true of other universally accepted physical conceptions. . . . The distortion of a moving body is not a physical change in the body itself, but is a scientific fiction. . . . Although these changes in the units of space and time appear in a certain sense psychological, we adopt them rather than abandon completely the fundamental conceptions of space, time and velocity, upon which the science of physics now rests. At present there appears no other alternative.[4]

Thus Tolman and Lewis, even if somewhat ambiguously, assured their readers that space and time might appear to be subjective but in fact were not. By taking this position, they indicated that they, too, were not ready to give up the a priori Newtonian space and time. But the problem of how to interpret the physical units of measurement had been raised, and it was not long before Tolman and Lewis each offered his own theory of measurement.

Bridgman was not a participant in the early debates over the meaning of relativity. His attention was taken up with the more mundane activities involved in establishing his career as an experimentalist. Even with the assistance of a mechanic, he found the work load increasing. Experimental findings had to be written up for publication. A treatise on acoustics had to be done for the Submarine Signal Company. He had too much work already, he said, to do abstracting for

Chemical Abstracts. Nor did he have the time to write an article for the *Journal of the Franklin Institute.* The weight of correspondence was also growing. In addition to the effort expended in procuring samples, he had to answer inquiries about his apparatus and experimental results. Independent inventors, as well as industrial engineers, were taking notice of his techniques. And the European physical chemists Alfred Stock, Gustav Tammann, and Andreas Smits were interested in information about the new allotropic forms of water and phosphorus.

Nevertheless, Bridgman was not destined to remain aloof from the revolutionary events already shaping his intellectual future. The death of B. O. Peirce in 1914 thrust upon Bridgman the responsibility for teaching the courses in advanced electrodynamics. For these he now had to prepare. The material he had to deal with was by no means unproblematic, since at the foundation of electrodynamics was an inconsistency, an asymmetry in Einstein's view, that rendered it irreconcilable with classical mechanics. Since special relativity provided the resolution to this difficulty, Bridgman had to study it. What he learned distressed him deeply and permanently transformed his outlook on physics.

Nevertheless, Bridgman did not immediately enter the debate on relativity. He came to it only indirectly through his disagreement with the theories of Tolman and Lewis. At issue was the ontological status of measured and derived quantities. Actually, the argument was not even about relativity, although Tolman and Lewis had made use of certain assumptions drawn from their own particular interpretations of relativity. It was about the legitimacy of dimensional analysis.[5] Bridgman objected to the idea, implicit in their theories, that the composition of the physical world, either material or rational, is pre-established. At an American Association for the Advancement of Science (AAAS) convention in 1953, Bridgman recalled that dimensional analysis was being expounded in such a way as to "cast doubt on the necessity of doing experimental work."[6] He might well have had in mind British physicist J. W. S. Rayleigh's remark that the dimensional method was underutilized, that he had "often been impressed by the scanty attention paid even by original workers in physics to the great principle of similitude [dimensional homogeneity]." "It happens not infrequently," Rayleigh had asserted, "that results in the form of 'laws' are put forward as novelties on the basis of elaborate experiments, which might have been predicted *a priori* after a few minutes' consideration."[7] Bridgman reacted to this unsatisfactory situation by entering the world of interpretive physics. At the time he could hardly have foreseen that the position he was taking would set him on a path toward subjectivity or that Einstein (the Einstein of special relativity) would be his mentor.

In the meantime, the dimensional analysis controversy revealed a generous amount of confusion about the meaning of relativity and measurement. The marriage of the two great concepts brought together by Newton—the mechanistic ontology and classical rationalism—was strained by the new physics. Einstein's abrogation of the traditional meaning of measurement had demonstrated that the relationship between mathematics and physical reality had to be reconsidered. The dispute over dimensions was just one manifestation of a general concern that would be stated with more precision and politicized by the logical positivists.

Overlapping with and confounding the dimensional analysis debate were ideas that it might be possible to discover a set of "ultimate rational units" (URU). This aspiration was premised on the belief that if the units of measurement were based on universal relationships, their arbitrariness could be minimized or even eliminated. In a sense, according to this view, measurement introduces an imperfection into a potentially ideal rational order. Knowledge is more perfect the fewer its ports to the empirical world.

In 1914, G. N. Lewis proposed his version of a suitable system of measurement, a scheme he believed would limit the degree of arbitrariness to the choice of one basic unit, say, of length (Planck had also discussed a similar possibility).[8] It was Lewis's goal to find relationships that would permit all quantities to be derived from one "dimension" or category of measurement. One such relationship, he thought, had been given by relativity theory when it showed that time was of the nature of length. He suggested that now only mass and temperature needed to be similarly reduced. It was his belief that if more than one set of such relations exists, they

will be dependent upon one another in a very simple way, and that if in the manner suggested we obtain the ultimate units of interval and of mass by the aid of two universal and fundamental relations, then all universal constants will prove to be pure numbers, involving only integral numbers and pi, just as we have seen that in geometry several different units of angle, area, and volume may be chosen, which differ only by such a factor. This we shall call the *theory of ultimate rational units.*[9]

Lewis's theory, like Planck's, was an affirmation of the belief that nature is intrinsically mathematical. URU eventually became caught in the crossfire of charge and countercharge generated in the debate over dimensions. For the moment, however, it remained relatively unnoticed in the background.

Also in 1914, R. C. Tolman announced in *Physical Review* that he had discovered a relativity principle of his own, a companion to Einstein's relativity principle. It was based on the universal property of relativity of size, and he called it the "principle of similitude." He stated it as follows:

The fundamental entities out of which the physical universe is constructed are of such a nature that from them a miniature universe could be constructed exactly similar in every respect to the present universe.[10]

Tolman claimed that the theory depended on knowledge of both relativity and electron theory. Although he did not explain this dependence, there can be little doubt that in Tolman's mind what he was doing resembled what Einstein had done. The "principle of relativity of motion," Tolman's name for Einstein's version of relativity, rested on the assertion that it is impossible to detect a difference between absolute rest and uniform motion. Therefore the laws of physics had to be formulated in such a way that they are unaffected by a change from one reference system to another that is in a state of constant motion relative to the first. The equations of transformation happened to be the Lorentz equations, originally conceived within the framework of Lorentz's electron theory. Analogously, Tolman asserted, it is impossible to detect absolute size. Therefore the laws of physics had to be indifferent to a transformation from a reference universe, the ordinary world, to one arbitrarily smaller. His equations of transformation were ones constructed on the basis of an arbitrary reduction of the unit of length.[11]

In the 1914 paper, Tolman applied his technique and derived the correct results for the properties of ideal gases, the electromagnetic field, and the electron, as well as the *hohlraum* (black body). In the case of gravitational attraction, however, he failed to achieve consistent results. Affirming his commitment to the principle of similitude, he took the negative result to mean "that the gravitational attraction between two bodies is not merely a function of the masses of the bodies and the distances between them, but must depend on something else as well, perhaps, for example, on the properties of some intervening medium." He believed that "the search for the true nature of gravitational action will now become an important problem of physics, and the Principle of Similitude will be a criterion for judging the correctness of proposed solutions." Thus, having demonstrated what he thought to be the universality of his "fundamental principle," Tolman concluded that he had discovered, "as a matter of fact, a relativity principle," that is, the "relativity of size." [12]

While criticism was quick to appear, none of it accused Tolman of misinterpreting the special theory of relativity. Instead, it was pointed out that Tolman had not come up with anything new, that he had simply restated the basic assumptions of dimensional analysis.

The first person to express this view was Edgar Buckingham, a physicist-engineer at the National Bureau of Standards. In the next issue of *Physical Review*, Buckingham made clear his opinion that Tolman had not discovered a new principle, and even if he had, its application added nothing to physics.

Now I do not know whether the developments set forth above have ever been published in just this form, but it is certain that they are merely consequences of the principle of dimensional homogeneity, which is far from being new or unfamiliar. . . . My feeling that Mr. Tolman's "Principle of Similitude" is not really new may, of course, be mistaken. But for the purposes to which he puts it, it is, at all events, superfluous. [13]

The principle of dimensional reasoning was, indeed, nothing new to physicists. However, Buckingham, like everyone else, conflated two versions of it—similitude and dimensional homogeneity. In practice this did not matter, but conceptually they were based on different premises.

The concept of similitude is sometimes attributed to Galileo, more frequently to Newton; for both it was a simple way to investigate the manner in which a change of scale affects the properties of physical systems. For example, does doubling the thickness of a supporting beam double its strength?

Dimensional homogeneity was introduced by Joseph Fourier in 1822 in his *Théorie analytique de la chaleur.* [14] According to Fourier, every physical quantity can be expressed as some combination of five fundamental units of measurement: length, time, temperature, weight, and heat. Velocity, for example, is length divided by time. The "dimension" of a unit is the exponent of its symbol as it appears in the "dimensional formula." Velocity, then, has the dimensional formula l/t. Thus, the dimensions of length and time are 1 and -1, respectively. The principle of dimensional homogeneity was the requirement that only terms of the same "total exponent," that is, those reducible to the same combination of fundamental units, may be combined in an equation. In other words, the operation of addition and the relationship of equality are valid only for objects of the same kind. Furthermore, every quantity has a unique dimensional formula, and the form of an expression is independent of the size of the units of measurement.*

Dimensional homogeneity became a very mischievous principle when some physicists began to think of equations as expressions of true material equalities. As time passed, "dimension" lost its original meaning as an exponent and began to be used in the sense of "element" or "quality." Mass, length, time, and sometimes temperature took on aspects of the old Aristotelian elements, earth, air, fire, and water. "Quantity" came to be used in the same sense as "substance." To inquire about the dimensionality of a quantity was to ask what kind of a substance it is—what is its composition or its "kind"? The following statement, published in 1892, by W. Williams, an assistant in the Physical Laboratory, Royal College of Science, captures the flavor of contemporary thought.

The dimensional formulae may be taken as representing the *physical identities* of the various quantities, as indicating, in fact, how our conceptions of their physical nature (in terms, of course, of other and more fundamental conceptions) are formed—just as the formula of a chemical compound indicates its composition and chemical identity.[15]

Furthermore, once this material interpretation of the mathematical formulations of physical quantities became common practice, a number of difficulties appeared that resisted resolution. These problems involved the interpretation of certain dimensional formulas seemingly devoid of physical meaning or apparently in contradiction of the general assumptions of the dimensional method. During the second half of the nineteenth century, much discussion among English physicists was taken up with questions of how many fundamental quantities there are, which ones are fundamental (primary or elemental) and which secondary or derived, and what might be the physical nature of such quantities as magnetic permeability and inductive capacity or the mechanical equivalent of heat. The interpretation of heat and temperature was especially troublesome. What was the physical nature of temperature, a variable that often appeared in equations? Should it be treated as a fundamental quantity to be added to the three mechanical quantities, or was it really itself mechanical, as the kinetic theory of gases seemed to imply?[16]

The effort devoted to the resolution of such issues illustrates the strength of the commitment to a concrete mechanical interpretation of

---

*For example, the dimensional formula for velocity is always l/t, irrespective of what units of time and length are used.

physical reality. The fundamental measuring units of mechanics were treated as chunks of world-stuff, the matter from which, in an orderly fashion, the universe had been built up. Dimensional formulas showed how the primary elements had been put together, and equations indicated when complex structures were equivalent. But whether big chunks or little chunks of elemental stuff were used in the construction, the combinatory proportions never changed. The problems encountered in interpretation were looked upon not as flaws in the mechanical worldview itself, but as a consequence of incomplete knowledge.

In the 1910's as the discussion was being reactivated, the basic issues were much the same, despite the intercession of special relativity. They were still essentially ontological concerns, arguments centered about the nature of physical existence and the constitution of the world. No one except Bridgman hinted that these might be physically meaningless problems, and even he took an approach that was, at the outside, quasi-metaphysical. At the same time, however, the arguments were implicitly as much about the symbols of physics and what they stand for as they were about the metaphysical makeup of the universe. Here again, the logical positivists would later delineate the issues in a clearer way.

Tolman acknowledged that he was aware of the Newtonian version of similitude, or "dynamical similarity," as it was also called, but claimed that his own principle of similitude was quite distinct, since it depended on knowledge of relativity and electron theory.

For particular kinds of dynamical systems a somewhat similar hypothesis was advanced by Newton but we shall see that any complete development of the consequences of our postulate is dependent both on a knowledge of the electron theory and the theory of relativity.[17]

Mrs. Tatiana Ehrenfest-Afanassjewa, mathematician and wife of physicist Paul Ehrenfest, echoed Buckingham's criticism. In the July 1916 issue of *Physical Review*, she offered her assessment.

An accurate analysis shows that Tolman's considerations possess at least a close connection with the reduction to a definite hypothesis of the conviction of the homogeneity of all the equations of physics, a conviction which is commonly used without any foundation. This is not the intention of the author, as appears from his third paper on the same subject, yet he really does nothing else but construct a system of dimensions of his own (indeed one that in some respects deviates from the C.G.S. [centimeter-gram-second] system), and he examines all equations with a view to homogeneity as regards this system of dimensions.[18]

In other words, Ehrenfest-Afanassjewa viewed Tolman's transformations to a miniature world as a change from one set of units to another. "Instead of imagining measurements to be made with the same units in two different worlds, we may conceive the measurements to be carried out applying two different sets of units to the same objects 'in the same world.'" Therefore, Tolman's principle of similitude amounted to

the hypothesis that all the equations of physics remain invariant (i.e. with constant coefficients) when the units of measurement are transformed in such a way that certain appropriately fixed relations exist between the factors of transformation. Tolman further indicates what these relations are to be—so that he fixes his own system of dimensions [system of units].[19]

Further, observed Ehrenfest-Afanassjewa, if among the fundamental equations (equations that define the fundamental units) for Tolman's system the equation of gravitation were included, it would have to be said that "the equations of physics are of such a nature that no model transformation exists in which all the universal constants have the same values as in our universe."[20] The way around that, she added, and this had been tried by Gunnar Nordström, Swedish theoretical physicist, was to maintain the model universe and "correct" all physical equations so that they satisfy the principle of similitude.

Tolman strongly objected to this critique. Though admitting to its logical possibility, he stated his belief that similitude did not determine what is ordinarily meant by a set of dimensions:

The dimensions of a quantity may be best regarded, I believe, as a shorthand statement of the definition of that kind of quantity in terms of certain fundamental kinds of quantity, and hence also as an expression of the essential physical nature of the quantity in question.[21]

In defending this view, Tolman reverted to the position taken in 1892 by Williams. But Tolman made an even stronger claim. In 1917 he published an article in the *Physical Review* entitled "The Measurable Quantities of Physics," in which he attempted to combine his dimensional ontology with the axiomatic mathematics of Bertrand Russell to show how physics could be made a rational deductive system and still retain the physicality of its objects.[22]

In constructing his synthesis, Tolman borrowed heavily from Russell's 1903 *Principles of Mathematics* and, in particular, adapted Russell's arguments for an absolute theory of quantity to his own purpose.[23] He began by designating his indefinables, that is, the fundamental physical quantities and the operations by which they can be related. He argued that the fundamental quantities should be a set of quantities with extensive magnitude, because these are directly measurable and "have an additive nature." He chose length, time, mass, the charge of the electron, and entropy. Derived quantities would then be constructed from the fundamental ones by applying the indefinable operations. For operations, Tolman selected multiplication, division, differentiation, and the two kinds of vector multiplication.

Tolman plainly stated his belief in the physicality of the operations. "It should also be particularly noticed that these five operations are to be considered as performed upon actual quantities, not upon numbers which are used to represent the magnitudes involved." The results of this procedure would be that derived quantities are represented by dimensional formulas, which "may be regarded as a shorthand statement of the definition of that kind of quantity in terms of the kinds of quantity chosen as fundamental, and hence also as a partial statement of the 'physical nature' of the quantity in question."[24]

Next, Tolman drew attention to what he considered a common error in dimensional reasoning. This gave him the opportunity to emphasize his interpretation of dimensional formulas again. Confusion has arisen, he declared, in understanding the meanings of "quantity" and "unity." It had become the practice to speak of the dimensions of a *unit* when the dimensions of a *quantity* are meant, and this was obscuring the physical significance of dimensional formulas.

In the meantime, Tolman and Bridgman had been corresponding privately about similitude. Bridgman wanted to know what could be meant by the hypothesis that it is possible to construct a miniature universe. Tolman answered that "you could take the actual substances, entities or materials, or whatever you wish to call them which exist in the actual universe and build out of them a small universe which would be an exact reproduction of our actual universe when experimented with by a miniature observer." Of course, admitted Tolman, that "does not mean that such miniature constructs actually do exist in our universe," but they could. "In fact," he confided, "I had some idea that isotopes of elements might be of this nature, but this does not seem to be borne out by spectrum determinations."[25]

In Bridgman's opinion all of this was unacceptable. He thought that Tolman's hypothesis implied the work of a creative intelligence and that it was therefore "opposed to the entire spirit of physics." He reasoned that if Tolman's principle were true, the fact that this universe with its particular elements exists rather than another "must be due to arbitrary selection."[26] Therefore, with the intention of demonstrating that the principle of similitude in its universal form could not be correct, Bridgman, in 1916, joined the debate.

Bridgman's article, published in the *Physical Review*, was a tightly focused analysis dealing with the pivotal issues already indicated by Ehrenfest-Afanassjewa. First, he demonstrated that the application of the principle of similitude is "precisely like that of the theory of dimensions, except that the former is much more restricted." Therefore similitude is not a relativity principle. He went on to show that the reason similitude gives correct results for expressions involving dimensional constants is that these constants just happen to "have such a special form that they are not changed in numerical magnitude by the restricted change of units allowed by the principle." In other words, Bridgman pointed out, Tolman had set up his system in such a way that the constants, except for the gravitational constant, which could not be accommodated, would not change. For a new situation with no outside evidence that a dimensional constant has this form, the principle of similitude would possess no more power than the theory of dimensions. Bridgman further recognized that all of Tolman's examples had been drawn from situations for which the result was known ahead of time. And finally, he reiterated Ehrenfest-Afanassjewa's view that Tolman had defined a new set of units.[27]

If Bridgman thought he had put the matter to rest, he could not have been more mistaken. In reducing similitude to a restricted application of dimensional reasoning, he had merely demonstrated, to his own satisfaction only, the nonuniversality of similitude, thus avoiding what he thought might be reason to reserve a place in scientific explanation for a divine creator. He had also shown that Tolman's units were arbitrarily chosen within a preselected domain, and therefore, as Buckingham had charged, similitude had no creative value. Overall he had just confirmed what Buckingham and Ehrenfest-Afanassjewa had already said. Nevertheless, this modest article marked the beginning of Bridgman's involvement in the philosophy of physics. It set him to thinking about how he might clear up the confusion about dimensions.

Tolman remained unconvinced. Nevertheless, he was still interested in Bridgman's opinion. In 1919 he wrote to Bridgman,

I am wondering how you feel about the principle of similitude these days. I still maintain my old heresies.[28]

To this Bridgman replied,

As for the principle of similitude, I am, if possible, more orthodox than ever. This summer, I spent some time trying to get straight in my own head some fundamental notions about dimensional reasoning, and the result of my lucubrayions [sic] was some ideas which possibly I may publish one of these days. I am sure when you see these clear and cogent remarks you will realize only to [sic] clearly the heretical deviousness of your ways.[29]

Bridgman was no less unyielding than Tolman. Again, in 1920, we find him chiding Tolman in the same Sunday School teacher tone.

[John] Hersey dropped into the laboratory yesterday, and said that you had been giving them a talk on Similitude, and that you were still as much of a heretic as always. It makes me very sad to see your obdurate refusal to come into the fold. I, on the other hand, have just been talking about the general matter of dimensional analysis, and am more than ever convinced of my position.[30]

It may or may not have been Tolman's intention to fuel a controversy, but that summer circumstances provided him with fresh encouragement. He described his experience to Bridgman.

I tried to find you when I was recently in Cambridge, but you were away on your vacation. While I was there I had my first chance to study Sommerfeld's "Atombau." It is fine stuff. I do not see how we are going to escape the Bohr atom in the case of monatomic hydrogen, singly ionized helium and doubly ionized lithium. The concordance of prediction and experiment is perfectly tremendous. I noted with pleasure that the complete expression for the Rydberg constant has such dimensions as to transform correctly in accordance with the requirements of the theory of similitude.[31]

Bridgman was not so sure.

I wish that you had gone a little more into detail about the Rydberg constant. You remember that when we talked in the spring, you agreed that if this constant has the dimensions of a time, that the principle of similitude could not apply to it. What have you done to it to make it change its dimensions? Perhaps you remember enough of the scriptures to know that even the tiger cannot change his spots.[32]

When Tolman tried to explain, Bridgman lost his equanimity. "I believe you will drive me crazy yet," Bridgman snapped.

The well known eel has nothing on you for high coefficient fo [sic] surface slip. . . . What are your rules for telling whether you are going to transform the constant according to your transformation equations, or whether you are going to leave it unaltered, as you did in the example of Rydberg's number? And do you relly [*sic*] think, if you admit dimensional constants to be played with, that there is anybody else in the world who will know how to juggle with them or that you will know yourself unless you know the answer beforehand?[33]

Then, referring to an unsolved problem discussed by Sommerfeld, Bridgman sarcastically added,

It may be that there is a new constant of nature involved, as unsuspected now as the quantm [*sic*] was twnety [*sic*] years ago. I recommend the principle of similitude to the rescue.

To Bridgman's attack, Tolman retorted,

Don't talk to me about eels. . . . I am very glad that you and Mrs. Bridgman have much more liberal ideas on politics than on the principle of similitude. After all the future of applied sociology is more important to the human race than the future of mathematical physics.[34]

In any case, Tolman did not give ground. He continued to apply his similitude principle and could not resist flaunting his success in achieving publishable results that might prove that similitude was not a sterile principle after all. Six months later, in March 1921, he wrote to Bridgman,

You will be interested to know that I have got an article on the Principle of Similitude and Entropy of Polyatomic Gases coming out in the number of the *Journal of the American Chemical Society*. Of course, under your influence I have come to realize that the principle of similitude is absolutely no good, nevertheless I still seem to be able to amuse myself with it.[35]

The ideas Bridgman mentioned to Tolman became, in 1920, a series of five lectures on the dimensional method given at the Graduate Conference in Physics at Harvard. In 1922 they were published by Yale University Press as *Dimensional Analysis*. Here we have the earliest application of the strategy Bridgman thought he had learned from Einstein. It was also his first step toward a repudiation of the Newtonian synthesis; for the present, he repudiated only one term—the mechanistic ontology.

Bridgman was clearly unwilling to admit that dimensions, or units, as the word "dimensions" had come to mean, were concrete physical things. He took pains to emphasize his disapproval of "the view that regards a dimensional formula as an expression of operations on concrete physical things."[36] This denial of the concreteness of physical entities was evidently the consequence of a drastic change from an earlier conviction, for subsequently he confessed to correspondent Arthur Bentley that there had been a time when he believed that things "really are."

Once I started a little essay with the statement [that?] my absolute starting point was the recognition that "things really exist." I cannot recapture now what I meant by that.[37]

The shift must have been accompanied by a severe and lasting disillusionment, for one senses a distinct feeling of resignation with respect to this disappearance of the absolute. Asked by the publishing firm of Macmillan to comment on a manuscript by Professor Tobias Danzig entitled *In Quest of the Absolute*, Bridgman wrote:

My chief criticism concerns the title and the preface. Certainly no one now imagines that he can reach the absolute, or in fact that there is any meaning in the concept, and I very much doubt whether as a historical fact the course of scientific development has been determined by a "gigantic effort of man to free himself of the anthropomorphism imposed on him by 'Nature.'" The preface gives the impression that the book will be devoted to this thesis, but in the actual working out it is not. It ought to be possible for the author to find some method of introduction as vivid and arresting as the one which he has chosen, which at the same time does not arouse false expectations.[38]

At the same time, Bridgman did not wish to embrace subjective idealism as an alternative. As he explained to Buckingham:

Now I am the last person in the world to take the attitude of Berkeley which neglects too much the contribution of the outside world to our mental processes. I always have vividly in mind that we are compelled to measure lengths with meter sticks in the way we do because of some characteristic of the things outside us. Now here is the difference between you and me. I think that having once made our measurements in the way which nature compels us to, this is an end of it, and from here on everything is purely formal.[39]

But if there is no meaning in the idea that there are "things" which "really exist," if the quantities measured by physics are not "things," and if they are not Berkelian ideas, what, then, are the objects of physics? To answer this question, Bridgman followed what he took to be the example of Einstein and turned to an analysis of measurement. The quantities of physics, he said, are the product of a method of measurement. They are numbers generated by acting according to a definite set of rules. Bridgman described the procedure and its outcome in *Dimensional Analysis*.

In dealing with any phenomenon our method is somewhat as follows. We first measure certain quantities which we have some reason to expect are of importance in describing the phenomenon. These quantities which we measure are of different kinds, and for each different kind of quantity we have a different rule of operation by which we measure it, that is, associate the quantity with a number. Having obtained a sufficient array of numbers by which the different quantities are measured, we search for relations between these numbers, and if we are skillful and fortunate, we find relations which can be expressed in mathematical form.[40]

Primary quantities "are the quantities which, according to the particular set of rules of operation by which we assign numbers characteristic of the phenomenon, are regarded as fundamental and of an irreducible simplicity." Again, "in the measurement of primary quantities, certain rules of operation must be set up . . . by which it is possible to measure any primary quantity directly in terms of units of its own kind."[41]

Furthermore, "the dimensions of a primary quantity . . . have no absolute significance whatever, but are defined merely with respect to that aspect of the rules of operation by which we obtain the measuring numbers associated with the physical phenomenon." Equations, being nothing more than expressions of equality "between numbers which are the numerical measures of certain physical quantities," have no physical significance either. Thus, for the application of dimensional reasoning, dimensional homogeneity is not, in general, necessary. Only the requirement that the absolute size of the unit should not make a difference was important.[42]

Secondary quantities are not obtained by some operation which compares them directly with another quantity of the same kind which is accepted as the unit (as are primary quantities), *but the method is more complicated and roundabout* [emphasis added]. . . . Now there is a certain definite restriction on the rules of operation which we are at liberty to set up in defining secondary quantities. We make the same requirement that we did for primary quantities, namely, that the ratio of the numbers measuring any two concrete examples of a secondary quantity shall be independent of the size of the fundamental units [within a given system of units].[43]

This is more than just a hint of the operational approach that Bridgman would present in the *Logic of Modern Physics* (1927). It is a preview and, as such, provides valuable insight into the semantic content of the word "meaning" as he later applied it. Bridgman said as much himself when he wrote in 1937 to Arthur Bentley, "I think my point of view with regard to dimensional analysis is very closely connected to my operational point of view, in fact I got started on the latter only after having worked through my ideas about dimensional analysis, although I did not express it in those words."[44] Clearly, in the context of dimensional analysis, an operation is an act of measurement. In a secondary sense, it is also a mathematical step—the execution of a procedural rule in the manipulation of mathematical symbols.

On this question of the physical concreteness of the fundamental units of measurement, *Dimensional Analysis* received disparate comments. For example, Ludwik Silberstein of Eastman Kodak's research laboratory appreciated what Bridgman was trying to accomplish: "Your abstinence throughout the book from 'mystical' elements and from hypostasy of the products of the human brain is a very welcome feature."[45]

John Hersey, physicist at the National Bureau of Standards, could not imagine that anyone could believe otherwise.

But you seem to imply that people in general may at times overlook the distinction and deceive themselves into imagining that you can operate on physical quantities in the sense of actual realities and not mere numbers. For myself the possibility that the word "quantity" in physical discussions could mean anything other than a numerical magnitude, never entered my head until reading your book.[46]

Nevertheless, the referee for Yale University Press felt differently, and George Parmly Day, head of Yale University Press, did not hesitate to tell Bridgman so.

In the report, I am advised, that the reader states that he can find only one point on which he would dissent from your position, and that is in your contention that not physical quantities, but only their numerical values, are susceptible of combination according to the rules of ordinary algebra.[47]

Reacting to this information Bridgman lightly countered, "I shall . . . keep my ear to the ground for any rumors as to the identity of the reader [could it be Tolman?] and will immediately attack him on the slightest suspicion."[48]

Buckingham, however, thought Bridgman's approach was too mathematical and subjective and that Bridgman was headed for trouble.

I think your own conception of the whole subject is too mathematical and too remote from the range of ideas that we get directly from our senses—I mean for your own good. In other words, it seems to me that you do not, in your own mind, ascribe enough importance to physical instinct or common sense, which can not in reality be replaced by any amount of logic. For example: to treat physical quantities impartially is to ignore a vitally essential fact, namely, that there are in reality only a very few physical magnitudes of which we can form any conception as measurable quantities. . . .

I spent several months, once, thinking about that subject and about what there really is in all our talk about "quantities" and about "measuring" them. But about all I got out of it, beside 150 pages of drool, was the conviction that when you mix gumption with metaphysics you only spoil the gumption and end up with a dirty and useless mess.[49]

To turn to another train of events, Bridgman was also on a collision course with the other member of the original pair of collaborators whose initiative had given rise to so much of this controversy over the meaning of measured quantities. With Tolman the disagreement had been about the nature of the objects of physics. With G. N. Lewis, the issue was the mathematical structure of the world. Bridgman's position that the equations of physics were not to be regarded as having any absolute significance meant that there could be no physical meaning to a system of rational units. In *Dimensional Analysis*, he characterized Lewis's theory as being "quasi-mystical."[50] It was too idealistic for Bridgman's taste. However, in taking his stand against Lewis's URU, he also moved in the direction of repudiating the second term of the Newtonian synthesis—the intrinsic mathematical organization of nature.

And, as if on cue, at the same time that *Dimensional Analysis* came out, G. N. Lewis revived his theory of Ultimate Rational Units in a paper published in *Philosophical Magazine*. According to Lewis's theory, "there is possible a set of units in terms of which all universal constants (and not merely those which are employed in defining the units) will be reduced to simple numbers." His justification for the assertion was his belief that "there is nothing nature abhors more than an arbitrary number which possesses no intrinsic meaning." "Every physical constant," Lewis claimed, "indicates some flaw in our scientific system," and "it becomes as useful a scientific service to eliminate these constants as to discover them."[51] With this in mind, he restated his argument that all physical quantities could be reduced to the dimensions of a power of length.

O. J. Lodge, the editor of *Philosophical Magazine* and the most conservative party to the debate, appended a disclaimer to Lewis's article:

I for one hold that an attempt to unify essentially different physical quantities, and obliterate the ratios or constants connecting them can only result in confusion. *A unit is not unity.* Length and Mass and Time and Energy and Momentum are not the same . . . and the arbitrary elimination of constants, or the attempt to mask them and replace them prematurely by pure numbers, is retrograde.[52]

For his part, Bridgman objected to the idea of mathematical simplicity in nature. He explained what he meant in a letter to Lewis.

Now to come back to the main question of U.R.U. I believe that the main difference between us is as to the extent to which we believe nature to be simple. You believe that all the laws of nature of wide applicability are simply connected with each other. By "simple" you mean simply expressible in terms of human mathematical analysis. It would seem to me very strange if our mathematics, which for many phenomena is such an extremely clumsy tool, should turn out to be so peculiarly well adapted for the correlation of all the broad universal relations.[53]

Lewis was more optimistic and pointed out that despite the ultimate unprovability of the rationality of nature, science is, after all, able to assimilate more and more experience into a rational framework." My dear Bridgman," he answered:

I was very much interested in your letter, and after reading it and considering our previous conversations I have come to the conclusion that our chief differences are due to the fact that you weld together your general philosophical or cosmic ideas with those of physics, whereas I, for the most part, keep them pretty well divorced. Speaking as a philosopher, there is no one who distrusts an ultimate more than I do. I suspect that a minute infusorial creature in the darkest depths of the ocean knows just about as much of the universe as we do. At best our science furnishes but a bit of an infinitesimal cross-section through nature, but in this bit it is surprising how much law and order has been discovered. We shall never know to what extent the apparent orderliness of nature depends upon our rigorous exclusions from observation of all the things which do not come within our little schemes. But granting the limited scope of our inquiries, it is really remarkable that we are able to assimilate constantly an increasing amount of new experimental material without any great increase in the complexity of our mathematical formulae and our laws.[54]

In the background, stepping up the intensity of the dispute, was the sharp-tongued physicist-philosopher Norman Campbell, lashing out at just about everybody involved. Earlier, in 1920, Campbell had published a book, *Physics, the Elements*, which he intended as a general critique of the fundamental principles of physics. In it, rather than considering the problem of measurement in the metaphysical terms of the ongoing debate, he clearly stated it as an epistemological difficulty.

The application of mathematical analysis to physical problems doubtless depends greatly on the introduction of measurement; but it raises some difficulties because mathematical ideas, being based upon "internal judgments," differ in their nature from physical ideas, based on "external judgments." Our view of the exact relationship of mathematics and physics is sure to be influenced by the view that we take of the exact nature of measurement; it will not do to show that we actually adopt certain processes; our conclusions will be uncertain unless we know that we could adopt no others.[55]

From this vantage point he criticized dimensional reasoning and ridiculed the idea that there might exist natural units of universal validity.

In January 1924, Campbell renewed his attack on rational units, arguing that Lewis's theory of Ultimate Rational Units was "demonstrably false." There are, he observed, available to Lewis "infinitely many sets of values assigned to his constants which would enable him to determine from them a set of units." Professor Lewis would have noticed this if he had not argued from the presumption that URU is true: "he apparently believes that by juggling with dimensions he can rid science of all 'arbitrary' numbers."[56] Furthermore, there is no criterion for judging a number to be simple. Nevertheless, after all this forceful argumentation, Campbell arrived at the weak conclusion that even if URU turned out to be true, it would be useless.

Two months later, in March, Campbell again turned his critical instincts toward dimensional analysis. He argued that the only conditions under which dimensional analysis is valid and useful are those such that the laws governing the phenomena are fully known and expressible in their differential form. In this case, "then all the formalities which various writers have elaborated are useless . . . all that is required is a sound appreciation of the meaning of similarity and knowledge of the rule, which is one of its obvious consequences, that all terms in a numerical law have the same dimensions."[57] Therefore, he insisted, the correct application of dimensional analysis rests on physical judgments rather than on logical analysis; that is, on the physical assumptions included in the decision that a particular law is relevant to the system under consideration and that the selected variables are indeed determinative.

Buckingham objected to Campbell's attitude. He felt that Campbell had exaggerated the superficiality of the dimensional method. Buckingham pointed out that Lord Rayleigh had used it to deduce new physical relationships, thus circumventing a certain amount of laboratory work, and while this was probably the application most interesting to Campbell, it was also the one most susceptible to error. "But," he continued, "the principal value of dimensional analysis is not for obtaining new theoretical results or for proving anything *a priori*."[58] Nevertheless, Buckingham argued, dimensional analysis is a valuable tool—and one of increasing utility given the recent and significant development of the physics of engineering and manufacturing.

Buckingham went on to draw a distinction between the experimentation technique of pure physics and that of technical physics. In pure physics, an experiment is so designed that "equations that describe it can be solved without great labour." In contrast, technical physics must find solutions for problems "as they are, with all their complications." It is in such situations that dimensional analysis proves its value. As for the trustworthiness of the results, Buckingham reminded the reader that "we are never absolutely sure of anything in physics till it has been proved by experiment, even if we have written down a differential equation; and that in almost any operation, avoiding mistakes is a matter of care, experience, and skill."[59]

But evidently Buckingham, who was an engineer, had not been accepted into the circle of physicists engaged in the controversy. His remarks went unacknowledged throughout the entire debate.

Meanwhile, Bridgman had asked his comrade under siege, Lewis, "Have you, or do you intend, to reply to Norman Campbell's article in the Phil Mag on ultimate rational units?"[60] Lewis replied, "I did send to the Philosophical Magazine an article called forth by Norman Campbell's effusion, and I shall be glad to have your comments on it."[61]

Lewis began his rejoinder to Campbell's "highly provocative paper" by pointing out the contradiction between Campbell's claim the URU is "demonstrably false" and his statement that if URU is true, it is useless. The fact that an infinite number of values may be possible, Lewis took not as disproof of URU "but rather [as] a statement that from a purely logical standpoint no proof of any physical theory can ever be obtained." As to the problem of simple numbers, Lewis could only express his faith in their eventual appearance as universal constants, reminding the reader that every scientist, to some degree, believes in the simplicity of natural laws.[62]

But the real basis for the undeserved criticism directed at URU he attributed to an unfortunate contamination spilling over from the controversy over dimensions. The idea that dimensions determine the physical nature of derived quantities is responsible for a resistance to "any suggested modification of the orthodox statements regarding fundamental and derived magnitudes." In Lewis's view, "there is nothing in natural laws which prescribes the choice of any particular measured magnitude, or of any particular number of such magnitudes which may be taken as fundamental."[63]

Again O. J. Lodge appended comments. He raised the question whether Lewis might "not [be] attaching undue importance to the numerical part of physical quantities, and trying to regard actual things as if they were or might be pure numbers. . . . Every physical quantity has a nature of its own, and it is the reverse of helpful to mask it."[64]

Then, in 1926, Ehrenfest-Afanassjewa took up the topic again, subjecting it to an extensive and confusing mathematical treatment. This time she appeared to favor the method of similitude.[65] Norman Campbell published a sharp reply. In his opinion, such "elaborate symbolism is not merely superfluous, but positively misleading. It concentrates attention on 'elegant irrelevancies' . . . and prepares the way for all the charlatanry with which the word *dimensions* has unfortunately become associated."[66]

Campbell went on to argue that the validity of dimensional reasoning is based on the physical assumption of similarity, which is equivalent to the mathematical statement of proportionality. Therefore, the mathematics is "so simple that it is hardly worth writing down; it certainly does not justify the elaborate conception of invariance under transformation."[67] Furthermore, he pointed out, most systems are not completely similar and really only need to be sufficiently similar for the immediate purpose. The judgment of similarity is a physical one and stands outside the domain of mathematics.

Bridgman could stand it no longer. He felt it was time for him to set things straight again. In the December 1926 issue of *Philosophical Magazine*, he spoke out.

The subject of dimensional analysis has been so much discussed in the pages of the *Philosophical Magazine* in the last few years that one would like to see it die a natural death, were it not that in the last two communications things have taken such a turn that an unchallenged acceptance of the views there presented may easily affect the use which the physicist makes of this analysis.[68]

Referring to his *Dimensional Analysis*, Bridgman took a stand against Ehrenfest-Afanassjewa's mathematical method of similitudes, arguing "that the dimensional method is much easier for the physicist to apply, and in many important practical cases is more powerful." And against Campbell, Bridgman argued that "it is certainly not necessary to assume that physically similar systems are possible in order to apply the dimensional method." He concluded by saying that if the application of dimensional reasoning really rested on such complex analysis, "few physicists would ever have the courage to attempt a dimensional analysis at all." [69]

Bridgman's remarks were notably unhelpful. They added nothing to what had already been said. The dimensional analysis controversy had no clear resolution, either. Whatever significant issues might have been raised about the nature of measurement were eclipsed by the appearance of the Heisenberg uncertainty principle. In 1931 Bridgman published a second edition of *Dimensional Analysis*, with no important changes. The theories of Tolman and Lewis were unproductive, and except for the persistent efforts of physicist turned author and philosopher L. L. Whyte, such discussions virtually disappeared from the formal discourse of physics journals. But the problem of dimensions did not vanish altogether. As a technique of physical reasoning, dimensional analysis remained alive in physics textbooks. As a philosophical issue, it reappeared briefly in the late 1940's. Dimensional reasoning also provided an exemplary reference for psychologists and social scientists attempting to establish their disciplines on a mathematical basis.

However, for this small group of American physicists, the controversy over dimensional reasoning was at least an instrument for a belated examination of some fundamental physical concepts. For Bridgman, it was a successful initiation into the field of interpretive physics. The generally favorable response to *Dimensional Analysis* encouraged him to expand his efforts along the same lines, the results of which culminated in the publication, in 1927, of the work that would so profoundly affect his later life, *The Logic of Modern Physics*.

Just as important for the historian of science, the dispute over dimensions is an indicator of what occupied the thoughts of contemporary physicists. It supplies a literary artifact that speaks the language of the age. As such, it provides a key to understanding the prominent philosophical concerns and a gauge to judge the flexibility of minds educated to function in the Newtonian universe, minds that had to adapt their beliefs to make room for the encroachment of novelty. It was not an easy task that Einstein set for this generation of American physicists, this unphilosophical group of men for whom science was supposed to simply discover the truth about the physical universe. For Bridgman, the task was especially difficult, and the adjustment proved to be a lifelong effort.

CHAPTER 5

The Shadow of Einstein: Relativity and the Operational Method

No doubt, those who are truthful in that audacious and ultimate sense that is presupposed by the faith in science *thus affirm another world than the world of life, nature, and history; and insofar as they affirm this "other world"—look, must they not by the same token negate its counterpart, this world, our world?—But you will have gathered what I am driving at, namely, that it is still a* metaphysical faith *upon which our faith in science rests—that even we seekers after knowledge today, we godless anti-metaphysicians still take our fire, too, from the flame lit by a faith that is thousands of years old, that Christian faith which was also the faith of Plato, that God is truth, that truth is divine.—but what if this should become more and more incredible, if nothing should prove to be divine any more unless it were error, blindness, the lie—if God himself should prove to be our most enduring lie?—*

—Friedrich Nietzsche, *The Gay Science* (1882)

■ ■ ■

PROBABLY NO ONE but Nietzsche could so dramatically have expressed the sense of despair following on the destruction of the very possibility that truth is absolute. For him this stark realization was a consequence of the profound epistemological implications of Darwin's evolutionary theory. Had Nietzsche lived to see the publication of special relativity, he might have interpreted it as confirmation of his prophecy. He would not have been alone. This fear was at the basis of an impassioned attack against relativity by both scientists and nonscientists in America. For those who had placed their confidence in the revelatory nature of scientific discovery, relativity was a betrayal, a scientific imposter. Jealously held values were threatened by the suggestion that common experience was not an adequate foundation for the edifice of physical reality built up by science. That such intuitive and elementary concepts as space and time should prove to have no absolute status made no sense within a body of knowledge supposed to be grounded in the experience of every individual. Relativity, its detractors charged, was sophistry, it was unintelligible, unnatural, anti-inductivist, subjective, antidemocratic. It was un-American and it was not science.

The full impact of relativity on American sensibilities was not felt until after World War I and the spectacular confirmation in 1919 of Einstein's prediction that light would be deflected by the gravitational field of the sun by 1.68 seconds of arc, precisely the value demanded by the general theory of relativity. In the years immediately following, an explosion in the number of books and articles on relativity took place in professional and popular literature—reporting, explaining, interpreting, attacking, defending—all concerned in one way or another with the possible meaning of Einstein's discovery.[1]

In the face of the counterintuitive character of the relativistic universe, philosophers and a number of prominent physicists (including Einstein himself) felt called upon to explain how the methods of physics could have found the world to be such a queer place and how the idea of physical reality could still have any intelligibility at all. They assumed the task of formulating the principle of relativity in a way that it might appeal to ordinary intuition and of translating the mathematical language of relativity into terms that could be understood by common sense.

This was hardly the intention of P. W. Bridgman when, in 1927, he made his contribution to this literature in a book entitled *The Logic of Modern Physics*.[2] To Bridgman's mind, these explanations were not dealing with the right problems. The difficulties entailed by relativity, he thought, were much more profound and far-reaching than acknowledged in these efforts to overcome the unintelligibility of its formal language. Indeed, Bridgman believed that some of these difficulties were part and parcel of this formal apparatus. But prior to them were the metaphysical and epistemological assumptions of relativity, and these, Bridgman felt, needed to be scrutinized. The fact that at the time much of what he sensed was only dimly perceived and inadequately communicated does not mean that his misgivings were unfounded. Bridgman was particularly annoyed with Arthur Eddington's interpretation of relativity, especially with what he saw as Eddington's absolutism: "Eddington's ecstatic contemplation of the absolute vision which is afforded by the new theory [general relativity] is pure tommyrot."[3]

The basic problem was the meaning of truth in a universe where the traditional notion of absolute knowledge had been rendered obsolete. To be sure, the damage had already been done by Darwin when he replaced the God of design with natural selection. The de-divinization of the cosmos implied a concomitant naturalization of knowledge. And it is not insignificant that the relativity debates coincided with a renewed attack by religious Fundamentalists against evolution and the scientifically sympathetic views of liberal Protestant modernists. But the difference was that relativity posed the problem with a more compelling concreteness as well as with a more refined logical specificity. What could once have been dismissed as philosophical speculation was now at the heart of a mathematically articulated physical theory whose validity could be experimentally confirmed.

Although Bridgman was not entirely convinced that relativity was here to stay because he believed that it contained some unresolved empirical difficulties, he thought that special relativity (most definitely not general relativity, where Einstein, in Bridgman's view, had abandoned his earlier empiricism) provided the key to the most urgent question— What can be known? (Or, perhaps more precisely, What cannot be known?) What is knowledge of, if it is not of the absolute, and what is the meaning of the individual terms of the language in which a physical theory is expressed? What do these terms signify? But the most important insight Bridgman believed that Einstein had given him, which lay for the moment undeveloped at the foundation of Bridgman's interpretation of physical reality, was that knowledge is of human, rather than of transcendental, origin.

At the same time, Bridgman was not prepared to embrace an anti-scientific irrationalism that would devalue science and undermine its cognitive authority. He was not willing to say that scientific knowledge had no objective content. Rather, its objectivity had to be founded on something other than a transcendental reality that has no empirically detectable aspects. Bridgman believed that Einstein had shown what this was. *The Logic of Modern Physics* was a treatise aimed at enunciating that alternative.

Bridgman offered *The Logic of Modern Physics* for publication with no pretensions of contributing to systematic philosophy. He described his work as intended for the general reader and other scientists, although perhaps philosophers might also be interested. Characterized by the publisher's referee as having a home-grown quality, the book was not a polished work in a literary sense. But it was intense and sincere, as Bridgman surely had intended. He felt that it was more important to express his ideas than to produce a polished philosophical dissertation. He even decided against changes suggested by Alfred Hoernle, a philosopher who advised him about the manuscript, "so the reader will find it [helpful] for his estimate of just how much of a damn fool the author is."[4]

As might be expected from Bridgman, *The Logic of Modern Physics* was a tribute to economy of style and directness of expression. He wasted no time getting to the point. In the opening paragraph, he immediately confessed:

Whatever may be one's opinion as to our permanent acceptance of the analytical details of Einstein's restricted and general theories of relativity, there can be no doubt that through these theories physics is permanently changed. *It was a great shock to discover that classical concepts, accepted unquestioningly, were inadequate to meet the actual situation* [emphasis added], and the shock of this discovery has resulted in a critical attitude toward our whole conceptual structure which must at least in part be permanent. Reflection on the situation after the event shows that it should not have needed the new experimental facts which led to relativity to convince us of the inadequacy of our previous concepts, but that a sufficiently shrewd analysis would have prepared us for at least the possibility of what Einstein did.[5]

This uncomfortably candid introduction immediately placed Bridgman's *The Logic of Modern Physics* in a category by itself. It promised no soothing rationalization of modern physics or inspiring exhibition of its metaphysical wonders: on the contrary, it drew the reader into an intensely personal attempt by a physicist to come to terms with the "disquietude" (Bridgman's word) he experienced when confronted with a conceptual upheaval at the foundation of physics. It led the reader on a search for something stable, an island of permanence, in the midst of the change that was so rapidly transforming physics. Thus Bridgman continued:

Looking now to the future, our ideas of what external nature is will always be subject to change as we gain new experimental knowledge, but there is a part of our attitude to nature which should not be subject to future change, namely that part which rests on the permanent basis of the character of our minds. It is precisely here, in an improved understanding of our mental relations to nature [whether he meant epistemological, psychological, logical, or even neuro-logical, Bridgman did not indicate], that the permanent contribution of rela-tivity is to be found. We should now make it our business to understand so thoroughly the character of our permanent mental relations to nature that an-other change in our attitude, such as that due to Einstein, shall be forever im-possible. It was perhaps excusable that a revolution in mental attitude should occur once, because after all physics is a young science, and physicists have been very busy, but it would certainly be a reproach if such a revolution should ever prove necessary again.[6]

Thirty years later, asked to comment on this work, Bridgman observed:

To me now it seems incomprehensible that I should ever have thought it within my powers, or within the powers of the human race for that matter, to analyze so thoroughly the functioning of our thinking apparatus that I could con-fidently expect to exhaust the subject and eliminate the possibility of a bright new idea against which I could be defenseless.[7]

At the time, however, the urgency of the matter overshadowed the possibility that what he was attempting to do might lie beyond the ca-pabilities of the human intellect. Therefore, in *The Logic of Modern Phys-ics* Bridgman set out to show where physics had gone wrong and to indicate how such error could be avoided in the future. The first step in his argument was to renounce still another Newtonian assumption, the principle of the uniformity of nature. Relativity, Bridgman asserted, has underscored what experience has shown us time and again, and that is as we cross new experimental frontiers, we are likely to encoun-ter phenomena of an entirely unexpected character. Because of this, he said, the idea that nature is necessarily a formally coherent unity must be given up. As a rule, new experience cannot be expected to fall under the same generalizations as the old.

Therefore, Bridgman proposed, the physicist needs to assume a new attitude toward the concepts used in physics. He can learn this by studying what Einstein did when he formulated the restricted (special) theory of relativity. In Bridgman's view, Einstein's approach exempli-fied the frame of mind proper to the physicist. His methodology incor-porated principles that could guide the interpretation of physics from a standpoint that would always maintain contact with experience. If adopted, these principles would guarantee that the physicist need never again be "embarrassed" by searching in nature for an unrealizable ideal such as Newton's absolute time. In other words, Bridgman was admonishing physicists to change their expectations of what can be known and advising them that Einstein's analysis of measurement was the key to making sure that physical knowledge would be truly empirical. Bridgman called the point of view thus characterized the "operational view."

Bridgman argued that Einstein had shown that it is necessary to be critical of the idea of absolute physical properties and scrutinize instead the method whereby such properties are concretely determined. If that is done, it is evident that no knowledge can be had of any physical property independently of a determinative act. Because it obtrudes the observer, this determinative act, which in physics is a measuring procedure, rules out the possibility of the absolute existence of the property being considered. In Bridgman's opinion, Einstein's analysis of simultaneity was fundamentally a recognition of this situation. Therefore, if an operational point of view had been adopted earlier, the discovery that simultaneity is not absolute could have been anticipated. Similarly, simple examination of the way time is measured would have shown that absolute time cannot be a meaningful physical concept. This idea, that there is no knowledge which does not depend on a human presence, was one that would come to play an increasingly prominent role in Bridgman's thinking.

In this context, Bridgman distinguished two senses of relativity, one perspectival and the other comparative. In the first sense, he said, relativity is "the merest truism" because all knowledge is built up from operations and therefore must be regarded from the perspective of the particular operations involved. In the second sense, because a physical state must be determined with respect to a standard, knowledge is relative to that standard. However, he added, "In saying that there is no such thing as absolute rest or motion we are not making a statement about nature in the sense that might be supposed, but we are merely making a statement about the character of our descriptive process."[8]

However, if what is measured has no absolute existence, then what does the physicist know after making a measurement? What is knowledge *of*? Bridgman decided that the only meaning he was able to attribute to a physical concept is that which derives from its determining operations. As the realities of physics, absolute or ideal properties, being experimentally out of reach, must be replaced by measuring

operations, which are empirically palpable. In short, knowledge is *of*
what the physicist does. Bridgman stated this principle, the cornerstone of his operational outlook, in its most extreme form:

In general we mean by any concept nothing more than a set of operations; *the
concept is synonymous with the corresponding set of operations.**

The epistemological justification that Bridgman offered for operational thinking was "if experience is always described in terms of experience, there must always be correspondence between experience and
our description of it."[9] The ground of stability or permanence that
Bridgman sought was to be found in experience, which is what it is no
matter what is said about it. At the very minimum, the physicist knows
what he experiences when he carries out a physical procedure. And to
emphasize the physical quality of this experience, Bridgman reminded
the reader that operations must be capable of actual experimental execution. Furthermore, he insisted that each concept, to be meaningful—
that is, to indicate something physically real—should be uniquely associated with an operation. One operation, one concept. At the same
time, he argued that the physical reality of a construct (a physical entity of secondary reality), such as stress, for example, is in proportion
to the number of independent operational routes that lead to it.

To illustrate the operational criterion for judging meaning, Bridgman
considered the concept length. What can be known about length? As
much as, and no more than, how to measure it. Thus by fixing the
procedure for the measurement of length, the meaning of length is
exhausted. And since "the operations by which length is measured
should be uniquely specified,"[10] if there is more than one way to determine length, there is more than one concept. Correctly speaking, each
should be given its own name.

And this is, in fact, precisely what Bridgman wanted to show to be
the case for the physical concept of length. Ordinary length (length in a
stationary system of reference) is measured by laying down a measur-

---

*Bridgman, *Logic*, p. 5. To this maxim, Bridgman added the qualification that "if the
concept is physical, as of length, the operations are actual physical operations, namely,
those by which length is measured; or if the concept is mental, as of mathematical continuity, the operations are mental operations, namely those by which we determine
whether a given aggregate of magnitudes is continuous." However, because he overwhelmingly emphasized physical operations in the development of his thesis, most readers failed to notice that he had even mentioned mental operations.

ing rod and simply counting the number of times this operation must be repeated. Bridgman characterized such length as "tactual" length. However, such a procedure cannot be used to measure length in distant or moving systems. Here it is necessary to take sightings and calculate a quantity we call length, making certain assumptions about the geometrical behavior of light beams. Now the length we are dealing with is optical in character, not tactual, and by virtue of what is involved in its measurement, it is not the same concept. Nevertheless, he pointed out, there is a "connection," and it is given by the transformation equations of relativity theory.

By concluding that Einstein's length and ordinary length were different categories, not only was it permissible to expect them to exhibit different properties, but the troublesome relativistic length contraction was changed into an attribute of something that was not really length at all, at least as ordinarily understood. The disturbing behavior of length at high velocities was removed by a change in the definition of length. Thus, Bridgman saved the traditional meaning of length. He saved empiricism by revising the way a definition is properly formulated—not in terms of ideal properties beyond knowledge, but in terms of the actual experimental procedure by which physical quantities are measured. Finally, Bridgman rescued scientific progress by a linguistic strategy that permitted new experience to be admitted cumulatively without necessitating radical reinterpretation of physical categories.

In the same way, Bridgman applied his operational reasoning to rescue the other fundamental physical quantities from relativistic distortion. He divided time into two kinds, local and extended, each of which was measured differently. Therefore, he argued, they could not be thought of as the same concept. Similarly, space and time could never fuse, he argued, because the operations for measuring them were distinct. Finally, mass and energy were of essentially different physical character because mass is a localizable aspect of physical reality and energy is a generalized property of a system. (Bridgman later changed his mind about this.) Bridgman reasoned that mass must be invariant because it was composed of an integral number of protons and electrons. He wrote in his personal notes:

When Rutherford succeeds in knocking hydrogen out of helium nuclei we may accept the ordinary relativity view as so much more convenient than any other that we need not worry again about other possibilities of interpretation. When

Rutherford does pull off this stunt, it will dispose of our idea that perhaps the idea of mass may be simply connected with the invariant number of protons and electrons and we shall have to search further for the phenomenological manifestation of this most important fact of structure.[11]

In this manner, Bridgman showed how examination of the determining operations can reveal an essential difference between concepts that had traditionally been treated as one and the same. It was this kind of reasoning that Bridgman wished to have applied to all concepts of physics and that exemplified what came to be known as *operational analysis*.

Bridgman thought of operational analysis as a tool for uncovering any unexpressed assumptions that might have deceived physicists into holding false beliefs, to prevent them from being caught off guard:

We must always be prepared some day to find that an increase in experimental accuracy may show that the two different sets of operations, which give the same results in the more ordinary part of the domain of experience, lead to measurably different results in the more unfamiliar parts of the domain. We must remain aware of these joints in our conceptual structure if we hope to render unnecessary the services of the unborn Einsteins.[12]

The operational standard was supposed to serve this prophylactic end by functioning as a criterion to distinguish what is physically real from what is not. It was intended to be an instrument for demystification, a method to be used to clear away any rubbish generated by the metaphysical imagination. It was an inquisitorial weapon wielded to expose unwarranted beliefs. Its application, Bridgman expected, would produce clarity of thought and a purified physics.

Operational analysis did not, as Bridgman hoped, absolve the physicist of the responsibility for making metaphysical judgments. Instead of providing a clear line of demarcation between the "real" and the "metaphysical," operations blurred the distinctions among metaphysics, epistemology, psychology, language, and behavior. As a consequence, *The Logic of Modern Physics* was afflicted with internal contradiction. On the one hand, for example, Bridgman asserted that operational knowledge of nature was the equivalent of nature itself; on the other hand, he claimed that nature is unaffected by man's categories. "Of what concern of nature's is it how man may choose to describe her phenomena," he cried, "and how can we expect the limitations of our descriptive process to limit the thing described?"[13] Indeed, because he had assimilated method to metaphysics, Bridgman often could not avoid treating operations, conceived of as units of arrested activity, as ontological realities.

The critical response to *The Logic of Modern Physics* was generally favorable, although varied. According to its publisher, it ranked "among the 40 most important books selected by the American Library Association." Most reviewers were sympathetic to the spirit of Bridgman's undertaking and agreed with F.E.B., of the *American Journal of Science,* who recommended it "to those who have been disquieted by the upheaval or revolution which has resulted from the recognition that the classical concepts so long unquestioned were inadequate to meet the present experimental situation." [14]

Many physicists thought that Bridgman had resolved the dilemma of how to interpret the new physics. His younger colleagues Edwin Kemble and Francis Birch, for example, recall their exhilaration at the prospect of being freed from mechanistic constraints. But R. C. Tolman had some reservations about the underlying attitude. He wrote Bridgman that although he had not read the book himself, he had heard that it was good. Nevertheless, he thought that Bridgman was being too cautious: "I get the impression that you stress the desirability of our proceeding in theoretical physics so that we shall not make mistakes; it seems to me, however, far more important to proceed so that we shall make progress even with mistakes." [15]

Some of the lay response bordered on the hysterical. A Mr. K. Liljegren wrote to Bridgman that "we are born with fundamental ideas which are deformed by brilliant men and it is dangerous to advocate that our fundamental concepts may be wrong." [16] And Mr. Redman from the law office of Redman and Alexander claimed that "Einstein is a faker. He is an unlicensed peddler of 'blue bosh.'" [17] Such statements were extreme and might appear to have been unsolicited. However, though Bridgman never violated the canons of propriety in his criticism of relativity, he could easily have encouraged such emotional expressions by his reference to Einstein as a "conjuror" who takes out of the hat only what he has already put into it.

Philosophers of science were not as easy to please. They sensed the difficulties that Bridgman's inner conflict had created. If they expected logical rigor, they found ambivalence and an excessive physical austerity. Bridgman seems to know what reality is, noted Harold

Jeffreys in his highly critical review, but does not believe in it.[18] Paul Weiss charged him with being "parsimonious with a vengeance."[19] L. J. Russell wrote that Bridgman had gone "whole hog" with his view and that "a purely operational view of the concept is impossible: and Prof. Bridgman shows it by not sticking to it."[20] Nevertheless, they all recommended the book as "important and enlightening" (Weiss).

Clearly Bridgman had touched a cultural nerve. Both professional academics and laymen responded with more than casual interest to *The Logic of Modern Physics*. Bridgman's examination of the meaning of modern physics found a waiting audience (this subject is treated more fully in Part III).

However, apart from the contemporary critical reception, two other historical questions need to be considered. How original is operational thinking, and how much does it owe to Einstein? In pursuing these questions, we will begin to discern some of the deeper sensibilities, personal and cultural, involved in the reaction to Einstein.

Explicit operational thinking, as a strategy formulated to avoid metaphysical commitment, had appeared independently at least twice in the work of prominent nineteenth-century scientists. In both cases the metaphysical status of the objects of knowledge was in doubt and therefore controversial. During the debates over chemical atomism, the chemist and mathematician Benjamin Brodie attempted to devise a chemical calculus in which the objects of chemical investigation—elements and compounds—were defined as operations performed on a unit of ponderable matter. Similarly, the mathematician Giuseppe Peano, whose intention was to found mathematics on physical experience, defined numbers (not without ambiguity) as operations. It is highly doubtful that Bridgman was familiar with either of these earlier manifestations of operationism.

On the other hand, it has been suggested that operational ideas were current in the Harvard-MIT community at about the time that Bridgman conceived his operational interpretation of physical reality. In particular, the MIT physicist William Franklin, with whom Bridgman was acquainted, espoused an operational approach similar to that formulated by Bridgman.[21] However, no documentation is available that would indicate an intellectual debt in either direction.

A more promising possibility is that Bridgman's operationism derived from some combination of William James's pragmatism and Mach's critical positivism (Bridgman's thought bears a remarkable similarity to theirs), together with ideas picked up from reading the mathematicians Jules Poincaré, W. K. Clifford, and the philosopher of science Johann Bernhard Stallo, all of whose works Bridgman had read, but at a very young age. Nevertheless, since there is no way to trace a specific influence, this conjecture is no better than saying that operationism was "in the air." It is not unreasonable to think that all of them were responding to a common intellectual ambience.

The question of Bridgman's debt to Einstein admits of a more definite answer. Bridgman implied that he was generalizing Einstein's reasoning when he designated operations as the medium responsible for the empirical content of the categories of physical thought. Was this an accurate portrayal of Bridgman's own thought, and was it a correct interpretation of Einstein's methodology?

A re-examination of the arguments in *The Logic of Modern Physics* reveals that conceptually, operational thinking has more in common with dimensional analysis than with relativity, although there can be no doubt that relativity was the nucleus around which operational analysis crystallized. The precipitating occasion was a symposium on relativity that was included as part of the Boston meeting of the AAAS in 1922. Bridgman claimed that the word "operation" was first explicitly used in this context, and the paper that he presented corroborates his recollection.[22] But although the subject matter was relativity, the language of the paper betrays his recent involvement with the interpretive difficulties of dimensional analysis. Bridgman later confirmed this observation. In his own view, the investigation of dimensional analysis was a prologue to operational analysis.

I had already come to grips with one comparatively minor detailed situation, that presented by dimensional analysis. Here, the bulk of professional writing on the subject afforded what seemed to me a pretty sorry exhibition of "metaphysics" in its bad sense. I had been able to think the situation through, separate the wheat from the chaff, and rid the whole dimensional situation of its metaphysics.

From this distance, I suspect that in writing the *Logic* I was really trying to do the same thing for the rest of physics.[23]

It was much more than just a preparation. Dimensional analysis provided a structural template for Bridgman's interpretation of Einstein's epistemology. The quantities, whose physical status was the subject of the dimensional analysis controversy, were translated in *The Logic of Modern Physics* into the "concepts" of physics, whose "meanings" were to be determined by the kind of operations—clearly, measurements—by which numerical values were assigned to them.

In fact, despite all his criticism of R. C. Tolman's principle of similitude, Bridgman also conflated relativity and dimensional analysis, but in a more organic way. Based on a subtly misconstrued analogy, Tolman had drawn what he thought was a parallel to the relativity principle, which turned out to be a variation of the dimensional method. Bridgman impressed one onto the other. Nevertheless, both interpretations were based on an elementalist, rather than a causal, metaphysics. Indeed, within the framework of his critique of relativity, the problem of causality was one that Bridgman was never able to resolve. He did not see that his essentially static conception of the universe stood in the way.

In *Dimensional Analysis*, Bridgman had denied that the fundamental measuring units could themselves possess physical reality. He also believed that he had freed himself from the idea that "things" really exist. Bridgman saw that Einstein had made clocks and measuring rods the empirical roots of special relativity. Here, it must have seemed, was a working model that retained the experiential foundation of physics while denying to time and length any absolute significance. Physical meaning was conferred by method. The 1922 paper indicates how Bridgman was thinking.

Now what appears to be E[instein]'s most important service, and the one whose effect is most likely to survive, is one which in its statement appears almost trivial, but which has nevertheless had farreaching effects. This is merely the insistence on the requirement that the quantities which we use in our equations have a meaning. It is obvious that in any exact formulation of a relation we are concerned with equations between numbers which are the numerical measures of various physical properties of the system under discussion. It is at once obvious, merely on saying it, that if the quantities which appear in our equations are to mean anything, we must be told the method by which we find the number measuring the property in any concrete case. In other words, we must be told the operations by which the numbers which enter our equations are determined. These operations must be physically realizable. It means nothing to talk about a length or a time unless the operations are specified by which concrete lengths and times are to be measured.[24]

According to Bridgman's reading of special relativity, Einstein had imposed the requirement that all terms, to be physically meaningful, must have operational correlates. This is something that Einstein emphatically denied. In 1949, referring directly to Bridgman's operational restrictions, Einstein wrote:

In order to be able to consider a logical system as physical theory it is not necessary to demand that all of its assertions can be independently interpreted and "tested" "operationally"; *de facto* this has never yet been achieved by any theory and can not at all be achieved. In order to be able to consider a theory as a *physical* theory it is only necessary that it implies empirically testable assertions in general.[25]

In the meantime, the elementalism Bridgman borrowed from dimensional analysis, coupled with his renunciation of the uniformity of nature, created some unique logical problems. His operational analysis failed to produce definitive results. Because he asserted that the quality of experience changes as the scale of measurement approaches the very large and the very small, and therefore the defining concepts of physical experience become different both in kind and number, he had to say something about what happens in the transition. Thus, he postulated a region where a "fusion" of concepts takes place. How the change could occur without passing through a phase in which the number of "concepts" or "building stones of our descriptions" was not an integer was a difficulty. There must be, Bridgman decided, a foggy area surrounding the mergers, since "the actual change in our conceptual structure in these transition regions is continuous, corresponding to the continuity of our experimental knowledge, whereas formally the number of concepts should be an integer."[26]

Critics of *The Logic of Modern Physics* did not know how to respond to such an idea. They could only repeat it in their synopses. They could neither approve nor disapprove of something that they did not understand. Nonetheless, it is clear that nothing of the sort could have come from Einstein. There is no sanction in Einstein's work for interpreting concepts of time or space as building blocks. The idea that operations are "building stones" of physical concepts could only have been a product of dimensional thinking, where a unit of measurement was treated as a kind of physical atom.

Bridgman's effort to pin physical knowledge to the ground involved him in a critique of relativity that occupied him to the end of his life. One of his very last literary undertakings, published posthumously in 1962, was a critical examination of special relativity entitled *A Sophisticate's Primer of Relativity*.[27] Relativity had lifted Bridgman to a level of self-consciousness that would never permit him to return to the familiar comfort that, in his words, draws one to mechanistic thinking with a longing "which has all the tenacity of original sin." But this longing must be overcome, he wrote; "just as the old monks struggled to sub-due the flesh, so must the physicist struggle to subdue this sometimes nearly irresistible, but perfectly unjustifiable desire."[28] However, his struggle with relativity was not only one of personal acceptance, but also one aimed at preventing physics from simply replacing one dogma with another.

Bridgman was especially disturbed about the central role Einstein had given light in relativity theory. There were several reasons for his discomfort. Relativity, he noted in *The Logic of Modern Physics*, was not compatible with quantum mechanics. Furthermore, quantum mechanics had shown that light possesses contradictory characteristics. On the basis of these two conditions alone, it appeared to Bridgman that it might not be safe to depend so heavily on something so poorly understood. But he also had metaphysical and logical objections. He thought that relativity implicitly assumed, in a holdover based on mechanistic thinking, that light was a material "thing travelling." Operationally, he protested, light was nothing more than "things lighted." Its path could not be followed through empty space, whatever that might be, and its presence can be known only by the illumination of an object.

Bridgman was also skeptical of the postulate of the constancy of the velocity of light and was troubled by the logical circularity of defin-ing distant time in terms of the velocity of light. He noted that time is "spread through space" using light as a vehicle, and at the same time the value for the velocity of light depends on how time is distributed. It bothered him that there was no independent way to measure time, no independently definable "clock." One reason for this difficulty, Bridgman asserted, was Einstein's "desire to think of light as a thing travelling, with a finite velocity," where "in all the deductions we in-evitably think of ourselves as an observer from outside, watching a thing that we call light travelling back and forth like any physical thing."[29] It was the desire to maintain this viewpoint, Bridgman thought, that in-fluenced Einstein to select the round-trip or "go and come" velocity, rather than the one-way, "here to there," velocity as the basis for spreading time through space.*

*Bridgman reasoned that if light is not a "thing travelling," it need not have a ve-locity in the usual sense of the term. A different concept of "velocity," which could be physically defined and which would possibly be a more primitive concept than either space or time, would avoid the problem of light having "the paradoxical property of bear-ing the same finite relation to each of two material systems which differ from each other by a finite amount (that is, the first postulate of relativity that the velocity of light is

Bridgman thought that Einstein's treatment of light as a "thing travelling" was a fundamental weakness in relativity theory, and therefore he questioned "whether Einstein's whole formal structure is not a more or less temporary affair." Indeed, he went so far as to suggest that "we ought at least to try to start over again from the beginning and devise concepts for the treatment of all optical phenomena which come closer to physical reality."[30] However, by the time he began work on *A Sophisticate's Primer of Relativity*, his final commentary on relativity, the validity of special relativity could no longer be questioned. Nonetheless, Bridgman was still puzzling over the definition of extended time and its dependence on the velocity of light. He was searching for an experimental check of the truth of Einstein's postulate, which could be made if he could devise a method for measuring the one-way velocity of light, since this would reveal whether, as Einstein assumed, the "go" time of light in a round-trip was equal to the "come" time. It came as something of a shock when his investigations forced him to the conclusion, in defiance of his operational expectations, that the one-way velocity of light has no operational meaning.[31] This meant that the truth of the principle of relativity could not be subjected to a direct experimental check and that Einstein's method for spreading time through space rested on a definition, or convention.

The idea that a true theory might be built around an arbitrary stipulation was repugnant to Bridgman, so much so that he attempted to give theoretical reasons (incorrect) why Einstein must have been constrained to only one choice. His arguments drew severe criticism from

---

3×10^{10} [cm/sec] in all reference systems)." Bridgman described the operational definition for just such a velocity, and conveniently enough, because it was allowed to take on an infinite value, it did not entail the relativistic distortions that Bridgman found so difficult to accept.

However, Bridgman pointed out, if this alternative is adopted, the idea that "velocity" is simply related to "go and come" (round-trip) time, as Einstein had assumed for his definition of simultaneity, must be given up. Applying the operational method just as he had for the concept of length, Bridgman argued that two different concepts of "velocity" could be distinguished. There is what he would later designate two-clock velocity (one-way, or "here to there," velocity), which is ordinary physical velocity, and then there is one-clock velocity (round-trip, or "go and come," velocity), which could conceivably allow an asymmetry with respect to reversal in direction. Einstein had conflated these two concepts, when in fact, because the operations involved in their determination are different, they are qualitatively distinguishable. Bridgman asserted that "which of these two treatments of velocity shall be adopted is to a certain extent a matter of convenience, determined by the sort of phenomena in which we are most interested and wish most to simplify" (*Logic*, p. 163).

the philosopher Adolph Grünbaum, who compared such a restriction to limiting the expression of physical quantities to only one system of units. The first postulate of relativity, Grünbaum wrote, would no more be violated if different conventions were chosen in different inertial systems "than would the following procedure: Using rectangular coordinates (or meters) in only one particular inertial system and using polar coordinates (or inches) in all other such systems." [32]

In addition, since time was the key to making sense of causality as well as of velocity, and since Bridgman perceived time as the weak link in the structure of relativity theory, he looked for a way to define causality that would not depend on the concept of extended time or the velocity of light. He suggested that cause might be directional or progressive in space without necessarily being concomitantly ordered in time, thus provoking further criticism from Grünbaum, who denied that this was logically possible.

A Sophisticate's Primer, like *The Logic of Modern Physics*, was an investigation of the boundaries between convention and what the physical world imposes upon the inquiring physicist. Its focus was Einstein's special relativity theory. Bridgman did not even consider general relativity, which he had long before dismissed as being hopelessly divorced from concrete experience. He had predicted that its approval by physicists "might someday become one of the puzzles of history." [33] Nevertheless, an examination of Bridgman's objections to general relativity, which were only to a lesser degree applicable to special relativity, reveals that Bridgman's argument with Einstein went far beyond what the prevailing positivist standard considered within the pale of science to issues that are deeply philosophical and moral.

In a sense, Bridgman felt betrayed by the Einstein of general relativity. The Einstein of special relativity had taught him that knowledge originates in human activity, that it is caused by human beings and is about human experience. With general relativity, Einstein seemed to be turning away from sensory ground to higher abstractions for knowledge. It was incomprehensible to Bridgman how anything so far removed from physical experience could qualify as truth.

Bridgman had carefully worked his way through the mathematical details of Einstein's 1916 paper, "Die Grundlage der allgemeinen Relativitätstheorie" and found "no real difficulty" with the mathematics. But he could not understand what it had to do with physics.

I still cannot see exactly what he is doing, and have no physical feeling for the situation. I am more and more convinced that all the talk about generalized coordinates is bunk, and that Einstein has been deluding himself by a metaphysical conception.*

It was from general relativity that Bridgman first learned that space and time had been stripped of the last vestiges of physical objectivity. Then, to compound the offense, in his own attempts to establish a physical basis for the covariance postulate, Bridgman ended up with nothing more concrete than a set of measuring operations: "Although the coordinate system may be arbitrary there is always an invariant scheme of physical operations independent of the coordinate system, the results of which cannot eliminate themselves from the final result."[34]

It looked as if measuring operations were the only permanent physical entities that remained, as he explained to Alfred Korzybyski, the founder of the General Semantics movement, in 1928.

The fact that you are able to set up an expression in generalized coordinates which gives the same length no matter what the coordinates seems to me to have a purely formal significance and to show that you haven't blundered in setting up your rules of calculation, but not to have any connection with anything "intrinsic." For you must remember that the physical operations by which you measure length are definite, fixed, "absolute" if you will, operations, which have no connection with a coordinate system and do not change when you pass from one coordinate system to another. It is therefore inevitable that when you describe an absolute set of operations correctly you get an invariant result.[35]

Special relativity, at least in principle, seemed to support this position by encouraging the interpretation of physical laws as descriptions of the relations between measuring bodies and clocks. In fact, in Bridgman's opinion, special relativity was for that very reason scientifically more respectable. Its postulates referred to scientific experience; they were grounded in what scientists actually do. General relativity, on the other hand, did not conform to the standards established

---

*Unpublished manuscript, Bridgman Papers. Bridgman's complaint regarding "all the talk about generalized coordinates" refers to the formal principle of general relativity or the principle of covariance that Einstein presented in "Die Grundlage" as follows: "The general laws of nature are to be expressed by equations which hold good for all systems of co-ordinates, that is, are co-variant with respect to any substitutions whatever (generally co-variant)." For this paper, see W. Perrett and G. B. Jeffery, trans. and eds., *The Principle of Relativity* (New York: Dover Publications, [1923]).

by special relativity. It seemed to be a return to the very absolutism special relativity had discredited.[36]

However, the distinction was not unambiguous, since Bridgman quickly realized that special relativity also contained many of the objectionable features that to his mind emasculated general relativity. Nevertheless, it was in commenting on general relativity that Bridgman most frequently voiced his dissatisfaction.

To begin with, the concept of "event"—the abstract, qualityless time-space conjunction that was the object of relativistic reasoning—was not, in Bridgman's opinion, a faithful representation of human experience. This seemed to him to be self-evident:

It seems perfectly obvious that the world of our immediate *sensation* cannot be described in terms of coincidences. . . . To justify the coincidence point of view we apparently have to analyze down to the colorless elements beyond our sense perception.[37]

This was something Bridgman thought unwarranted. It is not proper to speak of coordinates, he argued, without specifying "what it is that the coordinates are of."[38]

The more important problem, however, was "to find where the physics got into the general theory."[39] To Bridgman, this question was the same as asking What is the physical significance of the covariance postulate? The answer, he decided, is that there is none.

The assumption of covariance is itself sterile in physical consequences, as was pointed out by Kretschmann and admitted by Einstein. The physical content entered into the theory in other ways during the detailed working out, chiefly through the demand for mathematical simplicity.[40]

Bridgman rejected the "thesis [that covariance] is usually understood to have a physical content in addition to its purely formal content," because of what he saw as its metaphysical implications. The physical content, he explained, can be injected into the situation only by assuming the possibility that two different observers, each in his own frame of reference, can observe the same event. The word "same" introduces a difficulty. How do two observers know that they are observing the "same" thing?

The simplest way to make sense of what this could mean, Bridgman pointed out, is to assume a third observer in the background. And, Bridgman charged, "It does indeed seem that such a conceptual observer in the background is implied in the argument of Einstein, for we have already seen that a meaning is assigned to the event in its own right apart from the frame of reference which yields the coordinates." With this point of view built into it, general relativity includes the assumption of a "certain amount of pre-Einsteinian 'absoluteness.'"[41]

Bridgman characterized Einstein as a man who talks about his theory in the manner of one who believes that "there is a 'reality' back of all our multifarious experience," as one who "believes it possible to get away from the special point of view of the individual observer and sublimate it into something universal, 'public,' and 'real.'" The idea of "object," Bridgman argued, exists only in connection with the human nervous system and is palpably a human construction. His own analysis of how physical knowledge is gained had disclosed "the universal impossibility of getting away from the individual starting-point."[42] It is the individual who "knows." To the degree that relativity theory "postulates an underlying 'reality' from which this aspect of experience is cancelled out," Bridgman asserted, "it seems to me palpably false, and furthermore devoid of operational meaning."[43]

Here Bridgman was caught in the Nietzschean dilemma, whose resolution is ultimately a moral, rather than a scientific, judgment. As a physicist and man of common sense, Bridgman never doubted that there was an external reality, a real and objective world outside himself. But for human beings to elevate themselves to the standpoint of the divine, to assume that their imperfect intellectual instruments were capable of perfect and complete knowledge, was an arrogance that offended Bridgman's moral sensibilities. All physical knowledge, Bridgman insisted time and again, is approximate and inherently imperfect. And though he did not say so explicitly, Bridgman must have felt that with relativity, Einstein had exceeded the bounds of proper humility. "Man," Bridgman wrote,

has never been a particularly modest or self-deprecatory animal, and physical theory bears witness to this no less than many other important activities. The idea that thought is the measure of all things, that there is such a thing as utter logical rigor, that conclusions can be drawn endowed with an inescapable necessity, that mathematics has an absolute validity and controls experience— these are not the ideas of a modest animal. Not only do our theories betray these somewhat bumptious traits of self-appreciation, but especially obvious through them all is the thread of incorrigible optimism so characteristic of human beings.[44]

Indeed, in Bridgman's eyes, Einstein's "incorrigible optimism" probably bordered uncomfortably closely upon religious hubris. In contrast to Bridgman's denunciation of idealism, Einstein celebrated the inspiration of the Platonic promise and was not ashamed of its religious associations. He spoke of experiencing the satisfaction of feeling that God Himself could not have arranged the order of existence differently, of the "promethean element of the scientific experience," of "the particular magic of scientific considerations" as the "religious basis of scientific effort." In the creative principle provided by mathematics, Einstein found justification for admitting that "in a certain sense, therefore, I hold it true that pure thought can grasp reality, as the ancients dreamed."[45] Reflecting on the subject in a less personal way, he wrote:

While it is true that scientific results are entirely independent from religious and moral considerations, those individuals to whom we owe the great creative achievements of science were all of them imbued with the truly religious conviction that this universe of ours is something perfect and susceptible to the rational striving for knowledge. If this conviction had not been a strongly emotional one and if those searching for knowledge had not been inspired by Spinoza's *Amor Dei Intellectualis*, they would hardly have been capable of that untiring devotion which alone enables man to attain his greatest achievements.[46]

Neither did Einstein deny that he might have been motivated by an attempt to free himself "from the chains of the 'merely personal,' from an existence which is dominated by wishes, hopes, and primitive feelings." The world, "out yonder . . . which exists independently of us human beings and which stands before us like a great, eternal riddle, at least partially accessible to our inspection and thinking," he recalled in some autobiographical remarks,

beckoned like a liberation, and I soon noticed that many a man whom I had learned to esteem and to admire had found inner freedom and security in devoted occupation with it. The mental grasp of this extra-personal world with the frame of the given possibilities swam as highest aim half consciously and half unconsciously before my mind's eye.[47]

Bridgman's expectations were conditioned by a lesser estimation of humanity's proper aspirations. His God was not the god of Plato or of Spinoza, but the jealous God of Puritanism, with whom man's relationship was personal and submissive. This was the God whom Bridgman had renounced, but who still guarded his universal province. It was the God who was so incredible that Bridgman accepted resignation over hope, and chose the temporal world over the eternal. It was the God on whose account Bridgman was driven to conclude that

we can never get away from ourselves. . . . Not only is each one of us as an individual not able to get away from himself, but the human race as a whole can never get away from itself. The insight that we can never get away from ourselves is an insight which the human race through its long history has been deliberately, one is tempted to say wilfully, refusing to admit.[48]

It is unrealistic, indeed immoral, Bridgman believed, for humanity to seek comfort in the universal. Mankind is condemned to an imperfect this-worldly existence and must learn to accept it.

This was the insight that Bridgman felt Einstein had betrayed. But what if relativity, with all of the assumptions that Einstein had insinuated, were true? Might not that also mean that the divine is at least as enduring as science itself? This was not an admission that Bridgman was prepared to make.

PART III

Operational Reasoning and the Tyranny of Language

The Eagle soars in the summit of Heaven,
The Hunter with his dogs pursues his circuit.
O perpetual revolution of configured stars,
O perpetual recurrence of determined seasons,
O world of spring and autumn, birth and dying!
The endless cycle of idea and action,
Endless invention, endless experiment,
Brings knowledge of motion, but not of stillness;
Knowledge of speech, but not of silence;
Knowledge of words, and ignorance of the Word.
All our knowledge brings us nearer to our ignorance,
All our ignorance brings us nearer to death,
But nearness to death no nearer to GOD.
Where is the Life we have lost in living?
Where is the knowledge we have lost in information?
The cycles of Heaven in twenty centuries
Brings us farther from GOD and nearer to the Dust.
　　　　—T. S. Eliot, Chorus from *The Rock* (1934)

■ ■ ■

P. W. BRIDGMAN HAD correctly intuited that relativity was merely the tip of an iceberg when he expanded his operational analysis to include all of physics. Revolutionary change was rapidly rendering physics unintelligible even to the physicist, who was no longer sure what physics was about. The severity of the problem, Bridgman declared, made it permissible to violate commonly accepted sanctions and undertake a metaphysical enquiry. Moreover, as Bridgman also observed, the necessity for such an inquiry was not just the consequence of an upheaval of values following World War I, but derived from a more deep-seated scientific situation in the making for some time before the war.[1]

However, at the time Bridgman wrote *The Logic of Modern Physics*, he was not aware of the full extent of the problem he had broached and in which he was soon to become inextricably involved. Indeed, whether as cause or symptom, science was implicated in a cultural crisis of meaning, widely in evidence in philosophical discussion at all levels of rigor, amateur and professional.[2] The problem was both moral and intellectual, ideological and logical. Its philosophical roots can be traced to the deontologization of the modern universe and the loss of faith in the possibility of certain knowledge. To be sure, all ages have had their skeptics, but twentieth-century culture is particularly characterized by its philosophical skepticism and its concern with what is sometimes called the problem of knowledge. Not only was it heir to Darwinism and the downfall of the mechanical ontology in science, but trends in literary and historical criticism, together with developments in mathematics and logic, especially the discovery of non-Euclidean geometry and the introduction of mathematical formalization, all concurred in the verdict that the traditional ground of knowledge was no longer valid. Knowledge needed new justification, independent of its former ontological foundations.

The seventeenth-century Scientific Revolution had bequeathed to its immediate successors both an ontology and an ideology, a complex constellation of beliefs canonized in the Newtonian worldview. The mechanical ontology united the mathematical interpretation of nature and the ideologically more potent "experimental philosophy," which expressed, besides a cognitive agenda, the social aspirations of a rising middle class and tied science to democratic ideals and the myth of progress. Supporting this synthesis was the axiom that reason and experience would affirm the same truth, all within a Christian context of values. Thus whether God was a mathematician, architect, or clockmaker, He had created an orderly world that was universally intelligible. The categories which expressed this order, because they reflected God's plan, were absolute, true, and real, and the investigation of God's handiwork was a celebration of His universal intelligence.

By the end of the nineteenth century, no more traces of God's purposes were to be seen. There was nothing more than adaptation; the signs of Divine Intelligence turned out to be mere human ordering devices. The principles of physics were economical descriptions, useful pragmatic instruments, convenient conventions. Mathematical truth consisted of arbitrary formal arrangements of empty categories—flights of reason acting independently of experience. With the final collapse of Newton's absolute time-space framework, mankind was left not to transcribe the evidences of God's Intelligence but to contemplate the meaning of his own signs and statements.

The predicament was of more than academic interest. What happened in science mattered because of its cultural prominence. Science had become the model for all knowledge. If science did not produce genuine knowledge, its authority would be dissolved, and the ideological claims based on it would lose their legitimacy and force. Progress would be deprived of its intellectual rationale.

In this deontologized universe—that is, in a universe where traditional intellectual categories were no longer ontologically supported—new philosophical strategies had to be devised to establish and accredit the validity of these rational entities, to fix the locus of knowledge. The problem of *meaning* became a philosophical and cultural preoccupation. Attention shifted toward the role of language as the system of symbols and relations that functions as the common repository and social carrier of human knowledge. In a sense, linguistics became a surrogate metaphysics and semantics, a surrogate epistemology. In the words of philosopher Max Black, "Philosophical linguistics may be expected to provide nothing less than a pathway to the nature of that reality which is the metaphysician's goal. To this very day the hope persists that 'with sufficient caution, the properties of language may help us to understand the structure of the world.'"[3] The question was How do the categories of language become invested with meaning? What is the relationship between word and experience?

In America, where deeds were valued over ideas and where science was overwhelmingly dominated by the empiricist ideology, the strategy for accommodating revolution in science—not just its nature, but the fact of its occurrence—had to be compatible with empiricist doctrine, both philosophical and ideological. Bridgman's operational interpretation of physical concepts met these conditions. Not only was the reality of physics cast into an active and experiential mode—becoming rather than being—but its growth was presented as a progressive accu-mulation of experience rather than as the result of the revolutionary destruction of tradition. The key to this accomplishment was the de-mand that conceptual categories represent operational distinctions, the underlying charge being that thought had disregarded what is experi-entially distinguishable. Thought, or reason, the inventor of metaphys-ics, had corrupted truth and created rational categories that did not correspond to experience. Language had to be reviewed and brought into conformity with experience. Nonetheless, Bridgman admitted, it was questionable whether metaphysics could be entirely eradicated, "whether any picture that we can form of nature will not be tinged—sicklied o'er with the pale cast of thought."[4]

Bridgman's strategy of reasoning had wide appeal, doubtless as much because of its philosophical ambiguity as because of its sug-gestiveness and relevance to contemporary concerns. It was diagnosis and prescription wrapped into one package, knowledge and experi-ence reduced to common operations, word and object fused in action. It was much more and much less than Bridgman had intended.

The appearance of The Logic of Modern Physics in 1927 catapulted Bridgman to a new level of public exposure, now as a philosopher of science. It became the occasion for invitations to give lectures and write articles on the subject of scientific philosophy. His reply to such re-quests reflected a reluctance to undertake something that came to him only with great difficulty. He refused an offer of a one-semester lecture-ship from George P. Adams of the Department of Philosophy at Berke-ley because of his doubts that he had enough to offer for two courses, adding that "anything I produce along these lines is done very slowly, after many revisions, and with considerable nervous strain."[5] He turned down a request to write popular articles for the Appleton series, claim-ing that he did not write that easily. He expressed hesitation about taking on the logical positivist Herbert Feigl as an understudy because "nearly all my time is occupied with the many details of my experi-mental work."[6]

One offer he did not refuse was from George Pegram asking him to give a five-lecture course at Columbia University during the summer of 1928.[7] Afterwards he expressed disappointment over the experience. "I am not sure," he wrote to a correspondent, "that the lectures did any good. The class was small and not at all responsive. Some of them had quite inadequate preparation and were not prepared to profit by it." Besides that, the weather had been "torrid; one could not move without running a moisture."[8]

Bridgman had come only reluctantly to philosophy. He had no academic philosophical experience and did not profess allegiance to any particular scholarly tradition. His intention was neither to construct a formal philosophical system nor to persuade his readers that he had discovered something philosophically ultimate. He was trying to cope with revolutionary change in science. Bridgman was disturbed by the possibility that the foundations of science might not be secure, and his writing was first and foremost an attempt to sort out his own thoughts, a kind of thinking aloud. "I think that perhaps more than most I write to clarify my own ideas," he wrote. "I am fully as much interested in straightening out things for myself as I am in producing a systematic structure, which is doubtless not quite fair on the reader."[9]

The exploratory nature of Bridgman's writing, the fact that his ideas were not fully developed and rehearsed, together with his philosophical inexperience, account for much of its philosophical vagueness. Moreover, the situation was exacerbated by his publisher's unwillingness to accept his original title, *The Conceptual Foundations of Modern Physics*. This title undoubtedly conveyed the generality of his intentions, but Macmillan thought it too cumbersome. Two of Bridgman's other suggestions, *The Metaphysics of a Physicist* and *Reality in the Concepts of Physics*, indicated the content of the book more accurately. But the publisher wanted a snappy title to compete with a book such as Whitehead's *Science and the Modern World* and suggested *The Logic of Modern Physics*. After a rushed exchange of correspondence just prior to publication, Bridgman consented.[10]

The effect of the title change was to shift the emphasis away from Bridgman's original problem of what physics is about and how to decide what counts as physical reality (that is, what criterion to apply in admitting conceptual categories into physics) to the problem of how scientific knowledge is empirically created and how elementary categories are defined. Accompanying this shift of emphasis was a confusion of metaphysics, methodology, and linguistics. To be sure, Bridgman's assertion that meaning and method are synonymous reinforced this reading. But his was, as he often declared, supposed to be a "minimum" or "safe" position, meant to limit knowledge claims to what can be known here and now, without appeal to a transcendent reality. The most that a physicist can know with certainty is what he does.

Nonetheless, the word "logic" in the title called attention to the linguistic strategy Bridgman had employed to save physics from the revolutionary implications of relativity. It suggested that he was prescribing an empirical rule for the formation of logical categories and, at the same time, connoted the rigor usually associated with the discipline of logic. This was never Bridgman's intention, and he repeatedly and vigorously protested against such interpretations.

Similarly, the operational physical-conceptual synonymy suggested that he had found a monistic resolution to the troublesome sense-intellect dualism and that operational reasoning could bridge the epistemological gap between experience and knowledge. It is not clear whether Bridgman fully understood the nature of the difficulty when he wrote *The Logic of Modern Physics*. He probably did not. In any case, it appeared that he might have discovered a scientific solution to the "problem of knowledge," the intellectual aspect of the problem of meaning.

Here is where Bridgman's troubles began. He had forcefully affirmed the empiricist doctrine that knowledge is grounded in experience. As a general precept, the principle was sound enough but sufficiently vague to accommodate disparate interpretations. Among his interpreters there was a wide range of opinion on what constitutes "experience" and what characterizes the state or condition designated as "knowledge." On the one hand, writers on popular semantics, who believed that errors in the use of language were the cause of modern spiritual and psychological malaise, saw in operational reasoning the key to restoring proper linguistic usage and thus the unity of body and soul. Bridgman's dialogue with them is presented in Chapter 6. On the other hand, operational reasoning was taken as formula for creating scientific knowledge. This was the assumption of social scientists, the psychological behaviorists, and the logical positivists. Their appropriation of Bridgman's operational method and his reaction are discussed in Chapter 7.

However, a growing strain of mutual alienation marred Bridgman's relationship with his interpreters as he became increasingly wary of their cognitive goals. And when his confidence in the metaphysical capability of science was further shaken by Heisenberg's principle of indeterminacy, the metaphysical/epistemological anthropocentrism (knowledge is about and caused by human activity) inspired by Einstein's special relativity shrank to a radical subjectivism. Whatever encouragement Bridgman might have offered his interpreters was completely withdrawn. This turn of thought elicited no sympathetic response and marked the beginning of Bridgman's retreat into the isolation of subjectivity. It also marked the point where the problem of intellectual meaning began to shade into the problem of moral meaning.

CHAPTER 6

Wordless Truth: Deliverance from the Idolatry of Language

A theory of thinking which disregards mystical relations between the knower and the known and treats knowledge as a causal affair open to ordinary scientific investigation is one which will appeal to common-sense inquirers.

Words, as everyone now knows, "mean" nothing by themselves.
— C. K. Ogden and I. A. Richards,
The Meaning of Meaning (1936)

■ ■ ■

ONE OF THE FIRST to initiate extended correspondence with Bridgman about *The Logic of Modern Physics* was Scudder Klyce, a self-proclaimed philosopher, an antirelativist, and something of an evangelist for monism. Klyce, a graduate of the U.S. Naval Academy and a former engineer, had no inhibitions against preaching his doctrine to distinguished scientists. He made a practice of writing to them and had, he told Bridgman, selected Millikan as one in particular need of salvation. For that kind of "reconciler," Klyce felt the most profound kind of disgust, he said.[1] In Klyce's mind, arguing the coexistence of two different realms of knowledge, science and religion, each with its own validity, was tantamount to hypocrisy. It was at least the consequence of extreme ignorance. With Bridgman, he felt that he had found a kindred soul. Bridgman, he thought, had transcended the arbitrary and erroneous division between realism and idealism in his insistence on the primacy of experience and its continuity with idea.

In scientific circles Klyce was regarded as a religious fundamentalist and was refused membership in the American Association for the Advancement of Science because it was thought that his reputation might tarnish the image of science.[2] In fact, it was Bridgman who accepted the unpleasant responsibility for communicating the refusal.[3]

On the other hand, Bridgman did not take Klyce for a crank, just as John Dewey had not (at least at the time he wrote the introduction to Klyce's first book, *Universe*).[4] Much of what Klyce had to say, Bridgman felt, was "true and important."[5] Although he had reservations about Klyce's contentious manner of writing, Bridgman was willing to hear him out—up to a point.

Even before *The Logic of Modern Physics* was issued, Klyce had written Bridgman inquiring about it. He had heard that it was "undogmatic" and offered to exchange a copy of his *Sins of Science** for one of *The Logic of Modern Physics*. Klyce included excerpts from his book for Bridgman to look over. In *The Sins of Science*, Klyce expounded on what he thought was the error of modern intellectuals, namely, their failure to recognize the unity of all knowledge and the continuity of all existence—the coexistence of the Many and the One. He accused intellectuals, among whom he included both scientists and theologians, of the idolatry of words, of mistaking the words for the reality and the use of words for reason. He proclaimed the supremacy of experience and the superiority of common sense and pronounced it is intuition that grasps reality. Words are not knowledge, he insisted. Furthermore, language is inexact—necessarily so because it deals with the Many—and therefore it cannot be complete. Knowledge of unity or completeness, according to Klyce, is a mystical experience.

Furthermore, logic or reasoning, as Klyce explained in a subsequent letter to Bridgman, "*proves* nothing. . . . The observable or 'experimental' facts are the reality or truth." Logic "is nothing more than mutual agreements in speech."[6] Mathematicians, Klyce asserted, are especially guilty of self-delusion. They pretend to be realists when in fact they are idealists.[7]

Bridgman did not disagree entirely with Klyce's accusations, but he was suspicious of his religious sympathies. Thus in reply, he wrote:

My reason for this mistrust of the value of your book is not that I at all object to your dissatisfaction with many of the methods and attitudes of science, for I imagine that there are many points on which we can heartily agree in this matter, but because the extracts which you sent me seem to indicate that you

---

*Scudder Klyce, *The Sins of Science* (Boston: Marshall Jones Co., 1925). The book was widely reviewed, and judgments ranged from "great" and "substantial" to "insane" and "a joke."

have approached the whole question with a religious bias, and this it seems to me is a most dangerous thing to do, sufficient to cast suspicion on the whole superstructure.[8]

To this, Klyce answered that he had discovered that science is as dogmatic as religion, and it was his goal to eliminate all dogma. In his view sound science is also sound religion.[9] However, Klyce appeared unable to temper the anger that fueled his crusade against the dogmatism of science. This prompted Bridgman to caution Klyce that if he persisted in his "rabid and one-sided" attitude he would most certainly alienate the very audience he wished to address. Your "method of swearing at scientists has its grave defects," wrote Bridgman, and

no doubt it repels every sincere scientist from your book as it did me. The reason is that the picture you present of science and scientists as being a class apart from the common man, plotting against him, simply is not true to the facts. . . . A scientist is first of all a common man himself, and the attitude which science has come to adopt toward many questions is simply the attitude which the common man adopts when confronted with certain situations. And the fact that this attitude is not always correct, as you and I believe, is not evidence that science is trying to put over something false.[10]

In the same letter, Bridgman expressed his appreciation that Klyce had "been able in the rain to approve the point of view" in *The Logic of Modern Physics*, especially since "it looks to me now as though the book is not going to attract the attention from physicists that I wished it might." Klyce had an explanation for this. The reason that the "ordinary scientist shys off your book," said Klyce, is that "in it you have undertaken to balance ordinary scientific realism with laws or principles—i.e., with idealism."[11] Whether or not this was indeed the reaction of the ordinary scientist, and that seems rather unlikely, it surely reveals Klyce's bias.

The correspondence between Klyce and Bridgman continued with a relatively cordial tone for almost three years, although it is doubtful that Bridgman fully appreciated the seriousness of Klyce's mission. For example, in response to the receipt of Klyce's next book, entitled *Dewey's Suppressed Psychology*, Bridgman wrote, "I wish that I was able to have as much fun with myself as you are able to achieve with yourself."[12] Evidently Bridgman thought that Klyce's colorful language was an indication of lightheartedness. He was soon to find out differently. In January 1930, the exchange of letters came to an abrupt halt, leaving stranded a disappointed and very disillusioned Klyce. The falling-out had to do with an article Bridgman published in *Harper's* in 1929 entitled "The New Vision of Science."[13]

The circumstances leading up to its publication arose from Bridgman's increasing apprehension about the implications of quantum mechanics. The character of his concern is illustrated in the following words, written on March 18, 1928, to another of his correspondents.

Are you picking up any of the new quantum mechanics in Pasadena, and have you come across Tolman? In the intervals between soldering and wiping up the floor in the laboratory I have been thinking a little lately about Heisenberg's principle of indetermination. . . . If the physical basis for the theory turns out to be unassailable, I believe that this is the initiation of the biggest revolution in mental outlook since at least the time of Newton, much bigger than Einstein, for example. It means that the universe is forever bounded in the direction of the very small by becoming meaningless beyond a certain point. This sort of a limit to our activities is one which I did not suspect, and is one which is hopeless to try to surmount. There are tremendous possibilities in the idea, once it gets into popular hands. The advocates of "free will" on the one hand and on the other the old fashioned atheists with their "pure chance" will equally find justification, and probably the vitalists in biology will be equally comforted. In the mean time I find the greatest problem in the situation in learning how to deal intellectually with [a] situation that fades out on you by becoming meaningless. This being my present frame of mind I find it impossible to understand much of the present writing in physics, entirely apart from the mathematical difficulties.[14]

These sentiments were essentially the ones Bridgman expressed in the article. Moreover, the letter accompanying the manuscript was characterized by an unmistakable sense of urgency. "I would particularly like to get this before the popular audience of Harper's," he wrote, "because I believe that the consequences of the new discoveries are so important for everyone that all sooner or later, will have to make considerable readjustments to meet the situation."[15]

When *Harper's* accepted, gratuitously adding that they had been awaiting a piece from G. N. Lewis,[16] Bridgman communicated some second thoughts. The astronomer Harlow Shapley had told him, he said, that publishing this article was a very courageous act. And, "the more I think of it the more I am inclined to agree with him; I hope I am not irretrievably wrecking the Magazine."[17] He did not on that account withdraw the manuscript.

The thrust of Bridgman's message in "The New Vision of Science" was a worry that the limitations of scientific measurement as embodied in Heisenberg's principle of indeterminacy might unjustifiably be taken as meaning that scientific knowledge is not all of knowledge, thus wrongly leaving open the possibility that there might be a domain in which science has no authority. Science, he argued, had freed us from superstition by promulgating the idea that nature is orderly and not capricious and, being so, is therefore understandable. Quantum mechanics suggests that this belief might have been unfounded, that when the range of measurement is extended, nature is not understandable or subject to laws. This situation is worse than that caused by relativity. Now it looks as though the basic structure of the universe is discontinuous and the laws of cause and effect have failed. "The world," cried Bridgman,

is not a world of reason, understandable by the intellect of man, but as we penetrate ever deeper, the very law of cause and effect, which we had thought to be a formula to which we could force God himself to subscribe, ceases to have meaning. . . . The physicist thus finds himself in a world from which the bottom has dropped clean out; as he penetrates deeper and deeper it eludes him and fades away by the highly unsportsmanlike device of just becoming meaningless.[18]

However, in Bridgman's view, at least one mitigating principle could be abstracted from this revolutionary theory; namely, it is impossible to gain knowledge about something "without getting into some sort of connection with it, whether direct or indirect."[19] This seems so inevitable that it almost takes the sting out of the realization of quantum mechanics. By this, presumably Bridgman meant to say that knowledge cannot be gained by other than physical means; knowledge by revelation or contemplation, for example, is not possible. Thus empiricism is saved.

But if the scientist has been taken by surprise, the man on the street, Bridgman speculated, might find the quantum situation the occasion for a

veritable spree of licentious and debauched thinking. This will come from the refusal to take at its true value the statement that it is meaningless to penetrate much deeper than the electron, and will have the thesis that there *is really* a domain beyond, only that man with his present limitations is not fitted to enter this domain. . . . The man in the street will, therefore, twist the statement that the scientist has come to the end of meaning into the statement that the scientist has penetrated as far as he can with the tools at his command, and that there is something beyond the ken of the scientist. This imagined beyond, which the scientist has proved he cannot penetrate, will become the playground of the imagination of every mystic and dreamer. The existence of such a domain will be made the basis of an orgy of rationalizing. It will be made the substance of the soul; the spirits of the dead will populate it; God will lurk in its shadows, the principle of vital processes will have its seat here; and it will become the medium of telepathic communication. One group will find in the failure of the physical law of cause and effect the solution of the age-long problem of the freedom of the will; and on the other hand the atheist will find the justification of his contention that chance rules the universe.[20]

Bridgman's conclusion to this unsatisfactory state of affairs, characterized by Sommerfeld as pessimistic resignation,[21] was that new methods of education were needed to effect a major adjustment in thinking so that thought will spontaneously and freely conform to our knowledge and the actual structure of the world. Otherwise, there is nothing to do but accept the situation with "a certain courageous nobility." Ultimately, Bridgman predicted, man will admit that God's knowledge cannot be his own, that

in the end, when man has fully partaken of the fruit of the tree of knowledge, there will be this difference between the first Eden and the last, that man will not become as a god, but will remain forever humble.[22]*

Klyce was as furious as he was hurt. He wrote Bridgman a scathing letter denouncing the error of Bridgman's interpretation of Heisenberg. Klyce had believed, he said, that Bridgman accepted his "general idea that as an observed fact the universe is continuous." By this Klyce was referring to his principle of the coexistence of the particular and universal, and he was upset that Bridgman was asserting the existence of only the particular. He hoped that Bridgman would not be angry or offended when he, Klyce, pointed out just exactly how Bridgman's reasoning had gone wrong, which had to do with a confusion in the use of language.

One way to look at Bridgman's interpretation of the Heisenberg principle, Klyce said, is as "a new version of the old Greek paradox that

*In a lengthy review of this article in *Scientific Monthly* (Dec. 1929; pp. 506–14), A. P. Weiss wrote, "Bridgman is humble now, and there are a number of other physicists who feel the same, but this is because they are baffled by the complexity of the physical situation which must be fitted into a social system. Humility is the attitude of the conquered. Man triumphant has never been humble. The marvelous development of industry during the last half century has not developed humility in the business man. Even the physicist will again be proud when certain strained relations between subatomic and superatomic behavior will have been arbitrated" (p. 514).

a body can't move, because it can't move *where* it *is*, and it isn't anywhere else besides where it is. That is merely a way of stating the contradiction between the One and the Many—between an apparently-asserted discontinuous or separated single body and remainder of the universe." Bridgman, Klyce charged, was asserting the existence of only the Many and adding to that an incomplete principle of relationship, called "interaction," and then denying one or the other aspect of it—either the movement relation or the position relation.

At the foundation of Klyce's disillusionment was his realization that Bridgman was taking the stand that complete and perfect knowledge is impossible not just in fact but also in principle. To Klyce, there is nothing that is unknowable.

As Klyce went on, it is difficult to see how he could have expected not to offend Bridgman. "You say," continued Klyce,

that there can be no "meaning" to a finite but small enough phenomenon, because "the necessary analysis of the minimum interaction can never be made," "because of our fundamental dictum that things which cannot in principle be measured have no meaning." In the first place, all the "meaning" there is is in actually existing things—or, that meaning is those things themselves; or, that meaning is the *fact* which is the expression by words or gestures of these existing things. (In other words, when you insist that "meaning" does not involve "real existence" or insist that "meaning" is not simply a precise synonym of "really or actually existing things" then you are simply dogmatically asserting that "meaning" is merely a matter of *words* or of mathematics, and have slopped over to the side of the metaphysicians who think just the same thing. . .—*in short, your talk about "meaning" is verbal idolatry, pure and simple; you have quite abandoned facts or things or experiments, and have become a dogmatic word-slinger* [last emphasis added].)

Having hurled his ultimate insult at Bridgman, Klyce elaborated another point. "The point of this paragraph is that it is a fundamental principle that *only* things which can be measured have any meaning—any standing in science. If that is so, then just what is the 'consciousness' you use to observe, and then even to *measure*, your things? Can you *measure* that consciousness? If so, what size is it? And just what is its position? . . . Actually it is an observable continuity . . . a standard One." From another point of view, he went one, "it could be said that consciousness is a Many thing or a relationship, but that does not get you any where." The reason is, Klyce asserted, "because you end up with contradiction. For if you claim that consciousness can be measured, then you have said it is a scientific object. But it is in interaction with whatever is being observed, and therefore, according to the Heisenberg principle, either one or the other cannot have a meaning— and there you are back again with nothing."

Klyce concluded his attack by calling attention to the fact that Eddington, in his *Nature of the Physical World*, had come "to every one of those conclusions which you stigmatize by the severest terms in the English language," referring, of course, to Bridgman's warning against the licentious and debauched thinking that he predicted would result from the untrained mind of the common man, which might admit the "spiritual." Klyce thought that Eddington's logic was "not good" (and of course he had written to Eddington to tell him so), but Bridgman's was even worse. Both of them, Klyce exhorted, should study logic—his, naturally—which would induce them to understand the Heisenberg principle correctly, and then they should each publish the truth.[23]

Bridgman's reply was brief and restrained. He expressed his disappointment that Klyce did not approve of his article, reiterated his conviction that his statements were based on "the experimental situation," and declared therefore that the apparent irrationality of his position in the *Harper's* article did not bother him. "I am not particularly concerned with the precise verbal expression of the situation," he wrote, "so long as the experimental situation itself is clearly apprehended. Certainly it does not trouble me at all that the thing should appear to be irrational or illogical; I am not at all convinced that all experience can be brought within the forms of logic." He closed by emphasizing his disagreement with Eddington's views and assuring Klyce that he could not possibly overemphasize this disagreement.[24]

It was clear to Klyce that Bridgman had not understood the basis for his criticism, so he proceeded to compose a set of letters, none of which Bridgman acknowledged at the time, elaborating his original points. The tone of the letters became increasingly shrill. He accused Bridgman of not being willing or competent to carry out to the fullest extent Bridgman's own dictum that one should not compromise with experimental facts, because he had, in regard to the Heisenberg principle, drawn a self-contradictory conclusion, namely that "experiments prove that no experiments can finally be made."[25]

In the same letter, Klyce seized upon Bridgman's repudiation of logic, saying that it "simply amounts to another way of having your science commit suicide." The reason, in Klyce's view, is that "logic is nothing more than speech" and if experience cannot be brought within the forms of logic, as Bridgman had asserted in his letter to Klyce, "it then follows that there are some experiments that cannot be described, or communicated to others. So again, you in effect agree that such experiments are quite useless in what is customarily called 'science.'" In fact, said Klyce, if this is how you feel, "you should have made that book [*The Logic of Modern Physics*] consist in total of a statement to this effect;—'There is no logic of modern physics, and so far as I can observe the facts no need of one.'" Klyce could not have uttered a more prophetic statement.

There is certainly no denying that these charges appear to be the fist wavings of an enraged man. Yet it would be lacking in insight to dismiss Klyce's accusations as entirely off the mark. Klyce was merely magnifying a tendency in Bridgman's thought that would receive greater emphasis in his subsequent publications. Furthermore, the lack of philosophical rigor in Bridgman's writings made him an easy target for such a determined antagonist.

But Klyce did not stop there. The letter went on. He charged that Bridgman did not have the right to draw any conclusion from what is not observed. "If you do not know just what facts you see in certain experiments, then what right have you to say that those facts are such and such—that their opposites, which Eddington asserted, are mental debauchery, and other malisons? For all you actually know, Eddington may be right."

Klyce kept up his pounding of Bridgman, in the course of which he took the opportunity to upbraid Bridgman for pleading busy on pressure experiments—"You *published* an article in a magazine, and can't simply merely disclaim responsibility"—and attack A. P. Weiss's attack of the same *Harper's* article. You "orthodox physicists," he charged, "are so ignorant, and so excessively conceited, that none of you understood clearly what Mach was saying, let him die unappreciated." [26]

Four interminable letters later, Bridgman decided it was time to call a halt to this barrage. On January 5, 1930, he wrote to Klyce:

Such pertinacity as you display deserves at least the courtesy of a statement from me as to why I did not reply to your last two communications, and why I do not propose to reply to any more. Briefly, the reason is that life is too short for me to dig out exactly what you are driving at . . . it seems to me that the chances are in favor of your being wrong, and I will have to let it go at that. Altogether, I have spent a lot of time trying to find exactly what you are driving at . . . . I have decided that the probabilities are for me so greatly against you being correct that I cannot spend further time on the matter. . . . I hope you will be sport enough to accept it from me. [27]

Klyce did not consider what he was doing as sport. He charged Bridgman with being one of the know-it-all specialists who had invaded his field of general principles, who had taken "a few days off [to] publish a magazine article on the same subject, that is so defective and foolish that even a psychologist [Weiss] can show existing stupidities in it." Hence, he asserted, "I have no more time to waste on you than you have to spend on me." [28] But that was not true. Another letter followed immediately upon the last, in which Klyce accused Bridgman of not understanding what he himself meant, otherwise Bridgman would have understood what he, Klyce, was saying. [29]

The climax for Klyce came after he read Bridgman's address to the AAAS delivered in December 1929 in Des Moines. The speech, entitled "Permanent Elements in the Flux of Present-Day Physics," was published in *Science*. [30] In it Bridgman spoke of his reaction to the rapidity of recent changes in physics, confessing that "some of us might even in a moment of candor, admit a little resentment at our shortness of breath." Referring to "the electromagnetic theory of light, the special theory of relativity, the general theory of relativity, the quantum theory of Bohr, the matrix calculus of Heisenberg, the wave mechanics of Schrödinger, the transformation theory of Dirac and Jordan, the group theory of Weyl, and now the double quantization theory of Jordan and others," he complained that "these have come crowding on each other's heels with ever-increasing unmannerliness, until the average physicist, for whom I venture to speak, flounders in bewilderment." [31]

Revealing his disillusionment, Bridgman explained, "We used to demand that the ultimate goal of physical theories should be nothing less than the discovery of the underlying realities. To-day our demand for reality is much less insistent, in large part because we are much less confident that the ultimate reality, which we thought to be our goal, has any meaning." Citing Poincaré's insight that an infinite number of mechanical explanations might be adduced to deal with physical phenomena, Bridgman continued, "It is natural, therefore, to find that the demand that our theories reproduce reality is becoming replaced by the demand of convenience and simplicity." All of this means that one must be exceedingly cautious "in ascribing any finality to the details of the present mathematical theories." [32]

Applying this caution to contemporary theories, Bridgman asserted the need to look beyond mathematics to intuition to discover what is of "permanent significance in present developments." Out of intuition would be developed the art of conceptual experiment. "It should be possible to build up a formal structure in which the properties of photons and electrons and other elemental things, such as quantum interactions, are described in terms of conceptual experiments, and from simple properties deduce more complicated properties in much the fashion of Euclid." And after listing 21 questions whose answers might be grasped in terms of conceptual experiments, Bridgman admitted that mathematical theory might "at present" be the key to such answers, but concluded by declaring his belief "that this is merely a temporary phase and that ultimately we shall be able to demand that our theories be so formulated that we can answer these and other questions intuitively without recourse to formal mathematics." [33]

This was the last straw for Klyce. He was flabbergasted, he wrote to Bridgman, to see him "so irrational, so close to the verge of pathological insanity in one given brain area." He compared Bridgman's state of mind to that which

many clergymen have reached, when all of a sudden it dawns clearly upon them that the "religion" they have believed in is the bunk, and that they are left with no religion. . . . And your article is clearly to me at least a statement that you have lost *your* "religion," which was a belief in orthodox science, and are in a serious state of mind over the loss. . . . You are quite right in laughing at such mathematical physics. But your trouble is that you are still striving desperately to hold on to that former belief in such mathematical or quantitative physics. You laugh at your former religion, and yet you try to hold on to it, having nothing with which to replace it.

All the Heisenberg principle means, said Klyce, is

that in his mathematics Heisenberg has come plump up into a statement that no existing thing can have an absolute or mathematically classical boundary. . . . Or, another way of saying the same thing is to say that all quantitative science (physics, biology, and every other branch) can be only what you name statistical. What you call conceptual science can be and must be absolute, or KNOWN (statistics is an approximation: is uncertain: conceptual science is certain.) . . .

You yourself are badly worried over losing that *old* view—your previous religion. You are taking it all most seriously—in spite of your verbal bravado,

diagnosed Klyce. "I worked out that actual step years ago. I think that you are now too old to have the strength to work out the whole thing for yourself. . . . So it would be a good idea for you to get my *Universe* out of the Harvard Library and read it." [34]

This was the last letter that Scudder Klyce wrote to Bridgman. He died a week later, on January 28, 1933.

Klyce, for whom knowledge was the intuitive apprehension of metaphysical unity, stood on the threshold of modern sensibility. In his mind, the "problem of knowledge" was, strictly speaking, a false problem created by intellectuals, among whom scientists stood out as an egregious example, and whose "sin" is to judge partial knowledge to be all of knowledge. Klyce believed that the particular and the universal are not mutually exclusive or epistemologically divided, but coexist, and this coexistence can be intuitively grasped, or experienced, by an unprejudiced mind. He thought that in Bridgman he had succeeded in finding a scientist who had avoided this "sin of science," only to discover that Bridgman, too, had become a "word-slinger" and thus had shown himself to be an incompetent philosopher.

One might be tempted to brush aside this incident with Klyce, to dismiss it as accidental and therefore not particularly significant or revelatory, were it not that the story was played out again, with an uncanny similarity. This time the antagonist was a younger man whose terminology, reflecting a more modern "scientific" preference, was less directly metaphysical and more psychological, and whose moral precepts were not couched in the language of salvation but of therapy. Nonetheless, the prescription for curing the woes of mankind was the same—true knowledge. Only undistorted truth would free humanity from the darkness of sin (Klyce) or, equivalently, from the bondage of pathological thought.

Echoing the sentiments of Scudder Klyce that errors in the use of language were responsible for a miscomprehension of the nature of truth, but insisting on a more scientific mode of discourse, was Alfred Korzybski, another of Bridgman's early admirers. Korzybski, or Count Korzybski as he was often addressed, was a Polish engineer and man of affairs, a linguist and advocate of a discipline he called non-Aristotelian semantics (as in non-Euclidean geometry). An enormously charismatic man, possessing a worldwide following, he is probably best known as the author of the widely read *Science and Sanity*, published in 1933.[35] As the correspondence from 1927 to 1933 between Bridgman and Korzybski reveals, Bridgman gave generous support to the creation and publication of this work, not only by encouraging Korzybski but also by reading and criticizing drafts of various chapters. Moreover, when Korzybski was looking for a publisher, Bridgman put in a good word for him with Macmillan, which nevertheless did not publish it. (Korzybski founded his own publishing company.)

In addition, Korzybski is known as the founder and director of the Institute of General Semantics, established in Chicago in 1939. He was a fellow of the American Association for the Advancement of Science and a member of the American Mathematical Society, Chicago Society of Personality Study, Association of Symbolic Logic, New York Academy of Science, and the Society for Applied Anthropology. However, even though he held a place in a far reaching social network, he left behind a quickly fading intellectual legacy, his influence having rested to a large degree on his personal magnetism (and the susceptibility of contemporary sensibilities) rather than on the permanent merit of his ideas. This, despite the fact that two of his disciples, S. I. Hayakawa and Stuart Chase, wrote best-selling books based on his ideas.* On the other hand, the Institute of General Semantics is still in existence today and attracts reputable scholars to its conferences and workshops.

Sidney Hook has remarked somewhere that if Korzybski had "deliberately set about to obscure his primary insight he could not have proceeded any differently from the way he has or succeeded so well." That is probably true, but nonetheless something of his message reached its mark. His book was a popular success. For the present purposes, we may form some notion of the thought behind Korzybski's non-Aristotelian semantic program by comparing his ideas to those of Scudder Klyce.

In this connection, it is interesting to observe that when Bridgman pointed out to Korzybski that his ideas bore a distinct resemblance to Klyce's, Korzybski reacted by characterizing Klyce as insane, suffering from paranoia. "There is no doubt that he is gone, a classical example of paranoia with all the trimmings too. Sooner or later he will probably kill some 'persecutor' and will be locked up. . . . Great many geniuses

---

*S. I. Hayakawa, *Language in Action* (New York: Harcourt Brace & Co., 1941); Stuart Chase, *The Tyranny of Words* (New York: Harcourt Brace Jovanovich, 1938). Both men acknowledge a debt to Bridgman, Chase most explicitly. The concern about language and the attempt to purify it clearly reflects the anxiety of the era between the two world wars. It was part of an effort to regain control in a chaotic situation (political, moral, and scientific), to combat a feeling of helplessness, and to reassert the efficacy of rational thought and positive knowledge. Both authors argued that the world is "a madhouse of murder, hatred, and destruction" (Hayakawa) and attributed much of the cause to a manipulation of meaningless words, "peopling the universe with spurious entities, mistaking symbolic machinery for referents" (Chase). Society is falling apart because communication is impaired when words are used to manipulate rather than to enlighten. The cure is to be found in the study of language and in its subsequent purification, in discovering the true meaning of words and rejecting the "demonology of absolutes and high-order abstractions." (Chase)

have the paranoic make-up, but are not entirely gone. Klyce is. It is extremely dramatic." As to the similarity of ideas, Korzybski attributed them to "the age we are living in. The things are emotionally in the air, instinctively and intuitively. The same feelings have urged Einstein, you, Heisenberg, etc. myself and Klyce. . . . Klyce broke down but he still expresses and will express his feelings, which in many instances seem to be sympathetic to us, but they are ultimately incoherent and non-workable."[36]

Actually, it would be difficult to support the claim that Korzybski's system was any more coherent than Klyce's. Perhaps it was even less so.[37] It was his hope, he told Bridgman, to create a human "Principia" (as in Russell and Whitehead's *Principia Mathematica*). However, in his attempt to synthesize modern scientific findings, he ended up with a patchwork of concepts borrowed from various disciplines ranging from anthropology to mathematics. His work was a conglomeration of ideas that was certain to strain the imagination of anyone who might try to understand its primary thesis.

Except for a few minor points, the comparison with Klyce is valid. Where Klyce was unabashedly metaphysical, Korzybski thought he was being scientific. Klyce had taken Bridgman's *Logic of Modern Physics* to be a treatise on logic. Korzybski read it as one on psychology. But beyond that there was a noticeable similarity. What Klyce dubbed "verbal idolatry" Korzybski called "insanity," a state of mind characterized by the error of confounding words with objects.[38] Korzybski's "non-Aristotelian" logic, which was aimed at eliminating the universal "all" as a categorical attribute,[39] recalls Klyce's objection to using "One" (universal) words when "Many" (particular) words are demanded. Korzybski added the further refinement that a pathological condition resulted from the unconscious confusion of different semantic levels of language. In his view, ordinary language is severely pathological. "Objectification," he claimed, is an emotional disturbance.

Escape from this condition—ignorance for Klyce, insanity for Korzybski—was to be achieved by reaching for a truth beyond words. Language, both men assumed, is a kind of clothing that obscures reality and which must somehow be made transparent or closer fitting. Bridgman did not disagree. Truth for him was also wordless. "The event, then, is the starting point for us, and it is of necessity something that we think about. But we must not think about it in words, for the moment anything verbal gets into our thinking we descend from the level of the event."[40]

Whereas Klyce believed that truth was gained by mystical insight and consisted in the revelatory apprehension of the metaphysical unity of all knowledge, Korzybski spoke of a silent truth that was supposed to be an empirical absolute obtaining on "unspeakable objective levels,"[41] an experiential truth that must not be confused with words. It was "objective" insofar as perfect knowledge consisted in the exact congruence of the nervous system and "reality." Korzybski's therapeutic program included exercises for making the awareness of the difference between experience and language a permanent part of consciousness. It is not clear, however, how his exercises could produce the "objective" state of perfect knowledge or how this new level of consciousness could create the desired exact congruence between the nervous system and "reality."

Furthermore, just as Klyce had insisted on the primacy of experience, Korzybski also asserted the fundamentality of "empiricism." His "non-Aristotelian" semantics, he believed, was empirical by virtue of his rejection of the "is" of identity. Basically this meant that no event is exactly the same as any other.[42]

Korzybski's monistic tendencies were expressed through his desire to build a language that was "non-elemental,"[43] which meant that sense and mind, intellect and emotions, are not split (into separate elements, presumably). He believed that Einstein's time-space manifold provided the model.[44] As he explained to Bridgman, he regarded Einstein's theory as a "new monistic language closer to 'reality,'" where Einstein does not divide what should not be divided.[45] Bridgman appreciated Korzybski's judgment of the nature of Einstein's accomplishment, to the degree that he characterized it as being merely linguistic: "I was glad to hear that you ascribe only verbal significance to Einstein."[46] It was comforting to think that reality is still the same and that only the language used to describe it has changed. But Bridgman was beginning to sense some essential differences between his and Korzybski's viewpoints, if not the tangle of inconsistencies in Korzybski's writings.

Bridgman's major disagreement with Korzybski, he said, was with his "structuralism." Korzybski had written, "Since 'knowledge,' then, is not the first order unspeakable objective level, whether an object, a feeling; [therefore] structure, and so relations, become the only possible content of 'knowledge' and of meanings."[47] Bridgman objected. "I think you overstate your case in insisting that *structure* is the whole story. For example, a game of ball played with the hands and a game of ball played with feet, as do some Indian tribes, cannot be called the same in the experience of the players, although there may be a complete one to one correspondence between all the external motions and the muscular and nervous reactions of the players of the two games."[48]

Indeed, this parallels Bridgman's accusation that Einstein neglected the events themselves and concerned himself instead with only the coordinates.[49] The idea that knowledge should be constituted by bare relations would make it too abstract, too far removed from immediate experience, for Bridgman's understanding of what knowledge is. His self-examination was persuading him with increasing cogency that experience was concrete and personal, bathed in infinitely rich perceptual qualities; to abstract these away would be to "define away the truth."

The disagreement was more deep-seated than either man realized. Disillusionment was soon to descend on Korzybski, just as it had on Klyce, and in a manner almost identical. Again, the rift opened when Bridgman published an article responding to what he perceived as a crisis in the state of knowledge. This time the crucial article was concerned with the foundations of mathematics. Published in *Scripta Mathematica* in 1934, "A Physicist's Second Reaction to *Mengenlehre*" represented Bridgman's attempt to repudiate Cantor's set theory and, in particular, to discredit the diagonal proof by which the non-denumerability of the real numbers is established.[50]

Bridgman was disturbed by his recent discovery that the stability of the foundations of mathematics was being challenged by the paradoxes of set theory. Believing that he had satisfactorily resolved the "paradoxes" of relativity theory by operationally redefining the fundamental physical concepts, he thought he could do the same for set theory. He was guided by his maxim that "experience is not self-contradictory" and asserted that if mathematics were properly understood as an empirical science, no paradox could arise, because "impossible" mathematical objects (infinite sets) would not be admitted.

This effort involved Bridgman in a critique that superficially resembles the intuitionist program of the Dutch mathematician L. E. J. Brouwer, insofar as it placed primary emphasis on constructibility as a criterion for the admissibility of mathematical objects and denied the legitimacy of infinite sets as complete, self-sufficient entities.[51] Nevertheless, Bridgman's operational constraints were more restrictive and not firmly grounded on mathematically defensible principles. Indeed, his view of the nature of mathematical objects and proof elicited criticism from at least two prominent thinkers—the mathematician Abraham Fraenkel and the physicist Erwin Schrödinger. Their objections were, on Bridgman's subsequent inquiry, affirmed by the logician W. V. Quine. At issue were Bridgman's operational concept of number and his rejection of Cantor's diagonal proof of the non-denumerability of the real numbers.

Bridgman thought that the problems of set theory were a consequence of treating number as an object when it should be understood as a program. A number, he asserted, is not an entity or self-existing thing. It is a symbol standing for the rule applied in its construction. It is a formula for human activity. Nor is a point on a line, which point is supposed to correspond to a number, in any way a constituent of the line. "A line is not composed of points as the forest is composed of trees." Rather, a point on a line means no more than the rule specifying how it is determined. Bridgman argued that it is meaningless to "talk or think about an 'existence' of [numbers or points] in their own right." Furthermore, since a rule must be applied by a human being, it follows that the total number of rules is finite and any given rule can be applied only a finite number of times. Thus, he concluded, "we have no more reason to describe the points on a line as non-denumerable than the non-terminating decimals."[52]

Since, Bridgman continued, a non-terminating decimal is, in operational terms, a rule that is not self-terminating (it includes no condition indicating when to stop applying it), there could be no distinction between different orders of transfinite numbers. Moreover, he could not conceive of "different sorts of non-self-terminatingness." In other words, there cannot be different kinds of a non-condition. Non-denumerability is not a property present in degrees. The same operational reasoning led him to deny that the denumerability of the algebraic numbers could establish anything at all about the existence of the transcendentals, even though it can be shown that the reals, which consist of algebraics and transcendentals, are of a larger cardinality than the algebraics.

This argument rested on a particularly narrow concept of number, one that did not recognize the formal distinction among various kinds of number and did not accord with contemporary mathematical understanding or results. The philosopher, Georg Kreisel, in a letter to Fraenkel, remarked that Bridgman's misunderstanding of number was one natural to a physicist.

Why did I call Bridgman's attitude natural? In applications of numbers to measurements, of some continuous variable like the concept of length in contemporary physics, the difference between : rational and irrational, or : algebraic and transcendental, is spurious; and since in his discussion Bridgman only thinks of using numbers for measurements (not, e.g. for naming things), it is natural for him *to frame his definitions of concepts about numbers so that they are relevant in the application of numbers to measurements.* . . .

We might go further, and have to if we are to do Bridgman's problem justice: where is the difference between : rational and irrational, or : algebraic and transcendental, striking and interesting? For, it isn't in the physics of measurement. . . .

To counter Professor Bridgman's criticism we have to explain an application of the concept of decimals where Cantor's definition is interesting. An obvious one, of course known to B., but ignored in his argument, is the use of decimals as a *formal technique,* i.e. where one ignores their use as approximations, but thinks of *proofs* within a formal system. . . . *The diagonal argument, and with it the distinction between rationals and irrationals, is relevant to the systematic study of mathematical systems.* [Emphases in original] [53]

However, if Bridgman's idea of number was intelligible, and even perhaps challenging to the mathematician, the strategy he utilized in his repudiation of the validity of the diagonal method was not one that a mathematician could view sympathetically. Essentially it amounted to neutralizing it by redefining the concept of denumerability. We have seen that this is precisely what Bridgman attempted to do when he redefined the basic concepts of measurement because their relativistic attributes offended his physical intuition. In the present situation the goal of his approach was to eliminate the necessity to consider infinite mathematical objects.

Briefly, the diagonal proof of the non-denumerability of the real numbers is a method by which it is shown that "there is no rule for enumerating the real numbers between 0 and 1 exhaustively (i.e. no rule which pairs them off exhaustively with the positive integers)." [54] The procedure involves showing that given any sequence of decimal fractions, it is always possible to construct a real number not contained in the array. Or, in a more popular style, no matter which way you count the real numbers up, at least one is left out of the list. This proves that there are more real numbers than there are positive integers; that is, the real numbers cannot be put in one-to-one correspondence with the positive integers, which is, of course, what it means to be able to count them. [55]

This is the conclusion Bridgman denied in his article. He argued the proof was not compelling because it depended on the "impossible" assumption that the non-terminating decimals can be regarded as totalities that can be arranged in order.

For in the first place, what are these non-terminating decimals, which are supposed to be arranged in sequence, like apples in a row? "Non-terminating" is obviously only a polite way of saying "infinite," and such are not things like apples, for no one can present me with one which I may fit into its place. Operationally an infinite decimal means only a program of procedure. It is not possible to actually carry out the operations involved in the diagonal *Verfahren* [procedure]; the operations involved in producing the non-terminating decimal cannot be complete, so that it is not legitimate to postulate the performance of other operations, that is, arrangement in sequence, after the impossible completion of the non-terminating decimal.[56]

Instead, Bridgman offered his own formula, which supposed that a rule can be formulated such that any number constructed by the diagonal procedure can be located at a specifiable position in a finite array. This, he told Schrödinger, was what he, Bridgman, understood by denumerability. "The precise formulation was the outcome of some of my talks with the mathematicians this summer. There seems to be no agreement among the mathematicians; . . . I personally am very strongly of the feeling that this must be allowed; it seems to me that something very similar is involved in enumerating the rational fractions or the algebraic numbers."[57]

Schrödinger politely but firmly replied that he believed that Bridgman was completely wrong. "The concept of 'denumerable' is settled upon in a way which is not open to 'strong feelings' . . . the difference between denumerable and non-denumerable *Mengen* [sets] is a sound one. If you replace existing definitions by others, it is clear that existing propositions will no longer apply to them. But that is not an argument to shake them."[58] Schrödinger was careful to confine his criticism to the technical problem and made a point of reaffirming the affectionate relationship between the Bridgman family and his own. Nevertheless, he was emphatic in his insistence that Bridgman's argument was mathematically unacceptable.*

*Schrödinger continued, referring to Bridgman's formula: "You satisfy yourself . . . that you actually *can* indicate a series of sets S^1, S^2, S^3 . . . such that given an arbitrary definite set S' *and* given an arbitrary number m, however large, and comparing S' successively with S^1, S^2, S^3, etc.; you will always after a finite number of trials find out an S^n,

In contrast to the reasoned, even if critical, comments of Schrödinger, Fraenkel, Kreisel, and Quine, Korzybski's reaction was personal. "I have read attentively and several times your reprints, and find that your paper on Mengenlehre is one of the most severe blows I received for a long time . . . the trouble is that you did not read carefully my work." Korzybski attacked the paper as being "hopelessly antiquated to the point of being scandalous." He charged that "there is practically not one single paragraph which would give your operational methods one single scientific leg to stand upon," and exhorted Bridgman to study his book. "The crux of this matter depends on 'meanings,' 'existence,' 'all,' 'properties' and what not, which is a complete mess in the old [system of beliefs], and yet you have no consciousness at all that all these problems have been cleared up in my book, and this must be the foundation for any operational method." [59]

We have seen this response before. Klyce, too, had reacted by taking an article by Bridgman as a personal insult. And just as Klyce had deplored Bridgman's repudiation of logic, Korzybski attacked his lack of appreciation of generality in mathematics. "I have no intention to be satirical, but science and particularly mathematics depend on *generality* and you write from the point of view of personal experience the best you can. This paper should then be printed in a psychiatric magazine but not in a mathematical one."

Korzybski echoed Klyce's charge that Bridgman was clinging to an old set of values. "You curse out 'properties' and yet and although a 'physicist' you drag back unconsciously your 'properties' with the 'reality' of your 'object.' And so it goes on and on, and what can I do if scientists do not read, and hide behind a defense mechanism that their science is only their own pleasure and not a social activity, which may interest them but is beyond their own selfish pleasure."

You have something most important to offer to the world in your operational stuff but in your antiquated, most naive, self-contradictory, etc., form of presentation it is practically useless as a general discipline because it is thoroughly unworkable and based on *false to fact older fallacious principles which you somehow do not dare to face* [emphasis added].[60]

---

that coincides with S' *with respect to the admission of the first m numbers.* Excuse me, but I find this argument frightfully trivial. It simply states once again, what we all know, namely that the *rational* numbers are enumerable. It has not the slightest resemblance with the well known argument that proves the denumerability of the algebraic numbers."

Bridgman replied that Korzybski's letter was very disappointing and reminded him that he had supported Korzybski's work because he felt that the general subject was of "utmost importance." But that was not to have been construed as unqualified approval. "But this by no means implies that I can accept all, and I am sorry that you cannot accept this much from me, but must apparently have everything or nothing." [61]

By a stroke of improbable coincidence, the pattern of events was following exactly the same outline as it had with Klyce. Even the two-stage denouement was being replayed.

Korzybski, of course, did not give up, and continued to solicit Bridgman's support. But Bridgman had come to the end of his patience and informed Korzybski that he had already gone beyond "the elastic limit" [62] and could go no further. However, the final insight did not come to Korzybski until he read Bridgman's book, *The Nature of Physical Theory*, [63] the published version of the Vanuxem lectures Bridgman had just delivered at Princeton, "which no one," Bridgman complained, "seems to like at all." [64]

Korzybski most emphatically did not like what Bridgman had to say, although, he admitted, "this book made me understand a few things about you which have always baffled me." But, as a matter of fact, Korzybski had come to the realization that "from the preface to the conclusion your book is a deadly criticism of S+S [*Science and Sanity*]." [65] Still, he could not resist preaching once again against scientists who teach false knowledge and thus breed insanity.

Bridgman's reply is already familiar. "Now I have spent a great deal of time first and last on going over your ideas, as much as I am willing to afford in view of all my other interests, and the plain fact of the matter is that you haven't 'put it across' as far as I am concerned." In addition, Bridgman informed Korzybski, "You give too much the impression of a prophet with a mission." [66]

Even so, Korzybski's reaction was not unfounded. *The Nature of Physical Theory* was indeed subversive of Korzybski's program. It was, as he had correctly perceived, a thoroughgoing denial that the "cure" that Korzybski advocated could be achieved at all. His "non-Aristotelian" semantic therapy was predicated on the expectation that perfect correspondence between reality and the human nervous system was possible, which is another way to say that a state of complete knowledge can be achieved. This assumption, in turn, rested on the view that knowledge is of the general, abstract, or "structural." In Korzybski's words, "The old 'unknowable' becomes abolished and limited to the simple and natural fact that the objective levels are not words."[67]

In contrast, *The Nature of Physical Theory* was a concerted effort to argue that knowledge is a product of human activity and is therefore by its very nature limited and imperfect—incomplete. Bridgman believed that this is so by virtue of two undeniable conditions. First, the mind is incapable of transcending experience. Language, which is a human artifact, freezes a dynamic and changing experience into static and artificial units. But not only is any language or system of symbols unable to capture the fullness of experience, thought, although more complex than language, also has a different structure from experience. The laws of thought, which are said to give logic and mathematics their absolute status, are themselves human inventions, principles inferred from experience. Therefore they, too, are only approximate. Second, the future is in principle unknowable. As a consequence, truth is not eternal and absolute but depends on the state of human development. No extrapolation beyond what is within the reach of direct observation is to be accorded complete confidence. In any case, Bridgman argued, integrated principles are not necessary for practical activity, and elegance of representation is a utopian ideal.

His fundamental insight notwithstanding, Korzybski's attempt to unite the idea of knowledge as an abstract relational structure to a neurological/psychological state through a program of non-Aristotelian logic was riddled with self-contradiction. But that was not the basis for Bridgman's objections. His disagreement centered on the idea that knowledge consists of abstract relations rather than experienced qualities and that a total system of knowledge can be constructed.

Korzybski was not the only one who disapproved of Bridgman's message. *The Nature of Physical Theory* was severely criticized. Bridgman was accused of contradicting his own precepts, of reverting to mysticism, of being solipsistic, Bergsonian, and even of tending toward dialectical materialism.[68]

On the surface, Bridgman had developed a reasonably consistent, if extreme, empiricist stand. Having left behind the naive attempt in *The Logic of Modern Physics* to demystify scientific concepts by offering a demarcation standard that would preserve objectivity and cleanse science of idealization—a scheme that invited the interpretation that he was advocating a behavioral monism—Bridgman had settled into a strongly particularist position. It was distinctly anti-Platonist and antitheoretical and was quite clearly not monistic. He took the view that truth resides in the particulars of experience rather than in idealized generalizations. The jump from the operations of the laboratory to mathematical encapsulation, as he later expressed it, "is a jump which may not be bridged logically, and is furthermore a jump which ignores certain essential features of the physical situation."[69] For this reason, the intellect is bound to fail in its quest to put its straitjacket onto experience.

As a laboratory scientist, Bridgman was acutely sensitive to the fact that general principles in and of themselves are insufficient to produce success in a concrete working situation. They provide, as he asserted, only the rough outlines for the practical achievement of material goals. In a given experimental circumstance, any number of theoretically insignificant details may determine whether the program can actually be carried out. These mundane considerations are frequently swept aside in journal accounts, but they must be reconstructed by anyone who attempts to reproduce the results. Furthermore, the inevitability of error in measurement precludes sharp delineation of categories. For these reasons, Bridgman could, with justifiable confidence, claim that a concept is the imperfect representation of the experiential reality. Indeed, he went further in his argument and insisted that the metaphysical instinct urging the mind toward explanation was a fault to be resisted.

Had *The Nature of Physical Theory* been confined to arguments at this level, the worst criticism Bridgman would have had to suffer would have been for his failure to recognize the role of theory in making experiment itself intelligible. But what offended Bridgman's readers most was his foray into epistemology, the result of which was his conclusion that

There is no such thing as public or mass consciousness. In the last analysis science is only my private science, art is my private art, religion my private religion, etc. The fact that in deciding what shall be my private science I find it profitable to consider only those aspects of my direct experience in which my fellow beings act in a particular way cannot obscure the essential fact that it is mine and naught else. "Public Science" is a particular kind of the science of private individuals.[70]

If this looked like solipsism, he allowed (perhaps it was, since at the time he was not sure what was meant by solipsism), it was nevertheless a simple statement of what his direct observation had yielded.[71]

Bridgman made the reasoning supporting his position more explicit in a later publication, "Some Implications of Some Recent Points of View in Physics." Every intellectual enterprise, and therefore science, he explained, rests on an act of *understanding*—that is, it involves the mental process of "catching on"—and this is necessarily individual and private.

It is usual to ignore this private component and to think of science as an essentially "public" enterprise; in fact, the possibility of publicity is sometimes made a part of the definition of science. It is obvious that under this sort of a definition the much prized "objectivity" of science is ensured. Although this public component in science doubtless exists and is important, I believe that the private component is even more important. For any scientist worthy of the name does not accept the public body of scientific knowledge simply because it is public or has the authority of the consensus of his colleagues. . . . No argument presented to him by his colleague is accepted as valid unless it bears the impress of *proof*.[72]

Such proof, Bridgman argued, cannot be alive unless it is part of the understanding of an individual mind.

This position, Bridgman subsequently came to insist upon more confidently, is not solipsism as it is usually defined.* In fact, the traditional solipsistic interpretation of reality is itself a form of idealism, the very brand of metaphysics that Bridgman so thoroughly opposed. Bridgman had investigated the problem of knowledge and concluded that the locus of knowledge must be specifiable; therefore it cannot exist disembodied in a transpersonal ideal realm, in an unspecifiable somewhere. It must be empirically localizable, and introspection indicated to him that it resides in the individual consciousness. There is no such thing as a collective understanding because there is no collective mind. Furthermore, the trustworthiness of knowledge depends on an individual act of understanding, not on public consensus, which itself presents not the cogency of proof but only the authority inherent in social compulsion. In Bridgman's mind, this was a straightforward empirical observation.

---

* "It has been my experience that my point of view is very likely to be damned with the epithet 'solipsism.' What solipsism is in the popular view may be formulated in the statement; 'Only I exist and the external world is my construction.' This is usually felt to be so absurd as to constitute its own refutation. . . . I think it will be very hard to shake off the damning implications of the word 'solipsism,' and I wish that another more explicit word could be found for a point of view which is solipsistic only to the extent of

The Nature of Physical Theory proved to be the beginning of a permanent conflict between Bridgman and his philosophically minded colleagues. Even the friendliest of them could not assent to his subjectivism. This was not what they believed science is founded upon. Science is concerned with objective truth, not private intuition. It looked as if Bridgman had given up all pretense to the objectivity that his own operational denotation of physical concepts was thought to preserve.

Prodded by a desire to understand the foundations of physics, Bridgman had moved into the unfriendly territory of epistemology, and he was unfamiliar with the protocol. His statements were met with unsparing criticism. "His demand for operational criteria," it was said,

has no operational basis, and rests like the world-turtle of old cosmologies on a bottomless ocean of infinite regression. It is terribly unoperational to say "In the last analysis science is only my private science." What can an operationalist mean by "last analysis"? He definitely takes sides on a pseudo-problem, because he cannot eradicate in himself the very type of idealism he decries in Eddington and Jeans. His assertions as to the subjective character have no objective status at all on his own criteria. His subsequent treatment of the chain: experience, thought, language, has what appears to me to be a great disadvantage, and to him an advantage, of being rooted in a kind of mysticism of the internal with its introspective privacy and intuitional insight.[73]

No one was as sensitive to the change occurring in operational thinking as Arthur Bentley, another of Bridgman's interpreters, who has been described as "one of the most controversial political scientists the United States has produced."[74] Bentley, a Johns Hopkins Ph.D., is also known for his collaboration with John Dewey, with whom he coauthored the book *Knowing and the Known*.[75]

Bentley, along with Klyce and Korzybski, viewed Bridgman's operational treatment of the meaning of physical concepts as supportive of monism and believed it could be adapted to his own monistic social ideas in which "knowledge-language-experience-fact" was to be considered an uninterrupted continuum. Bentley's social analysis, being methodological rather than theoretical in intent if not in content, had much in common with Bridgman's operational analysis. Bentley was therefore highly sympathetic to Bridgman's protests against the ten-

---

feeling the need for an analysis of what one means when, on the private level, one talks about the public or external world" (P. W. Bridgman, *The Way Things Are* [Cambridge, Mass.: Harvard University Press, 1959]).

dency to read his attempts at achieving analytic clarity as philosophy.

Bentley's intellectual influence on Bridgman is most direct in the article on *Mengenlehre* (set theory). The debt is far greater than Bridgman's offhanded reference to Bentley's work would suggest. It is doubtful that Bridgman would have written the article had it not been for Bentley. Not only was it Bentley who first alerted Bridgman to the possibility of a mathematical critique, it was he who first suggested the strategy of redefining denumerability.[76]

The correspondence between Bridgman and Bentley constitutes a running commentary by Bridgman on Bridgman, as Bentley probes for insight into the origin and meaning of operational thinking. Such close questioning made Bridgman somewhat uncomfortable, he admitted, since "it makes me afraid that I may say something that isn't quite right and that I'll get jumped on."[77]

In the course of this exchange, we learn that the arguments Bridgman advanced in *The Nature of Physical Theory* (the Princeton Lectures) were intended as a counterweight to overconfident theorizing. "In the first place," wrote Bridgman,

with regard to the background of the Princeton lectures, I had constantly in mind the audience that I hoped to address, containing such mathematicians as Weyl and such physicists as Einstein and von Neumann, whose cocksureness I have always found profoundly irritating. The point that I was trying to ram home was that there is no justification for cocksureness.[78]

The need for humility is a recurring theme in Bridgman's writing and reflects his human, this-worldly philosophical attitude. We have seen it at the basis of his attack on Einstein. It is indicative, as well, of the cautious temperament that held him captive, fearful of overstepping the bounds of certainty. And if operational analysis retained any thread of consistency throughout Bridgman's career, it was as an instrument to tie knowledge firmly to human activity, to keep it from escaping into an abstract transcendental realm. He thought the aspiration to achieve a unified worldview to be a fault, an indication of a desire to escape from reality.

In contrast to Klyce and Korzybski, Bentley made a sincere effort to understand Bridgman's thought and frequently compared Bridgman to C. S. Peirce and John Dewey. Furthermore, he did not demand allegiance to a unified philosophical system, as they had. An unusually abrasive and contentious person, however, Bentley often overwhelmed

Bridgman with his abstruse rhetoric. Typically his style was vicious and deprecatory, and he made good use of his pernicious talent to encourage Bridgman's antagonism toward positivistic philosophies of science, probably because he perceived them as being inimical to his own views. He had selected Henry Margenau and Rudolf Carnap as particularly deserving of his invective. To Bridgman, Bentley wrote, "How Margenau does hate you."[79] Yet for all of his unpleasantness, Bentley was an astute critic, well acquainted with the contemporary literature.

Thus he was not for a moment deceived by Bridgman's change of direction. Bentley had made it his business to keep track of the state of operational thinking and was merciless in his criticism of everyone, including Bridgman, who did not seem to mind. In Bentley's judgment, the "new" Bridgman had adopted a hopeless position, which he characterized with an evocative metaphor in a letter to Bridgman: "I cannot agree with your remark in your last letter that the 'later' Bridgman is not so bad. Every additional examination I make shows him up worse— the Bridgman-Hyde, I mean, not the Bridgman-Jekyll."[80]

However, Bridgman's vision was confined to his own individual perspective, and for that reason he did not understand the basis for the criticism. His ambitions, he claimed, fell far short of what was imputed to him. Thus in answering Bentley's criticism of his subjectivism, he wrote:

I think that you think that I am trying to do more than I am in my insistence on the "private," just as most people thought I was trying to do more in my operational analysis than I was. I am only trying to say that any faithful description is self-centered. . . . It may very well be that I am giving you "not bread but a stone." If it should turn out that this is the case it would mean merely that the solution of your problem is not primarily a question of faithful description.[81]

His readers, he held, had simply misread his intentions. But beneath the surface, a new awareness was coming to Bridgman. It was one that he had not yet expressed in a public forum or even in his private correspondence. Nevertheless, it must have been discussed among his family members, for his daughter, Jane, drafted a letter to Bentley in which she charged him with distorting her father's ideas. "Father is the man," she pointed out,

that thought up Operationism, I think logically he is the one to go to find out how it works out. Seems to me that's the only honest way to approach the book. Say you don't like the way it works out, if you must, but for heavens sake don't say that shows that Father hasn't grasped the Operational method (well, you didn't, but you made me feel almost as though you had—) because it's his method, not yours.[82]

The letter was not sent, but the protest would have been too late and, in any case, futile. The operational method Bridgman had presented in *The Logic of Modern Physics* no longer belonged to him. It was a public object now, with a life of its own. It had become Operationism and was even being used in arguments against Bridgman himself. What had started out as a private idea had been transformed into an external artifact, subject to new interpretations and put to uses outside Bridgman's control. This was nowhere more apparent than in the operationist movement in psychology, and in the assimilation of operationism into logical positivism, both discussed in the next chapter.

Bridgman's frustration is reminiscent of that experienced by C. S. Peirce, whose pragmatism was elaborated by William James into a form Peirce could not sanction. Peirce reacted by giving his own philosophy a new name—pragmaticism. Perhaps Bridgman should have followed Peirce's example and renamed his project "operationalysis" or something equally unpronounceable in order to distinguish his idea from its interpreted forms. But that would also have misrepresented Bridgman's standpoint, for the operational method had undergone considerable change in meaning since Bridgman's original presentation. Moreover, the "old" operational analysis had become publicly appropriated, objectified, and thus rendered inauthentic. If the new operational method were to be named, it would become similarly an object and thus a new untruth.

CHAPTER 7

The Positivists and the Behaviorists: Defining away Private Experience

Reality and deceit are equally possible, and . . . deceit can clothe itself in the same appearance as reality. Only the individual himself can know which is which. . . . I can lay hold of the other's reality only by conceiving it, and hence by translating it into a possibility; and in this sphere the possibility of a deception is equally conceivable. This is profitable preliminary training for an ethical mode of existence: to learn that the individual stands alone.

. . . With respect to every reality external to myself, I can get hold of it only through thinking it. In order to get hold of it really, I should have to be able to make myself into the other, the acting individual, and make the foreign reality my own reality, which is impossible. For if I make the foreign reality my own, this does not mean that I become the other through knowing his reality, but it means that I acquire a new reality, which belongs to me as opposed to him.

—Søren Kierkegaard,
Concluding Unscientific Postscript (1846)

■ ■ ■

BRIDGMAN HAD CONCEIVED the operational method of *The Logic of Modern Physics* as a simple, practical way to sort out what was physically real from what was not. However, as we have seen, it did not retain that meaning for either Bridgman or his early sympathizers, if for the latter it ever had that meaning in the first place. In its evolution a bifurcation occurred, one that, philosophically speaking, could not have been more radical. Bridgman's retreat into subjectivity was contrary to what everyone else thought a scientific philosophy should accomplish.

This became emphatically clear in his relationship with the logical positivists and the behavioral psychologists.

At first glance, there seems to be little in *The Logic of Modern Physics* to suggest that Bridgman would arrive at so uncompromising a position, one so far removed from the mainstream of scientific philosophical thought. Although the seed of subjectivism is noticeably evident upon a retrospective reading, nevertheless, there was ample reason for the positivists and behaviorists to interpret *The Logic of Modern Physics* not only as being favorable to their goals but also as being an independent statement of the same desiderata—the elimination of metaphysics and a guarantee of objectivity in science. Especially attractive was Bridgman's methodological exegesis of physical meaning: "The concept is synonymous with the corresponding set of operations." Positivists and behaviorists seized on this dictum and made it the basis of the "operational definition," a device supposed to ensure the empirical validity of formal objects (concepts). However, each group had its own idea of how operational definition fit into the construction of scientific truth. For the logical positivists, the operational definition was a link between formal and factual propositions. For their part, the behaviorists treated operational definitions as directly constituting scientific concepts. The reason for the difference will emerge from the following discussion.

By all accounts the man at the center of all this interpretive activity was Herbert Feigl, a member of the Vienna Circle of logical positivists and the first to introduce their ideas to American scholars. He was responsible for bringing Bridgman into the positivist fold. He was also the agent who brought Bridgman's ideas to the attention of the behaviorists. Feigl's mission was immediately taken up by S. S. Stevens, at that time a student of the Harvard psychologist Edwin Boring.[1] Feigl himself was more interested in promoting the philosophy of logical positivism, into which Bridgman's operational method was soon absorbed.

According to Feigl, he had learned of Bridgman's *The Logic of Modern Physics* in Paris in 1929 from an American student from Johns Hopkins University named Albert Blumberg. In Feigl's opinion, "Bridgman's operational analysis of the meaning of physical concepts was especially close to the positivistic view of Carnap, Frank, and von Mises, and even to certain strands of Wittgenstein's thought."[2]

Having just recently also become acquainted with pragmatism, Feigl sensed that in America he would find a favorable philosophical atmosphere for the promulgation of positivist ideas. Therefore, he applied for a Rockefeller Foundation fellowship and then wrote to Bridgman asking that he be allowed to work under Bridgman's auspices at Harvard, doing research on the logic of scientific theories.[3] After setting out his qualifications, Feigl elaborated:

It has been a rare pleasure to me to find a physicist who at the same time has so interestingly contributed to the logic of physics. It would be of great advantage to me to be able to continue my work under your auspices. The desirable further training in pure physics would thus be combined with the unusual opportunity of having my more general results subject to the competent and understanding scrutiny of a physicist.[4]

Bridgman's reply affirmed a philosophical compatibility but cautioned Feigl not to expect too much.

I have your letter and that of Professor Schlick about your desire to work here next year on fundamental problems in physics, and the copy of your book has also just arrived. I have not yet had time to completely read the book, but I have looked at it enough to get your point of view and what I have read is after my own heart. I may say at once that I would be very glad for you to come next year, if you really care to do so.

But I think that I ought to say a few words of introduction of myself to you in order that you may not come with false expectations, or be disappointed when you get here. My work on fundamental questions in physics has been entirely outside my formal acedemic [*sic*] activities. My book was written during my sabbatical leave; nearly all my time is occupied with the many details of my experimental work, and in particular at present I am engaged in writing a book collecting my experimental work on high pressures of the last 20 years. I do not offer any courses of instruction on such topics as would be suggested by my book, . . . and any discussion of fundamentals which I give is entirely incidental.[5]

Feigl arrived in the fall of 1930. He quickly became the leading publicizer in America of the Wissenschaftliche Weltauffassung (scientific worldview) of the Vienna Circle. Soon after his arrival, in the spring of 1931, he and Albert Blumberg published an article in the *Journal of Philosophy* entitled "Logical Positivism: A New Movement in European Philosophy," which in Feigl's words, "started the ball rolling" toward making logical positivism a widely discussed philosophy in America.[6]

The article by Feigl and Blumberg set the stage for the difficulties that followed. The formal similarities between the linguistic edifice the positivists were attempting to construct and the pattern underlying

Bridgman's operational assumptions made them mutually reinforcing. The problems appeared when it came time to state the concrete basis for the entire production. It was at this point that solipsism threatened to sabotage the positivist program. And it was essentially over this issue that Bridgman parted company with the logical positivists, for he saw no reason to try to escape from it.

There is no question that in *The Logic of Modern Physics* Bridgman had presented an argument that had strong affinities with the positivist approach. Both were characterized by the desire to banish metaphysics from legitimate knowledge by introducing a criterion of empirical meaningfulness; according to this criterion, concepts of various orders are treated as having been built up or constructed from simpler, empirically pure elements. For Bridgman, as we have seen, these elements were operations. For the positivists, they were "atomic propositions" or "reduction sentences." By a process of analysis or disassembly, one should be able to distinguish empirically valid expressions from those that are simply "meaningless" because they cannot be linked, for Bridgman, to operations or, for the positivists, to experience (that is, the "given"). This process was supposed to allow the products of unchecked imagination to be identified and consequently eliminated from valid discourse. "The truth or falsity of propositions," Feigl and Blumberg assured their readers, "is ascertained by comparing them with reality." Further, the "isomorphy of the system of language and system of facts would hold . . only when all molecular propositions had been analyzed into their constituent atomic propositions."[7]

Both Bridgman and the positivists believed that by invoking experience (respectively, operations or the given), they were establishing a metaphysically neutral standard for truth and securing a self-evident foundation for knowledge. The given, however, was open to criticism in a way that operations were not, for the given suggested a presentation to the consciousness of a passive subject, whereas an operation was regarded as a publicly observable activity of the investigator. As it turned out, this distinction could not be maintained, but for the moment, it was not seen as problematic. Thus the accusation of the philosopher P. A. Schilpp, that "the 'given' is the point where the logical positivists turn mystic,"[8] was not a charge to which operational analysis seemed susceptible. It was only after Bridgman pressed his analysis further into the foundations of knowledge and ventured into epistemology as he did in *The Nature of Physical Theory* that his ideas drew similar criticism. In the meantime, operations appeared to hold a privileged position, being neither mentalistic (subjective) nor objective ("thinglike").

Feigl and Blumberg, to be sure, were not unaware of the problem of the given, and they tried to minimize it by certain dialectical strategies, which Schilpp labeled "a combination of *ex cathedra* dogmatism and philosophical oversimplification."[9] The basic problem, to explain how what is experiential and therefore subjective can be transformed into something public and objective, was treated along two analytical axes—first, as a question of the relationship between the factual and the formal, and then as one of passing from individual knowledge to public knowledge. It was precisely around these two questions that Bridgman's disagreement with the logical positivists was centered.

Taking their cue from recent developments in mathematics and relativity theory, Feigl and Blumberg, speaking for the Vienna Circle, drew a sharp distinction between formal and factual truth. The positivists were very much influenced by the school of mathematical formalism, which regards mathematics as a formal system of signs with its own internal rules of combination having no reference to the empirical world, and the success of general relativity, which showed how such a formal system could be interpreted or applied to experience. Thus they proposed that language be regarded as a formal system, governed by logic, which in turn was to be understood as a set of syntactical conventions. Complex linguistic expressions, then, were to be considered as compounds of simpler atomic propositions constituted by symbols whose empirical content was correlated with experience by means of "applicational definitions." "This application of purely formal patterns to facts is effected by means of concrete definitions which lay down the empirical meaning of our symbols and thus enable us to raise the question whether certain symbol-patterns actually fit the facts that have been gathered by observation and experimentation." Feigl and Blumberg identified these definitions with the "operational definitions" they attributed to Bridgman.*

*See Albert Blumberg and Herbert Feigl, "Logical Positivism, a New Movement in European Philosophy," *Journal of Philosophy* 28, no. 11 (May 1931): 288–89. Feigl and Blumberg continued: "The method of implicit definitions thus lays down the purely formal structure of concepts; in order to apply them to the description of the formal characters of experience 'applicational' definitions are required which correlate these formal concepts with empirical content. Reichenbach, in particular, has stressed this distinction between the 'Verknupfüngsaxiomen' [connection axioms] which implicitly define the

Language, then, was a structural network of symbolic relationships to be fitted to experience by adjusting either the internal arrangement of symbols or the applicational (operational) definitions. The underlying assumption that "knowledge consists of propositions concerning the formal characters of experience" [10] was not, as we have seen, shared by Bridgman. His goal, he said, was descriptive and multiplicative: "The first task of the operational approach is usually to recover the full complexity of the primitive situation." [11] This meant to him dismantling a conceptual edifice to discover what concrete truth it contained. Bridgman, Feigl was to remark later, "was not a formalist, and I suppose he felt that Carnap and other logicians were doing a rather far-fetched and artificial job of all too exact reconstruction." [12] Feigl did not overstate Bridgman's position.

However, Bridgman's objections were even more subversive than a superficial repudiation of formalism, and they struck at the very point where logical positivism was vulnerable. For while there was an existing model—relativity—which provided the positivists with an example of a successful treatment of the relationship between the empirical and the formal, it did not include guidelines for treating the problem of the gap between individual and public knowledge. The mathematical framework was useful only for handling abstractions, including "experience," which itself was used in the sense of a general, idealized concept. When it came to accounting for how knowledge can be empirical in a concrete way, Feigl and Blumberg slipped temporarily into subjectivist thinking. The structure, but not the content of experience, they argued, is what is communicable. But at the same time, the ground floor of knowledge, the immediately given, "is private and non-communicable." [13] This, as S. S. Stevens later exclaimed, "from the mouth of a Logical Positivist!" [14]

Feigl and Blumberg did not see the situation as a problem. They pointed out that experience includes both the physical and the psychological and invoked Bertrand Russell to assert that the difference is merely one of types. The language of one can be translated into the language of the other. This state of affairs, they argued, was not equivalent to solipsism. According to Carnap, it was instead "methodological solipsism," a doctrine that allows one to acknowledge that "the elemen-

---

concepts and the 'Zuordnungsdefinitionen' [applicational definitions] which apply them to experience. In physics, the 'Zuordnungsdefinitionen' takes the form of what Bridgman terms 'operations'" (p. 289).

tary experiences which are at the basis of the edifice are, if we speak from a higher lever, 'my' [subjective] experiences." Nevertheless, at the same time—and here is where the philosopher W. M. Malisoff detected some "black magic"—it can be "shown in detail how on this basis the objective (i.e. intersubjective) world can be constructed through a series of increasingly complicated combinations of the given." [15] The assumption is, of course, providing that such a translation—from psychological to physical—can be carried out, that "physical" is the same as "objective."

By the time he wrote *The Nature of Physical Theory*, as we know, Bridgman had rejected the necessity of taking this final step of inventing an intellectual device for projecting his subjective experience into the public domain. However, it appears that he did not fully appreciate the disparity between his position and that of the positivists until it was pointed out to him by Arthur Bentley in the course of his own attempt to discredit Carnap: "These Carnap people love to give you patronizing pats on the head; your name has appeared as consorting with them in conventions; you occasionally use a phrase or two of their patter (although never without taking all the poison out of it in your next sentence); while I want to exhibit your operations and their positivism as at opposite poles in modern knowledge." [16]

Bridgman's reply was characteristically candid.

It is true that I have always felt more sympathetic to the Vienna Kreis than to any other bunch of philosophers with whom I am familiar. This is partly because I have been told so (Feigl was at Harvard for a year) and partly because what I have found in my little superficial reading, and I am positively illeterate [sic] in the amount of reading that I do, has seemed to me eminently sensible, such as their emphasis on the importance and existence of "pseudo-problems" and their impatience with metaphysics. But a year and a hald [sic] ago Carnap sent me a copy of his little book on grammar, and I had to read it intelligently enough to make a self respecting acknowledgment. I couldn't make it fit with the way I like to think, and I have been left wondering ever since. [17]

There was no longer any doubt in Bridgman's mind when he addressed the 5th International Congress for the Unity of Science, held in Cambridge, Massachusetts, September 1939. In a speech entitled "Science: Public or Private", he drew attention to his differences with the logical positivists. [18]

Depicting the publicity (public nature) of science as "being in accord with the intellectual fashion of the times of emphasizing that all

our activities are fundamentally social in nature" and as "overdefini-
tion," Bridgman charged that a "streak of verbalism or metaphysics"
was at the basis of this claim.[19] He was willing to allow that the publicity
of science might be a convenience insofar as it relieved the individual of
the chore of having to certify everything for himself, or that it could
help to safeguard against someone else's unreliability, but in and of itself, publicity is not necessary for science. Authentic science is an activity carried out by an individual. Science is not what is written in
books. Books contain only marks for guiding activity.

The process that I want to call scientific is a process that involves the continual
apprehension of meaning, the constant appraisal of significance, accompanied
by a running act of checking to be sure that I am doing what I want to do, and
of judging correctness or incorrectness. This checking and judging and accepting that together constitute understanding, are done by me and can be done
for me by no one else. They are as private as my toothache and without them
science is dead.[20]

Responding rhetorically to the charge that this view might be
trivial, Bridgman drew on the example of the syllogism to argue that
any formalism is incomplete and requires for its meaning a "text" of
background understanding, practically never made explicit, which is
brought to it by the individual. This text includes the meaning of the
symbols and their range of application. It is within this text that commonly accepted but unexamined assumptions may also be found, and
they cannot be discovered by virtue of the publicity of the formalism
but rather by the insight of an individual into the inadequacy of his
text, that is, his realization of the inadequacy of the unstated background assumptions. For example, paradox is a consequence of inadequacy in the text and cannot be guarded against. But when a particular
paradox is discovered, it is

in the first instance discovered by some individual, who thereby discovers the
inadequacy of his private text, which may of course also have been in part publicly accepted. It is, I think, particularly evident that advances into new territory are the result of private activity and therefore of private science.[21]

Bridgman was arguing that public acceptance of scientific knowledge in no way guarantees its truth; furthermore, to hold that science
must be public is to overdefine it—to impose a superfluous requirement. Science happens in the understanding of an individual, and
without this "catching on" it has no life. This activity takes place in the
individual mind, which has no counterpart on the public level. Nor

does public consensus contribute to new insights or discoveries, since a state of agreement is by its very nature a static condition. It cannot move itself to a new awareness. Only the individual is capable of discovering new conditions, new relevancies, new perspectives. Only the individual is capable of the reflection that creates science.

At the time Bridgman did not, however, ask how one can know that he himself is not deceived. When he did address the question shortly afterward, he did not deny that mistakes could be made. This in his view was simply part of the human situation. He certainly did not believe that method could rid science of this possibility. In fact, he did not think there was a method unique to science. All advancement in knowledge is the consequence of what he called the "method of intelligence," and that consisted in "doing one's damnedest with one's mind, no holds barred." [22] Therefore, "the so-called scientific method is merely a special case of the method of intelligence, and any apparently unique characteristics are to be explained by the nature of the subject matter rather than ascribed to the nature of the method itself." [23]

Bridgman's dissatisfaction with the positivist program arose from his lack of sympathy for attempts to schematize or reduce the production of knowledge to a formula either in advance or in retrospect. Nor was his operational analysis intended to give aid to such constructions. On the contrary, his operational method, a method of disassembling formal systems, disclosed to him the artificiality of the positivist attempt to build a linguistic continuum from subjective experience to public objectivity—to obscure the disjunction between the private and public. "It is in the general scientific tradition, and in particular, I think, it is one of the presuppositions of the Unity of Science movement to ignore these dualities. . . . The validity of this assumption has never, I believe, been adequately examined." [24]

Bridgman did not, however, refute the continuity of the physical-psychological dimension so prominent in Feigl's early exposition. Indeed, he criticized the physicalistic turn positivism had taken. For Bridgman, it was illegitimate to leap from private physical experience to public objectivity. Private experience includes both the psychological and the physical, he believed, but the public-private dichotomy cannot be erased by the mere assertion that the physical is equivalent to the public. The latter view, he charged, can be achieved with "sufficient virtuosity," but only at the expense of "complication, artificiality, and clumsiness." [25]

Such efforts, as Feigl admitted, were for Bridgman intellectual artifices outside the empirical domain, not even comparable to the pencil and paper operations (mathematical) Bridgman allowed in the creation of derived or higher-level concepts. The point that Bridgman was trying to make—that private experience cannot be defined away—was lost on the philosophers, whose interest was the rational reconstruction of scientific knowledge. It was not that the positivists failed to appreciate the notorious solipsistic sink that inevitably threatens empiricist systems or that they were confident that they had successfully dealt with it. But they were aspiring to a goal Bridgman thought utopian and humanly unattainable. He preferred to take his chances with the imperfect accomplishments of concrete human activity, and this meant facing up to the fact that truth is a very individual matter. The logical positivists, he charged, had chosen to ignore the ultimacy of this condition.

In asserting the existential ground of truth and what it revealed about the politics of knowledge and language, Bridgman was pressing beyond the limits of the positivist program. His concern was with the ultimate cogency of proof, its coercive power, and with that, the social meaning of a view of knowledge that founds its authority on the collective acceptance of truth. He was offended by the idea that the coercive power of knowledge should be invested in the community; to him it rightly should be reserved for the individual. Compounding that affront, it seemed to him, was an arbitrary act underlying the principle of the publicity (public nature) of science, an act that in effect defined away the truth of individual experience. The traditional use of language, he pointed out, serves to reinforce the oversight. Science should more correctly be reported in the first person.

This is a far cry from the interests of the logical positivists, and Bridgman, in essence, accused them of being scientific and linguistic ideologues, promulgators of inauthentic knowledge. In an essay published in 1940 entitled "Freedom and the Individual," he followed up the implications of his insights in more dramatic language. "The last thing," he began,

that the average human being wants to be made to see is that as a matter of fact he is already inescapably free. This obviously does not mean free economically or politically, for these forms of freedom have not yet been generally attained and are still the object of passionate endeavor. What is meant is that inner freedom in virtue of which every individual leads his own life eternally free from his fellows within the walls of his own consciousness. . . . All the intellectual machinery which an individual receives as his heritage and with which he strives as best he can to adapt himself to his environment is a tissue of rationalization inspired by fear lest he see that he is really free.[26]

Bridgman went on. When an individual realizes his freedom he discovers that he has "no *right* to the interest or even the sympathy or understanding of others." There is an irreducible and eternal separation between "my" sensation and "your" sensation that language traditionally obscures. An individual can give authentic meaning only to his own experience. The experience of the other is inferred by imaginative projection. There is only one situation where this condition is reversed, "in which I experience what is yours and never what is my own. This is with respect to death." The reason is because death is not, strictly speaking, an experience, but its cessation.[27]

Extending his insight, Bridgman argued that science, which is a living enterprise, depends on proof, which in turn comes down to the conviction of the correctness of the proof. This judgment can be made only by an individual. Proof is validated by the process of understanding.

The feeling of understanding is [as] private as the feeling of pain. The act of understanding is at the heart of all scientific activity; without it any ostensibly scientific activity is as sterile as that of a high school student substituting numbers into a formula. For this reason, science, when I push the analysis back as far as I can, must be private.[28]

However, Bridgman had discovered something even more significant. In the privacy of understanding, in the radical subjectivity of knowledge, he encountered the existential condition of individual freedom.

The public level is tremendously important, and most of our individual and social living is done on this level. Our language is so constructed that we are almost forced to talk on this level. . . . But always beyond the public level, waiting for a deeper analysis, is the private level. It is on the private level that I realize my essential isolation; here is my awful freedom that I can hardly face.[29]

"My awful freedom that I can hardly face." There is no mistaking the existential impulse behind this terrible realization. This awareness was for Bridgman concrete and real. It was not an abstraction of the intellect. Over and again he stressed the essential nature of his isolation. He would not even go so far as to assert definitively the isolation of his fellows from others. For "to say that my fellows are isolated from me or from each other involves an extension and alteration of meaning like that we have already analyzed."[30] Thus in the final analysis, as Bridgman explained to the sociologist Hornell Hart, private meaning is totally incommunicable, unintelligible to another. "There is only one proper privacy, namely, my own privacy. . . . Only I can talk about it. You can not talk about it."[31]

If the positivists believed that the rationality of science was the antidote to political totalitarianism, which they saw as having been made possible by obfuscatory metaphysics, Bridgman was exposing the oppressive potential of even a scientific ideology. There was no place in his value system for conceptual despotism of any kind, including scientific. The social compulsion implicit in the ideal of public knowledge offended his individualistic priorities. However, he was caught in a conflict because of his loyalty to science as his life's work. Therefore, rather than submit to an interpretation of science that would subordinate the unique to the general, he insisted that science was also essentially private.

By adopting this standpoint, Bridgman not only gave up the comfort offered by the warmth of community, but the possibility of certainty as well. He was alone in an indifferent world.

It is not that the world is really neither beneficent or malign but instead neutral; it is that it is meaningless to think of the world in terms of beneficence. We are trying to apply an intellectual category that is inapplicable; we are trying to do something with our minds that cannot be done. . . . It is hard to admit that there are no certainties, and the probabilities with which I would fain replace them cannot have the meaning I would desire.

I stand alone in the universe with only the intellectual tools I have with me. I often try to do things with these tools of which they are incapable, and I have often been misinformed and have delusions as to what they are capable of, but nevertheless it is my concern and mine only that I get an answer.[32]

Against such odds, Bridgman saw that for someone like himself, there is nothing to fall back on but fortitude. Nevertheless, he hoped that future generations could be educated to accept this isolation more naturally, so they would not have to engage in heroics in order to give up a belief based on false expectations.[33]

To a group of philosophers intent on formalizing the construction of scientific theory and consolidating the authority of science, Bridgman's existentialist ideas must have seemed irrelevant indeed. There is no indication that they evoked any significant response. For the positivists, Bridgman was the author of the operational method, a technical detail that found its place in the framework of the rationalization of scientific theory.

The fate of the positivist criterion for delineating scientifically permissible concepts closely paralleled that of Bridgman's operational criterion for valid scientific knowledge or meaning. Both were too restrictive and had to be relaxed. The positivists had demanded empirical verification as the absolute demarcation between scientific and metaphysical claims, and Bridgman had insisted on operational specification. To many, the positivist designation of meaning as the method of verification and the operational specification of meaning as a set of operations were the same thing, although the first referred to statements and the second to objects. Be that as it may, the positivists had to weaken their criterion of verifiability to "confirmability," and Bridgman had to extend the range of operations to include not only physical measurement but also "paper and pencil" operations.*

The operationism that suffered this attenuation was not the operational method that led Bridgman to discover the privacy of knowledge and the dichotomy of language. Rather, it was the positivist version of operationism in which operations played a mediating role between experience and formal categories, and this version, Bridgman lamented, was a monster to which he had only a historical connection. Listening to others talk about operationism, he declared, it seemed to him that

I have only a historical connection with this thing called "operationalism." In short, I feel that I have created a Frankenstein, which has certainly got away from me. I abhor the word *operationalism* or *operationism*, which seems to imply a dogma, or at least a thesis of some kind. The thing I have envisaged is too simple to be dignified by so pretentious a name.[34]

Nonetheless, it was this form of operationism that was the subject of discussion in a seminar devoted to its clarification during a conference held by the Institute for the Unity of Science in 1953. This meeting took place in cooperation with the American Academy of Arts and Sciences and the National Science Foundation, as part of the AAAS Christmas session in Boston. The title of the conference was "The Validation of Scientific Theories." With regard to operationism, the issues raised were limited to those fitting into the positivist framework, and except

---

*"Paper and pencil" operations were mentioned, but not emphasized, in *The Logic of Modern Physics*. However, since Bridgman's focus was overwhelmingly on physical operations, as a practical matter he did not consider paper and pencil operations until many years later.

for the very briefest acknowledgment by the philosopher Adolf Grünbaum, no mention was made of Bridgman's concern with the privacy of scientific knowledge.

The greater part of the discussion took place against the background of the positivist model of scientific theory as an interpreted system of signs, and interest focused on the problem of where operations belonged in this idealized scheme. It was generally agreed that operationism was not a systematic philosophy, but that operations nevertheless were intimately involved with establishing the empirical validity of scientific generalization. Neither the commentators nor Bridgman could be more specific than that.

Henry Margenau held that operational definitions mediate between observations and constructs; Gustav Bergmann claimed they were in-use definitions of the form, if A then B, where A is something the investigator does and B is what is observed. Carl G. Hempel argued that as a verification theory of meaning operational standards were too restrictive and had to be relaxed for the same reasons that the verifiability/falsifiability requirements of the logical empiricists had to be reduced to confirmability. R. B. Lindsay took sides with a conventionalist interpretation of scientific theory, complaining that operationists appeared to be searching for a "true" representation of reality. Grünbaum felt that operations belonged to pragmatics, not to the logic or semantics of physics. Only Raymond J. Seeger argued that truth is something more than literal.[35] In all of this, Bridgman felt like an outsider.

The operational method had become absorbed into positivism, and by virtue of this assimilation, it was now being treated as a step in an attempt to create a new dogma, a methodological dogma that placed the burden of cognitive certainty on methodological norms. This was not Bridgman's idea of the operational method. It was a surprise to him, he said, "that, since the publication of my book [*The Logic of Modern Physics*], so much of the concern of others has been with abstract methodological questions suggested by the endeavor to erect some sort of philosophic system."[36] There is nothing normative about operational analysis, and it can be carried out by anyone upon any conceptual entity. So far as Bridgman was concerned, operational analysis was an instrument for demystification, a technique for finding out what a situation was really about. It was a way to uncover deceits and misconstruals, either deliberate or innocent, and thus to empirically purify knowledge. It was not a link in a formal constructive system.

Either way, however, the fact of human knowledge had not been shown to be any less mysterious or miraculous. The positivists could no more explain how language becomes invested with shared meaning than Bridgman could explain the private experience of understanding, how the individual "knows" when proof is cogent. Neither could give an account of how knowledge appears to human awareness. No amount of methodological analysis could create knowledge that was not already given or justify why formal systems can replace the subjective with the objective. And no amount of existential denial could explain the efficacy of scientific theory. To approach such an explanation would have required the articulation of a full-fledged metaphysical system, something both Bridgman and the positivists agreed was, at best, superfluous and, at worst, deceptive and harmful.

However, Bridgman had thrown down the existential gauntlet, challenging the validity of the positivist epistemological program. His analysis had yielded an entirely different picture of the meaning of experience and the nature of knowledge. Existentially, Bridgman had discovered it is only possible to "experience" subjectively, and even more significantly, that truth itself is a subjective act. At the same time, nonetheless, he had cut himself off from the possibility of participating in a meaningful transpersonal order.

Nowhere was the faith in scientific methodology more evident than among the behavioral psychologists. Because of their rebellion against philosophy and their impatience with introspectionist psychology, they had renounced the traditional subject matter of psychology. Mind and consciousness were declared private and inaccessible and therefore not the proper domain for scientific study. A growing interest in animal studies and the acceptance of the stimulus-response model as a behavioral paradigm encouraged the expectation that psychology could be founded on the basis of the study of behavior alone. The banishment of mentalistic subject matter was regarded as equivalent to the renunciation of metaphysics, and the substitution of behavior as the object of psychological knowledge was seen as a necessary step in advancing psychology to the status of a natural science. In addition to redefining the subject matter of psychology as publicly observable activity, the behaviorists, heeding the empiricist injunction against metaphysics, accepted the constraint that scientific generalizations should be expressed as correlations between observables. In principle, this limitation confined their theoretical expressions to statements of functional relationships between observable behaviors.

The ideals of the early behaviorists were not, however, met in practice. Mentalistic categories were not so easily eliminated. Therefore the second generation of behaviorists still sought methodological guidelines that would carry them toward their avowed goals. In the early 1930's the behaviorist program received a vigorous boost from the aggressive positivist campaign carried out by members of the Vienna Circle, but it was Bridgman who provided the nucleus around which the operationist movement crystallized. The positivist spirit sustained the behaviorists' efforts to break away from philosophy and lift their science to the status of physics, but it was inadequate to meet the methodological needs of this new discipline. Positivism failed to provide the prescriptive directives the behaviorists sought. Psychologists enthusiastically welcomed Bridgman's operational method for determining the meaning of a concept because it appeared to provide a formula for creating valid conceptual categories. Just as important, however, it gave immediate sanction to their inductivist methodology and, what was better, by none other than a prominent physicist.

Psychologists became operationists. The lead was taken by S. S. Stevens, a Harvard graduate student who at the time was working under Edwin Boring. Stevens extolled the virtues of operationism, announcing that finally there was a method for introducing rigor into psychology. Other psychologists followed suit, and in the mid-1930's operationism became the touchstone for neo-behaviorist theorizing.[37]

Boring made this preference for Bridgman's operational procedure explicit in the critical comments he wrote on an early draft of Stevens's first paper on operationism. He advised Stevens to say that

psychology for many years used to appeal to the positivistic principles which formed its philosophical frame of reference. The rule of positivism is this: when in doubt or controversy, go back to the experience which is the empirical foundation of the matter in question. Nowadays, however, this simple conception of empirical science seems a little naive, and at any rate experience is too far away from modern scientific fact to be readily within reach. *Fortunately there is an adequate modern substitute for positivism at hand*. It is what we may call operationism, a scientific method for validating concepts which has been advanced by Bridgman, the physicist. Operationism consists simply in referring any concept for definition to the operations by which knowledge of the thing in question is had.[38]

The positivists had a mature body of physical theory—relativity—upon which to base their analysis of how method certifies theory. The behaviorists had no such theory to reconstruct; instead they had only the certification process and a program built around it, a program that essentially transformed method into content. For the positivists, whose paradigm was an interpreted formal system, operational definition was an intermediary between formal and factual statements. The behaviorists, on the other hand, were proceeding on an inductivist model of science and without the benefit of hindsight. They rejected the analytic-synthetic distinction.[39] Instead, operational definition was treated as being constitutive of a psychological concept, and at the same time, a guarantee of its scientific validity. (On the latter point, the behaviorists followed Bridgman, whose attention was directed toward explicating the meaning of concept as object, rather than the positivists, whose concern was with the meaning of a proposition or sentence, that is, an assertion.)

In one other important way, Bridgman's operational approach was favorable to behaviorist assumptions. It is at this point that its provenance in dimensional analysis becomes significant. In a striking manner, dimensional analysis was to physics what associationism and elementalism were to psychology. Just as physical reality was supposed to be composed from elements called "fundamental quantities," psychological reality was made up of the attributes of sensation. Combinations of mass, length, time, and perhaps temperature created the higher-order physical entities treated by dimensional analysis. Analogously, quality, intensity, duration, and so on were the elemental attributes of the sensations studied by nineteenth-century psychology. Again, Boring betrayed the substratum of presupposition.

In order to understand the systematic role of the dimensions of consciousness we must turn to physical science, always the model for psychology.

As far as possible the realities of physical science are described by reference to the c-g-s system. Ideally the centimeter, the second, and the gram provide sufficient terms for the description of any physical event. . . . Physics went far throughout a long period with its three dimensions, and a young science like psychology may do well to emulate it.

We must not let ourselves be confused by this use of the word dimension. It is plain that each of these fundamental dimensions may contain dimensions within itself. Euclidean space, which physics has used so successfully, is tridimensional. Time seems to be unidimensional, but, when we come to describe an acceleration in c-g-s terms, we find that it must be expressed in terms of centimeters per second per second or cm./sec.^{2}, so that conceptually we have a squared or bidimensional time. We shall need the same degree of freedom in establishing a set of conscious dimensions.

The dimensions of consciousness are the immediate successors to the old attributes of sensation. Modern psychology certainly needs four dimensions: quality, intensity, extensity, and protensity. Titchener's fifth dimension, attensity, has in the view of the present author, become unnecessary.[40]

Historically and intellectually, Boring's role in the formation of the link between behaviorism and operationism was that of a mediator rather than an active contributor. (Indeed, he was not a behaviorist, but a follower of the psychological tradition of mentalism.) His argument, however, calls attention to the noticeable parallels between dimensional analysis and psychological associationism and elementalism.

Behaviorism did not discard the associationist and elementalist assumptions when it redefined its subject matter; it merely took its reference elements to be actions rather than perceptions. Similarly, Bridgman rejected the hypostatization of the fundamental physical quantities of dimensional analysis, but he retained the architecture. Like the behaviorists, he substituted actions—operations—for "metaphysical" categories but kept the basic structure. To Bridgman, physical concepts were built up from operations instead of from fundamental quantities. Bridgman's operational universe was morphologically similar to the behaviorist one because both were predicated on the same elementalist assumptions.

Besides methodological and structural compatibility, Bridgman's operational message conveyed a further promise, the promise that psychology could be unified and stabilized around a consensus founded on operational principles. The antirevolutionary attitude that reverberated throughout *The Logic of Modern Physics* had an especial appeal to psychologists, whose discipline was divided into conflicting schools of thought. It was with this theme that Stevens began his proselytizing.

The revolution that will put an end to the possibility of revolutions is the one that defines a straightforward procedure for the definition and validation of concepts, and which applies the procedure rigorously in a scrutiny of all fundamental concepts in psychology. Such a procedure is the one which tests the meaning of concepts by appealing to the concrete operations by which the concept is determined. We may call it *operationism*. It insures us against hazy, ambiguous and contradictory notions and provides the rigor of definition which silences useless controversy.[41]

Or, as he wrote a few months later, "Operational definition provides for a progressive evolution rather than a *volte-face*." [42]

Unfortunately, operationism was never able to fulfill its promise. In principle it was supposed to induce consensus among psychologists by providing an unambiguous starting point for psychological theorizing. However, there were as many operationisms as there were operationists. While there was general agreement that definitions should be operational, there was no clear idea of what operations were or what it meant for a concept to be operationally defined.

In Stevens's version, all psychological operations were reducible to the fundamental operation of "discrimination." "Complex operations are always reducible to more simple discriminatory acts, and all 'explanations' consist of detailing a complex set of operations in terms of simpler ones." Discrimination was Stevens's substitute for what was formerly called "experience" or the "immediately given." [43] "Discrimination or differential response is the fundamental operation. . . . By Discrimination we mean the concrete differential reactions of the living organism to environmental states, either internal or external. Discrimination is, therefore, a 'physical' process, or series of natural events, and all knowledge is obtained, conveyed and verified by means of this process." [44] Stevens's elaboration of the role of discrimination as the foundation of all knowledge involved him in numerous epistemological tangles, the most obvious being that he had not escaped the subjectivism that operationism was supposed to replace. But he was not the only one.

Operational ideas multiplied. Boring was willing to allow "certain operations of introspective report, which adequately imply the differentiation of the perception [of perceived duration]." [45] J. A. McGeoch understood "the notion of operational definition as a technique of building concepts in physics. . . . It is the procedure actually used in building concepts employed in laboratory work." [46] D. McGregor called operationism "the belief that an entity is adequately defined only in terms of the specific operations involved in its observation" and added:

We must continue to work with operationally defined concepts and in terms of discrimination, for these notions are fundamental to the logic of all measurement. In such terms, however, we shall be able to understand much more exactly than ever before the nature of these "psychological" magnitudes. If, in the process, we seem to lose the distinction between psychological and physical magnitudes, between psychology and physics, we can console ourselves by remembering that this fusion is the inevitable result of the operational technique which defines innumerable entities by reference to innumerable operations instead of seeking first to impose the Cartesian dichotomy of mind and matter upon all reality.

Furthermore, he asserted, "Psychological measurement, understood in operational terms, is a *fait accompli*. It is physical measurement." [47]

These are programmatic prescriptions, indicative of various methodological perspectives. A straightforward example of the operational principle as practiced is given in E. C. Tolman's article, "Operational Analysis of 'Demands.'" Tolman proposed that a behavior could be represented as a function of "intervening variables," which can then be isolated in "standard experiments." Thus, in analyzing demand for food, he set up the following relationship: $B = f(M, S)$ where B, the number of food crossings, was taken as a measure of the persistence of the food-seeking behavior, M was the time since feeding, and S was the kind of food. In the standard experiment, each intervening variable, M or S, would be varied independently, "so that behavior is directly correlated to intervening variable." "That is, under these conditions, $B = I$" (I is either M or S).[48] Overall, then, the behavior could be expressed in the desired functional form and was thus "operationally defined."

The goal of the methodological machinations of the operationists was to gain scientific objectivity in psychology. More specifically, they believed that by making psychological knowledge public, they could assure objectivity. Here, again, there was no clear concept of exactly what that meant. Stevens made an attempt to clarify the situation, in an exposition that only illustrates the difficulty of the problem.

The systematic application of the operational procedures to the facts and concepts with which men concern themselves generates a body of knowledge which we call science. An essential characteristic of all facts admitted to the body of scientific knowledge is that they are *public*. Science demands public rather than private facts. "Objectivity" in science is attained only when facts can be regarded as independent of the observer; for science deals only with those aspects of nature which all normal men can observe alike. Scientific knowledge has, therefore, what we may call a social aspect.

Science is a thing agreed upon by members of society. [Stevens added a footnote emphasizing this point: Not only is science a social convention, but, as Carnap and others have shown, language itself, including the rules for its use, is a convention based upon social sanction. From the social usage of words, there is no appeal.] . . .

The moral for the psychologist is that his science is not built up out of unique experiences which are private and known to him only. Unless the psychologist can report his experience in such a way that others can verify it, we are left with no dependable operation for establishing it. Operationism requires that we deal only with the reportable aspects of experiences. Not only must the experience be reportable; it must be actually reported verbally or otherwise. Operational psychology knows precisely nothing of unreported consciousness.[49]

This passage illustrates the spirit of operationism particularly well. It also reveals the profound ambiguity that enveloped the concept "operation" and the subtlety of the relationship between objectivity and publicity in science. How, for example, is objectivity achieved by requiring that experience be "reportable"? Is the report a valid operation? Given a report, what knowledge is thus established? Whose report is being considered, the report of the experimenter or that of his subject? Stevens asserted that "only the reported part of 'experience' can ever get into science; the rest is meaningless since we have no operations for dealing with it. Moreover, the reported part manifests itself as some form of reaction, and gets into science because it is verifiable, repeatable, and consequently agreed upon by other scientists. Thus it satisfies the criteria of 'objectivity.'" His reasoning led him to conclude that "what other observers cannot verify is not knowledge."[50]

The problem that naturally follows upon this assertion is that of self-knowledge. How is the psychologist to regard his own perceptions? Stevens approached this question by introducing the idea that self is to be treated in the same manner as "the other."

We cannot, of course, put up with the psychology of a particular man; we must have a psychology of men, and be able to bring the "mind" of any person into the scientific subject-matter. Consequently, psychology regards all observations, including those which a psychologist makes upon himself, as derived from "the other one" and proceeds upon this basis to formulate operations which define consciousness in general.

The psychologist can, Stevens argued, "observe a phenomenon and then, taking his observations as data, he can in the role of experimenter treat them as 'objective,' verifiable facts and draw conclusions from them." It is difficult to reconcile this statement of Stevens with his claim that "nothing gets into scientific psychology which does not appear in the form of overt changes in all or a part of the physical system which we call the subject, for only such changes can be dealt with by means of rigorous, univocal, repeatable operations."[51]

Perhaps this was one of the reasons that Bridgman's judgment of

Stevens was rather unflattering. In a letter to Bentley dated May 4, 1936, he wrote:

Your somewhat flippant treatment of Stevens was most refreshing. Stevens is very young—only three years away from his PhD and is regarded as the "white hope" of the psychology department here, one member of which has told me that Stevens is the ablest student he has ever encountered. Stevens probably in consequence is very sure of himself and probably has a swollen head. He certainly has done some important things in the physiology of hearing. He has talked with me at length about a couple of his papers before publication and professes to be most enthusiastic for "operational ideas" and does, as you say, "give me a fine salute in the dark," but I simply cannot make him see that his "public science" and "other one" stuff are just plain twisted. I have also discussed with him his "basic act of discrimination" without making much impression, and I have rather washed my hands of him.[52]

But the matter was not so easily put to rest. Bridgman and the psychologists who had adopted operationism as the watchword for scientific authenticity in psychology would eventually be drawn into a public confrontation over the meaning of his operational ideas.

In the meantime, criticism that portrayed operationism as a relatively minor part of science, if it had any applicable meaning at all was beginning to mount. The psychologists R. H. Waters and L. A. Pennington published an ambivalently critical article in which they denied that operationism was the instrument to bring about unanimity in psychological thought. While affirming its usefulness, they charged that operationism was nothing new in science. It was merely another name for methodism. Furthermore, they pointed out, the operational method as originally presented by Bridgman would not reduce the number of concepts employed in scientific discourse but multiply them. (Every operation would correspond to a separate concept.) Moreover, operations were not relevant to either theory formulation or theory selection, and the idea of experimental error is meaningless in an operational context since there is no standard against which operations themselves could be validated. Nevertheless, despite the implications of their criticism, Waters and Pennington still thought it important to declare that in psychology "introspective methods are operational in character" and that inference and generalization are valid operations![53]

If operationism was thus "being threatened by its friends," as Stevens observed, a review article he wrote in 1939 did little to add to its strength. In this article Stevens told his readers that operationism is not positivism, behaviorism, monism, dualism, or pluralism. It is, he said, a set of induced methodological principles that are valid because of their survival value in science making. These principles, as he listed them, were the same as those he had presented in his earlier papers except in one respect. He now accepted the positivist formal-empirical dichotomy. Nevertheless, he had not modified his enthusiasm for operations as the basis for the public nature of science and social consensus as a criterion for valid scientific knowledge. "An operation for penetrating privacy," he declared, "is self-contradictory."[54]

The sharpest and most perspicacious criticism came from psychologists Harold Israel and B. Goldstein in 1944. In a paper published in the *Psychological Review*, Israel and Goldstein attacked psychological operationists for understanding neither scientific method nor Bridgman's intentions. Characterizing the operationist practice of defining concepts in terms of "either the operations performed to produce phenomena or in terms of the events resulting from phenomena," they accused operationists of neglecting to define psychological concepts in the first place.

In the usual way of thinking, the concepts of operational psychology are not defined at all; the phenomena for which the concepts stand are not identified. Concepts merely appear in statements or equations specifying functional relationships, and they mean nothing more than these functional connections.[55]

Operationists, according to Israel and Goldstein, were blindly attempting to reconstitute scientific method by substituting "a new kind of meaning, purely functional meaning," for "meaning as ordinarily defined and for meaning as specially defined by Bridgman's testing and measuring operations."[56] Insofar as psychologists have accomplished anything, they said, it was because operationism did not interfere with their work. In that case their method is no different from that advocated by the behaviorists.

It is only when the simple measuring operations cannot be performed or when they yield ambiguous results that the operationists resort to their use of functional connections as defining operations. By the use of this device they presume to establish the existence and determine the identity of hypothetical phenomena which are otherwise unidentifiable and indistinguishable.[57]

No other critic came so close to suggesting that operationism might be a giant hoax, or at least a cover-up for scientific impotence. Boring, realizing the seriousness of Israel and Goldstein's assault, quickly organized "a symposium on operationism in order to clear up some of the disputed points."[58] Papers were solicited from Bridgman, psychologists B. F. Skinner, Carroll C. Pratt, and Harold Israel, philosopher Herbert Feigl, and Boring himself. Stevens and Tolman, according to Boring, were prevented from participating because of the war. The papers, together with rejoinders, were published in the *Psychological Review*.[59]

As might be expected, no resolution or consensus issued from the symposium. Each contributor argued his own interpretation. And if the psychologists expected a precise clarification of the operational method from Bridgman, they must have been disappointed. Bridgman's exposition was vague and unsatisfactory. The thrust of his statement was that operational analysis is a way of insuring that the objects of which we claim to have knowledge are the consequence of performable action rather than mere verbal inventions. All events are unique; membership in a category is a pragmatic assumption, not an absolute necessity. Understanding means knowing what to do to make the same thing happen again. Bridgman certainly did not lay down any guidelines other than exhortations to strive to attain maximum awareness of conditions surrounding an event. "Any method," he asserted, "of describing the conditions is permissible which leads to a characterization precise enough for the purpose in hand, making possible the recovery of the conditions to the necessary degree of approximation."[60]

As has been observed by the psychologist Sigmund Koch, Bridgman's operational method had evolved into a general hermeneutic. (For Koch's views, see Note 37 of this chapter.) It had lost its original focus on measurement as the scientific criterion for physical reality and become diluted into something vaguely understood as "human activity." That is not to say that operational analysis was interpretively irrelevant, but it had become so broadened that Bridgman found it difficult to convey what it meant for him.

What the symposium did accomplish was to expose in psychological operationism the classical limitations of empiricism as a complete philosophy of science. Of particular sensitivity was its vulnerability to solipsism,* especially since operationism was supposed to be a way to

---

*Referring to the conflict between Bridgman and the logical positivists (see above in this chapter), the problem may be characterized as follows: There is no formal procedure (proof) which can show with certainty how "what is experiential and therefore subjective can be transformed into something public and objective" along either of the possible analytical axes—as a relationship moving from the factual to the formal (particular to general), or as a relationship moving from individual knowledge to public knowledge (subjective to objective). The first difficulty gives rise to the problem of induction (see

bypass the problem. The intrinsic nature of psychology only exacerbated the difficulty. The behaviorist approach, commonly equated with operationism, was, of course, to repudiate the possibility of an authentic science of mind by ignoring the mind altogether. Psychology was redefined as the study of behavior, although this strategy does not actually suffice to eliminate the fundamental epistemological problem created by empiricist assumptions. Nevertheless, not all the contributors to the symposium were strict behaviorists, and there was a wide range of opinion about how knowledge becomes public and objective. Underneath it all, though, ran the belief that operational definition was in some concrete fashion the methodological key to scientific objectivity.

Boring added no new insights. He was now essentially the spokesman for Stevens's views. Publicity (the public nature of science), empiricism, and operationism, in his eyes, were all part of the same package, an opinion wonderfully expressed by the following circumlocutions. "Since science is empirical and excludes private data, all of its concepts must be capable of operational definition." Or, "Science does not consider private data. If a datum is public, the operation of its publication is statable. There can be no scientific data that can not be operationally defined."[61]

Harold Israel thought the problem was too difficult for psychology and supported this judgment by pointing to Boring's confusion. He focused his critique on a metaphysical point. The argument that two sets of operations are equivalent if they yield the same numerical results, he pointed out, "involves detaching the concept of quantitative value from its operational meaning and assigning to it the status of an absolute property, quantity of a kind which transcends the methods by which it is determined." Furthermore, Israel argued, the question What is an operation? has not been satisfactorily answered, although the foundations of operational theory are quite clearly elementaristic.[62]

Carroll C. Pratt tried to face up to the epistemological question directly. No operational definition of experience is possible, he declared. "Every scientific observation starts life as a bit of private experience. . . . The escape from solipsism can only be made by a leap of animal faith." Pratt handed the problem back to the behaviorist: "The initial data of behaviorism are no more public than are the data of introspection."

chap. 9); the second, to the problem of solipsism. Carroll C. Pratt's comments (see below) are to the point.

Operationism, as he understood it, was scientific method and, when practiced correctly, included the specification of the conditions that must be created in order to reproduce a reported experience. But the operational construction of the conditions does not define the experience, nor is it the same thing as the experience. "The experience is private and indefinable. It can only be pointed to." "Neo-introspectionists" such as Boring and Stevens stand in danger of mistaking "the pointer for the thing pointed to," whereas the behaviorists nearly throw out the baby with the bathwater. Still, Pratt's leap of faith did not really explain how an empirical science is possible. In fact, he seemed to be committed to a formalist way of thinking in his insistence that scientific hypothesis is circular. He summarized his belief in the statement that "science is a vast and impressive tautology."[63]

Herbert Feigl represented the logical positivists. He had refined his categories considerably since the publication of his early papers, and his discussion of operationism was more mature and qualified. Nevertheless, he, too, felt compelled to address the problem of the experiential basis of knowledge. "Private, immediate experience as such is only the raw material, not the real subject matter of science . . . it is not a construct but that small foothold in reality that any observer must have in order to get at all started." Attempts to convert this phenomenal experience into something scientifically useful can be of little significance. Therefore, Feigl asserted, it is preferable to adopt "a strictly physicalistic or behavioristic approach right from the start." In this way the "traditional metaphysical pseudo-problems of solipsism, the mind-body puzzle, etc.," can be eliminated.[64]

B. F. Skinner could not have agreed more. Science, he said, is the behavior of scientists, and the meanings of terms "emitted" by the scientist are to be found in the conditions that stimulate the verbal response. Verbal response—the use of a particular term or set of terms—is fixed or set by a reinforcing verbal community upon whom supposedly the same stimuli are acting.[65]

The requirements of this hypothesis led Skinner to a treatment of private experience that drew him into a prolonged debate with Bridgman. For in order to be consistent with his stimulus-response-reinforcement model of language behavior, he had to consider the language descriptive of private experience as the product of social reinforcement. Accordingly, these private events are inferences supported by "appropriate reinforcement based upon public accompaniments or consequences," and being conscious, which is nothing more than "a form of reacting to one's own behavior, is a social product." [66]

Skinner's radical behaviorism, with its ruthless obliteration of the uniqueness of individual experience, touched Bridgman's most tender philosophical nerve. He bristled. His attention was captured by Skinner's naked denial of so self-evident a truth as the experience of one's own consciousness. He heard virtually no other voices. Thus, his rejoinder to the symposium papers was a concentrated protest against Skinner's thesis. Indeed, this was only the beginning of an intellectual confrontation that lasted for the remainder of his life.

First, Bridgman reminded his audience that he opposed the commonly held assumption that the most important aspect of science is its public nature. From the operational point of view, he argued, "a simple inspection of what one does in any scientific enterprise will show that the most important part of science is private." Here he was referring to his argument about the private nature of scientific innovation and the coercive power of proof. Second, he asserted, operational analysis justifies the distinction between one's private thoughts and what thoughts might be imputed to him. "The operations which justify me in saying 'My tooth aches,' are different from those which justify me in saying, 'Your tooth aches.'" That is to say, a difference of operations implies a difference in the realities. "Your" reality is an abstraction for "me," Bridgman is arguing, but "my" reality is not an abstraction for myself. They are different kinds of knowledge, and one is not justified in ignoring the intimate in favor of the abstract. "Simple observation shows that I act in two modes. In my public mode I have an image of myself in the community of my neighbors, all similar to myself and all of us equivalent parts of a single all-embracing whole. In the private mode I feel my inviolable isolation from my fellows and may say, 'My thoughts are my own, and I will be damned if I let you know what I am thinking about.'" [67]

In addition, Bridgman indicated that he was also thinking of the social and political ramifications of such a view of knowledge. The prerogative of the social group in ratifying scientific knowledge was for him much more than just an epistemological matter. It was a threat to the self-sufficiency and integrity of the individual in society.

The whole linguistic history of the human race is a history of a deliberate suppression of the patent operational differences between my feelings and your feelings, between my thought and your thought. . . . All government, whether the crassest totalitarianism or the uncritical and naive form of democracy toward which we are at present tending in this country, endeavors to suppress the private mode as illegitimate, as do also most institutionalized religions and nearly all systems of philosophy or ethics.[68]

In his crusade to protect whatever is unique and individual, Bridgman pictured the operational method not as a technique for building up general conceptual categories in science but as a way to recover what has been lost in the process of generalization—of regaining the uniqueness of original experience. He emphasized this point in the conclusion of his rejoinder. "The operational approach demands that we make our reports and do our thinking in the freshest terms of which we are capable, in which we strip off the sophistications of millennia of culture and report as directly as we can what happens."[69] This meant that scientific reports should be made in the first person in order to avoid the unwarranted impression of universal validity.

As it turned out, Bridgman's confidence in the authenticity and reliability of ordinary experience was later shaken when he was introduced to the Ames demonstration in which an optical illusion impressed upon him the inadequacy of the senses in apprehending a total experience.* He wrote to Ames that he thought the work was of revolutionary significance because it revealed the possibility of a pre-perceptual order of experience.[70] But whatever revolutionary implications it had in Bridgman's mind, it did not lessen his resolve in the debate with Skinner over the significance of private experience. Skinner's behaviorism continued to challenge his thoughts. In private memoranda, in personal discussions with Skinner, Stevens, and Boring, Bridgman invoked his operational creed against what he believed was Skinner's evasion of the dual aspect of knowledge. It seemed to him that Skinner was simply not being true to the facts.

"What is Skinner trying to do?" he asked himself.[71] "Skinner's main point seems to be that I can best understand myself by standing outside myself and thinking about myself in public terms—he thinks that our forms of thought are strongly molded by social influences during

---

*"Adelbert Ames, Jr. (1880–1955), was an American trained as a lawyer who became interested in perception. He devised a number of fascinating demonstrations, mostly designed to show that under many circumstances our perceptions are not veridical representations of external reality, that is, are to some extent illusory" (Ernest R. Hilgard, *Psychology in America: A Historical Survey* [San Diego: Harcourt Brace Jovanovich, 1987], p. 151).

our infancy. To which my reaction is that this is all right as a program, but that it is poor description of what I as a matter fact now do. I think of my pain in its own terms, not in terms of the pain of the psychologist defined in terms of the observed behavior of the other fellow." [72] Knowledge of intimate private experience, as opposed to the abstract private experience of the psychologists, Bridgman hypothesized, might even be [beyond] the limitations of science itself. Intimate private experience has the property of quality, and that, he recalled, was also what he felt was missing from the reality of Einstein and Eddington, who claimed that science "is concerned only with structure, with the intersections of the four dimensional mesh of world lines or with pointer readings. I have always violently objected to this as incomplete, for the experience that is being described cannot be reproduced given only the pointer readings or the intersections of world lines—we must also know world lines of what. There is no quality in intersections of world lines." [73]

There is little profit in searching Bridgman's ideas for philosophical depth. He did not develop his thoughts systematically. His writing is exploratory and addressed primarily to himself except for the image of Skinner hovering over his thoughts. In 1953, he wrote, "This is going to be one more attempt to make my peace with the problem of public versus private sufficiently so that I may get on with other matters. . . . In thinking at this problem I shall probably continually have Skinner in the back of my head, imagining that I am discussing or arguing with him." [74] Nevertheless, certain points in Bridgman's philosophical explorations stand out because they appear so often.

The first is a keen sense of isolation from his fellow man, of the "unbridgeable gulf" separating "me" from "you." This is because "you" cannot know "my" thoughts in the same way as I know them. They are two different universes of knowledge. Acknowledgment of this difference, Bridgman believed, was basically what he was "fighting for." To deny it not only was contrary to the revelations of the operational method but also was to participate in an epistemological deception.

Second, this separateness should be valued and emphasized. There should be a way to capture in language the self-referential meaning of private experience, a way to indicate that certain words, such as "conscious," "understand," "feel," "think," "know," "remember," have different meanings when they are used in the first and second person.

Third, the reason that separateness should be preserved is a fear that the collectivity might swamp what is unique and particular. Bridgman was offended by the idea that public meaning could be prior to private meaning. He recoiled against any rationalization that would increase the power of society over the individual or grant the universal priority over the particular.

As an unyielding champion of the individual, Bridgman perceived Skinner as the representative of the stand most directly opposed to his own. He believed that Skinner was deliberately turning his back on a significant segment of reality that the operational method makes perfectly obvious. He felt this was the reason that "although I can apparently translate what Skinner says into my language it would appear that Skinner cannot translate what I say into his language."[75] In other words, from Bridgman's standpoint, Skinner was treating a partial truth as the whole truth.

This breach was never closed. Bridgman's operational method was not Skinner's idea of operationism. On May 10, 1956, Skinner wrote to Bridgman, "My efforts to convince you of the possibility of extending the operational method to human behavior have long since suffered extinction."[76] Bridgman noted a similar frustration. Attached to the collection of manuscripts on which the above discussion is based is a note-card which simply states: "These various reflections on Skinner were shown to him after his letter and lunch meeting, June 1958, and presumably read by him; later returned by registered mail with no comment." The "unbridgeable gulf" between himself and Skinner could not have been brought home to Bridgman more objectively.

Independently of similar contemporary stirrings and without knowing it by a name, as he pressed the meaning of "experience" to its utmost limit, Bridgman fell upon existentialism, the very antithesis of traditional metaphysics. In doing so, he showed to what degree the "empiricism" of the positivists and behaviorists was but another brand of metaphysical aspiration, and their antimetaphysical rhetoric, mere lip service to a scientific ideology. Furthermore, Bridgman had achieved his insight within the framework of scientific, rather than religious or ethical, enquiry, although it would be wrong to assume that there was no ethical aspect to his position. Indeed there was, and as we will see in Chapter 11, its outcome was bleak and despairing.

Within the community of scientists and scientific philosophers, Bridgman had become the lone spokesman for a radical existential subjectivism, a philosophical position that made no sense as a scientific philosophy and that critics mistook for solipsism. In fact, Bridgman's "new" operational stand had much less affinity to philosophical empiricism than to contemporary Protestant neo-orthodoxy. Ironically, his intellectual bearings were pointed along lines more similar to the existentialist theological epistemology of Reinhold Niebuhr than the positivism of Rudolf Carnap or even the pragmatism of William James. Indeed, as noted above, in his quarrel with Einstein, Bridgman's objection to relativity finally rested on a moral judgment ringing with Puritan overtones.

As his struggle with relativity has illustrated, Bridgman was never able to reconcile the theoretical side of physics with the empirical. Klyce, Korzybski, the logical positivists, and the behaviorists were no match for Einstein, and Bridgman was painfully aware of this fact. Bridgman could attribute the attempts of the former to erect a system of knowledge to an excess of missionary zeal or to philosophical artificiality, but he could not do the same with Einstein. Relativity, by its very success, defied the ultimacy of his radical subjectivity.

Nonetheless, Bridgman's epistemological inquiry had led him to discover the subjectivity of knowledge and, in this subjectivity, the grounds for a claim to cognitive freedom. At the same time, he realized a radical existential human freedom, a state of freedom wherein he could choose to accept or reject social and political authority. However, it did not occur to him that this same condition of radical freedom might also be a state of moral freedom—that he might be free to embrace the universal—and that loneliness and alienation might themselves be conditions of enslavement. So heavy was the burden of Adam's sin.

PART IV

Measurement and Communication

My work on the 1930's led me increasingly to probe the question of a world somehow suspended between two quite distinguishable systems and ways of life. The crucial and perhaps climactic stage of that battle . . . was fought in the 1920's and 1930's.

. . . Any study of the culture of abundance begins with the obvious cultural consequences of the new communications. It is not simply that these inventions made abundance available to many and made possible increasingly effective distribution. Consciousness itself was altered; the very perception of time and space was radically changed.

. . . No other culture expended so much of its energy and resources discussing and analyzing communication and its problems. . . . Some observers began to wonder whether anyone could really communicate at all.
    —Warren I. Susman, Culture as History (1973)

■ ■ ■

IF BRIDGMAN'S scientific colleagues ignored his "new" philosophical position or if the behaviorists and positivists found it incomprehensible, it was not without a scientific rationale. To be sure, operationism had found a receptive audience, but in Bridgman's view, it had been thoroughly misinterpreted. Part of the difficulty may be attributed to the vagueness of his original statements in *The Logic of Modern Physics*, and part to his change in attitude, a change that did not interest his academic interpreters, however disturbing it was to Klyce, Korzybski, and Bentley. The latter were well aware of Bridgman's new meaning, especially its moral implications, but they had little, if any, influence on the mainstream of scientific philosophizing. Nevertheless, they, as well as the academics, were committed to defending the traditional idealistic view of the meaning of science, whereas Bridgman believed that modern physics had shown this to be an impossible objective.

However, Bridgman's operationism harbored serious internal contradictions, and these contradictions, when made explicit, will help to explain both the relevance of Bridgman's operational thinking to contemporary intellectual challenges and the reason why Bridgman failed to live up to the expectations of his critics. In addition, once the scientific problem with which Bridgman was struggling is specified and his interpretation of its resolution clarified, Bridgman's existentialist mood and his radical views on the subjectivity of knowledge and the dichotomy of language will become much more understandable. Bridgman's position was by no means arbitrary or frivolous but was grounded in a subtle interpretation of scientific principles.

Indeed, to probe into what was philosophically and morally at stake carries us to the heart of the scientific method—to the science of epistemology, to the questions of how knowledge is possible in the first place and, concomitantly, what knowledge is available to humanity, and furthermore, what the answer to either question has to do with morality. At center stage in physics was the problem of measurement, and while it was in quantum mechanics that the question became most urgent because it appeared there in its most irreconcilable form, it had already been introduced, as we have seen, by relativity theory. However, it had been foreshadowed even earlier in thermodynamics.

What was this innovation, this new feature that made modern physics so daring, so demoralizing, that was centered upon an activity as ordinary as measurement? Why was Bridgman's operationism so attractive, yet so out of touch with contemporary scientific sensibilities? Why was it that Bridgman's conception of operationism was not applicable to the interpretation of relativity, or, as we shall see, to quantum mechanics, or even to thermodynamics, the so-called phenomenological science, which, contrary to common expectation, also failed the operational test? There are several ways to characterize this novelty. One might be to say that modern physics had discovered the need to account theoretically for the act of measurement itself. Another, more precise and also more useful for the analysis of Bridgman's meaning, is to emphasize the discovery of an epistemological separation between subject and object, a separation that must be mediated by some communicating agent. Thus, we may characterize the "revolutionary" feature of the new physics as a transformation of the epistemology supporting the concept of measurement from one based on direct intuition, where the distance does not count, to one dependent on a process of communication, where the behavior of the mediating agent must be accounted for in physical theory.

Indeed, the thrust of this transformation would be to propel twentieth-century physics out of the classical world of balanced forces and permanent qualities into the modern world of ongoing process and continual change, a vision that, besides creating a host of problems for philosophers to debate, would subsequently give birth to new communications sciences such as information theory, cybernetics, and artificial intelligence. The twentieth century would leave behind the quaint notion of the universe as a clockwork and its crude successor, the universe as steam engine, and redefine it as a vast network of interaction and communication. Operationism was a product of this transition from the old to the new, and it suffered from the ambiguities characteristic of and indigenous to an era of change. However, Bridgman chafed under the discomfort of ambiguity and sought a clear resolution, a resolution that in the end, we will see, eluded him—but just barely.

As we make our way through this transformation of scientific meaning, we shall witness the struggles of a mind gifted with exquisite intuition that was held back by the tenacity of inculcated belief and an ambivalence toward changing scientific standards, a mind whose full power of insight was choked by traditional scientific presuppositions and a personal overreaction to an orthodox Congregational religious upbringing, a mind whose expressive impulse was blunted by difficulty with the demands of verbal communication. Bridgman's son-in-law, the mathematician Bernard Koopman, once remarked that in a way Bridgman knew too much and, at the same time, knew too little, that he was in the habit of discovering what professionals in mathematics and logic had already known for a long time, where the problem was not to expose the difficulty, but to solve it.[1] The next two chapters will seek to elaborate Koopman's observation, to uncover the intuition behind operationism and to reveal its implicit assumptions, to find out what Bridgman "knew" that no one else knew, or at least, what he was trying to know, but did not quite know, and what stood in his way. It seems fair to say at the outset that for the most part, contrary to what might appear to be the implication of Koopman's remarks, Bridgman was probably not motivated by a desire to expose intellectual difficulties for the benefit of the community at large. Rather, his interest was his own clarity of mind.

This part begins with a brief review of the challenges presented by quantum mechanics and suggests that operationism was inadequate to deal with them because of its unacknowledged foundation in classical epistemology. It then goes on to argue that this was also the reason why Bridgman was confounded by relativity theory, why he could not comprehend its full significance. Finally, the clock is turned back to consider the transitional nature of classical thermodynamics with respect to the epistemology of measurement in order to engage Bridgman's treatment of thermodynamic concepts as an instrument for probing into the meaning behind his operational judgments.

CHAPTER 8

Quantum Mechanics: Signals from Beyond the "Here and Now"

The only thing harder to understand than a law of statistical origin would be a law that is not of statistical origin, for then there would be no way for it—or its progenitor principles—to come into being. On the other hand, when we view each of the laws of physics—and no laws are more magnificent in scope or better tested—as at bottom statistical in character, then we are at last able to forego the idea of a law that endures from everlasting to everlasting.

—John Archibald Wheeler, *Law Without Law* (1979)

■ ■ ■

THE FALL OF physical truth from Newton's ideal metaphysical loft to the theater of human action was for Bridgman a shock and a grave disillusionment tantamount to a loss of faith. He had once believed that the goal of science was to discover truth, plain and simple, metaphysically absolute.[1] Modern physics shattered that belief.

Originally, Bridgman had hoped to use operational analysis to salvage some remains of extrinsic objectivity while adjusting to scientific modernity, to the "disquieting" realization that science is an artifact of human ingenuity, that truth is formulated in human terms. Indeed, no insight made a greater impression on him than his realization that knowledge had been demoted from the height of absolute metaphysical perfection to the imperfect domain of human activity. Upon this point he insisted, and by postulating operations as the elements of physical reality, he believed he had made a successful transition from the old physics to the new, from the metaphysical to the human, from the passive to the active.

But somehow, operationism fell short of achieving its purpose. Somehow, it failed to capture the essence of modernity. Bridgman's operational reinterpretation of special relativity had provided him with a temporary psychological and rational crutch that enabled him to cope with the breakdown of his belief system. Nevertheless, he made no attempt to employ the same strategy in dealing with quantum mechanics, which, as we have seen, he denounced as being even worse than relativity in its philosophical implications. Something was missing; operational analysis did not serve Bridgman when it came time to confront quantum mechanics.

This curious situation demands an explanation, especially since the categories of quantum mechanics quite naturally invited an operational interpretation. Indeed, the vocabulary of operationism is commonplace in discussions of quantum mechanics, although further research might be necessary to determine whether this usage should be attributed to Bridgman or to Heisenberg.[2] In any case, other physicists did not hesitate to invoke operational language when speaking about quantum mechanics. As it turned out, even though operationism was formulated in response to special relativity, it was much more useful to physicists for rationalizing the concepts of quantum mechanics than those of relativity. The innovations of relativity were not, after all, so radical as to require the drastic operational reorganization of physical experience that Bridgman had proposed. Bridgman's redefinitions of the mechanical properties of relativistic objects were, at best, superfluous or irrelevant and, at worst, simply wrong. Except for characterizing to a limited extent the epistemological standpoint of special relativity, operationism was not otherwise especially useful to physicists for interpreting relativity.

However, the same was not the case for quantum mechanics, where interpretative difficulties were rampant, and fresh enigmas appeared at every new turn of theorizing. Here, both method and concept defied understanding in traditional terms. Established physical principles were brought under scrutiny and their validity questioned. The new concept of reality which was emerging was alien to the sensibility of ordinary physical intuition. As Heisenberg described the situation:

Almost every progress in science has been paid for by a sacrifice, for almost every new intellectual achievement previous positions and conceptions had to be given up. Thus, in a way, the increase of knowledge and insight diminishes continually the scientist's claim on "understanding" nature.[3]

This situation, as Bridgman often repeated, called for a re-education of physical intuition so as to cultivate new forms of understanding, forms that would make quantum mechanical reality readily accessible to intuition. Exactly what reorientation he had in mind or at what stage of scientific advance this re-education should be introduced is not clear. What is clear, however, is that theoretical physicists were trying out new modes of representing and interpreting what experimental results had forced upon them.[4]

Early on, for example, in an effort to preserve the classical intuition regarding the electromagnetic field, Bohr and Dutch theoretical physicist H. A. Kramers suggested giving up the principle of energy conservation as an exact law, a proposal that turned out to be unnecessary. There was talk of the existence of mysterious physical states possessed of "intermediate" reality, which may or may not have been what were later formalized as probability functions. Eventually, as the theory matured, physicists had to make sense of the paradoxical and irreducible wave-particle duality of light and matter. They had to confront Heisenberg's philosophically controversial uncertainty (indeterminacy) principle (the precision to which one of a pair of attributes or conjugate properties is measured limits the precision to which its partner can be measured) as well as Bohr's dualistic complementarity principle (only one aspect of the wave-particle potential—the one that is being measured—can be actual at a given time; even though the two aspects are mutually exclusive, both are necessary for a complete characterization of microscopic reality). A wave-particle can present itself as a wave or a particle, but not both simultaneously; which aspect is manifested is determined by the experimental setup. And as if these problems were not enough, there were more.

The mathematical formalisms of quantum mechanics created parallel difficulties. Heisenberg's matrix mechanics seemed at first to be physically unintelligible, and even sophisticated mathematical physicists were unacquainted with its formal properties.* Schrödinger's wave equation had a more intuitive appeal to the physical imagination because analogies from acoustics could be invoked, but the kind of re-

---

*E. H. Kennard, for example, wrote in "Quantum Mechanics of an Electron," *Journal of the Franklin Institute*, Jan. 1929, p. 50, "We shall also omit all mention of the earlier 'matrix' form of the theory, which is mathematically equivalent to the other [Schrödinger's], but is much more abstract and to many physicists repellent." At the same time, he affirmed Heisenberg's (not Bridgman's) version of operationism: "The great guiding idea of Heisenberg has been that in constructing our theories we ought to be very clear in our minds as to what quantities can really be made the object of physical observation; only observable quantities and such others as can be calculated from them by definite rules can be said to possess physical significance."

ality the wave function described was problematic—the motion of what? Whatever it was, it was not the familiar motion of a vibrating violin string or stretched membrane, the most easily pictured classical analogues. In addition, the formal equivalence of the Heisenberg and Schrödinger versions contradicted the classical assumption that a true scientific language should be unique. Did not two languages imply two realities? Judging that this theoretical situation was marred by "an unresolved dualism," Bridgman joined Einstein and physicists Boris Podolsky and Nathan Rosen in pronouncing quantum mechanics an incomplete theory. However, as we shall see, Bridgman argued that incompleteness—that is, if it meant nonuniversality—was not necessarily something intrinsically undesirable in a physical theory, although duality, meaning a mixture of separate theoretical languages, was an undesirable compromise indicative of human intellectual shortcomings.

When a statistical interpretation of the wave function was formulated by the German physicist Max Born, it provided a rationally, if not necessarily physically, satisfactory substratum of meaning. The Schrödinger waves could now be understood as representations of probability amplitudes corresponding to the potential outcome of measurements made on microscopic systems—as probability waves. The British mathematical physicist James Jeans was inspired to call them "waves of knowledge." However, the statistical interpretation raised even more questions about the kind of reality physics deals with. What could be the physical meaning of probability as an object of physical knowledge? What is it that the physicist ends up measuring, or "knowing"? Is probability intrinsically built into reality as an objective feature, or is it, as classically regarded, a consequence of human deficiency—either ignorance or error? In other words, is probability something objectively real, or does it symptomize (hopefully temporary) imperfection in our knowledge? Moreover, if probability is all that we can know of microscopic reality, does that not imply that deterministic laws are ultimately superseded by chance? It is for this reason Bridgman judged that quantum mechanics had created a situation worse than that caused by relativity. We recall the alarm that Bridgman expressed in his *Harper's* article upon realizing this possibility—the prospect that "the world is not intrinsically reasonable or understandable," and that cause and effect therefore have no meaning.[5] Furthermore, Bridgman must have regarded the incorporation of probability into quantum mechanical reality (the same difficulty also pertains to statistical mechanics) as a repudiation of the very goals of science—the establishment of irrefutable, certain truths—and their replacement, at best, by the predictions of a practical but clumsy procedure and, at worst, by nothing better than the prognostications of gamblers. Bridgman did not neglect to remind his readers that the rules of probability were originally worked out in reference to the question of proper betting procedures in games of chance.[6]

Similarly, since the indeterminacy (uncertainty) principle was the heart of the theory, it seemed necessary to give up the expectation that "exact" knowledge can ever be the goal of science. If all measurement involves at its most refined level an interaction between the object and the measuring instrument, an "uncontrollable" (Bridgman's word) interaction whose effect cannot be precisely known beyond the limit set by the Heisenberg principle, then in principle exact knowledge is impossible. Bridgman translated this to mean that science would have to lower its expectations, that in the end, "man will not become as a god, but will forever remain humble."[7]

We will consider some of these problems in a more specific context shortly. For the moment, it is sufficient to note that the interpretive difficulties were formidable and demanded a revision of ideas about physical reality and that the most significant departure from classical physics was the need to account for the intrusion of the act of measurement into the field of knowledge.

Measurement was the focal point of Bridgman's operational interpretation of reality. Indeed, operationism was predicated on the idea that the observer or measurer actively participates in determining the nature and content of physical truth. On this premise, measurements determine constitutively (but not interactively) the entities (objects or concepts) of physical reasoning. The vagueness of this concept is characteristic of operationism. Its rationale will be clarified shortly. Bridgman's later discussions (stimulated, probably, by his analysis of thermodynamic concepts) of the role of paper and pencil operations (by which he meant, as he explained to Carnap, operations "in which there is a large element of mathematical calculation") did not signify a change in his criterion for physical reality.[8] It only shifted the emphasis by calling attention to the nonphysical aspect of scientific activity he had already included, but not elaborated, in *The Logic of Modern Physics*. Essentially, Bridgman's operational claim was that physical concepts, or objects, are constructed from operations, both physical (measurements) and paper and pencil (computations), although he attributed a lesser reality to the outcome of the second. Given, then, that the essential ingredient of operationism was the determinative role of the measurer, we may well wonder why Bridgman did not attempt to apply his operational reasoning to quantum mechanics or, at least, why he did not adduce quantum mechanics as vindication of his operational method. The answer is subtle and qualifies the question, but it is not obscure.

In fact, Bridgman was not unaware of this possibility. He acknowledged that in certain formal respects quantum mechanics could be called operational, although in his view the characterization was superficial. Referring to the mathematical procedure, the calculus of operators, that was the formal machinery of quantum mechanics, Bridgman commented that

there is a very close formal connection between the theory and the things that we do, that is, it is formally a thoroughgoing operational theory. This end is achieved by labelling some of the mathematical symbols "operators," "observables," etc. But in spite of the existence of a mathematical symbolism of this sort, the exact corresponding physical manipulations are often obscure, at least in the sense that it is not obvious how one would construct an idealized laboratory apparatus for making any desired sort of measurement.[9]

How would one, Bridgman went on to ask, set up the laboratory apparatus to answer such questions as "Can e [charge of the electron], or m [mass of a subatomic particle], or h [Planck's constant] be measured separately with unlimited precision by a single experiment, or may they be measured simultaneously in a single experiment?" Furthermore, in terms of what apparatus does a given "observable" acquire meaning? We have now had to give up the idea that theoretical terms always correspond to something in experience. But, he intimated, how do we know with reasonable certainty that any of them do? This situation justifies a "certain amount of disquietude," if simply because it has not been thought through thoroughly.[10]

The implied criticism here is that there is no trustworthy (that is, immediate physical) standard to judge the validity of quantum mechanical concepts. Quantum mechanics is a mathematical model, which is to say that it is primarily a paper and pencil affair. Furthermore, Bridgman observed, the lack of agreement between Einstein and Bohr on the fundamental meaning of quantum mechanics only serves to confirm the untrustworthiness of conclusions based on quantum mechanical considerations. Furthermore, since no mathematical theory is unique, there might still be ways to save "the old concept of reality." Perhaps yet undiscovered mathematical models are compatible with both quantum mechanical and classical notions.[11]

Similarly, Bridgman admitted that Heisenberg's matrix formulation "appears to meet the demands of operationism," but again he withheld full endorsement.[12] Heisenberg, like Bridgman, had been influenced by Einstein's discussion of simultaneity.[13] He realized that the electronic orbital motions of Bohr's atom were not susceptible to empirical observation. Therefore, he insisted that the objects of theoretical manipulation be "observable magnitudes," such as the optically measured frequencies and intensities of atomic spectra rather than the hypothetical positions or momenta of unobservable particles, electrons, circling the atomic nucleus.[14] Heisenberg's matrices were sets or arrays of numbers representing these observable magnitudes.

Bridgman described Heisenberg's interpretation as a mixture of the old requirement that "every step in the mathematical manipulation of the equation had its counterpart in some feature of the physical system" and "the dictum of the operationist that only those physical concepts have meaning which can be defined in terms of physical operations, which means in particular that no quantitative physical concept has meaning unless it corresponds to something measurable."

Since every equation of physics essentially deals with numbers, Heisenberg demanded that only those quantities shall enter the equations which are intrinsically measurable. Bohr's atom was an example of the other sort of thing, for there was no physical evidence for the existence of the discrete orbits, but only of the frequencies emitted when the electron made the "jump" (for which again there was no physical evidence) from one orbit to another. The frequencies are of course measurable, and Heisenberg's form of wave mechanics did as a matter of fact use only these frequencies, without introducing the superfluous notion of orbits.

Nonetheless, and here is Bridgman's qualification:

I have always wondered, however, whether perhaps this requirement of Heisenberg was not formulated after the event as a sort of philosophical justification for its success, rather than having played an indispensable part in the formulation of the theory.[15]

In other words, in Bridgman's view, Heisenberg's operationism was not a constructive principle; operational thinking played no part in creating the theory, and therefore the putative operationism is a philosophical rationale added after the fact.

However, from a moral standpoint, if not an operationally technical one, Bridgman regarded quantum mechanics an advance over previous theories. It was exemplary insofar as it did not embody aspirations to universal scope, to a god's eye view, which was the error of relativity. (Bridgman rarely missed an opportunity to slap Einstein for being so presumptuous.) By its very nature, he argued, because quantum mechanics is probabilistic and consequently cannot be applied to individual occurrences, it is an incomplete theory; it "is intrinsically incapable of application to the entire universe." Indeed, Bridgman's pet objection to probability theory was that it cannot be applied to an individual event. On the other hand, Bridgman pointed out, imagine what "would be involved in a theory which was comprehensive enough to embrace in a single coordinating point of view the entire universe. Such a theory would have to exist in the mind of the theorizer; but his mind is also a part of the universe, so that the theory would have to involve a theory of the theory, and so on in never ending regression. A complete theory is thus intrinsically impossible."[16] The "incompleteness" of quantum mechanics was therefore not a strike against it. Furthermore, Bridgman added, quantum mechanics has the considerable virtue of recognizing the inevitability of error; a special kind of error, to be sure, one that results from the "lack of sharpness when we try to measure simultaneously position and momentum," nevertheless one automatically included in the theory itself.[17] Moreover, to the extent that quantum mechanics recognizes, indeed incorporates within its theoretical boundaries, the act of observation, it participates in the spirit of operationism. Thus, in a "qualitative" way, quantum mechanics reproduces experience more faithfully than any preceding theory.

However, Bridgman warned, a closer examination shows that all is not quite what it seems, and these advantageous features may have been secured by paying an operationally unacceptable price, in particular by introducing probability as the unanalyzable basis for the theory. For example, the "error" or uncertainty of quantum mechanics is clearly not what is ordinarily thought of as error on the macroscopic level of experimentation, and its order of magnitude, determined by Planck's constant, h, "is so far beyond present experimental possibilities that it is at present without physical meaning to inquire whether the hypothesis of a finite inherent error in the measurement of certain elemental things is 'correct' or not." As a matter of historical fact, h was "not introduced to ensure the presence of error as a qualitative fact" and only more or less incidentally turned out to provide for it. If, in fact, the observer is ever really to be successfully included as part of the system, errors—meaning human blunders, not irreducible quantum-level interactions—will have to be explained on the classical level.[18] In this respect, Bridgman was suggesting that the "error" attributed to quantum-level interactions is an "ideal" or theoretical error, as opposed to actual macroscopic human inconsistencies, and the latter are what science must eventually explain if the activity of the observer is to be a genuine part of physical theory.

Nevertheless, the fact that placing probability at the heart of quantum mechanics requires the abandonment of determinism, Bridgman asserted, "puts no intolerable burden on our powers of intellectual adaptation." What bothered him even more, he said, was the mixture of classical and quantum-scale concepts, where the unanalyzables show up against a background "not describable in terms of the ordinary concepts of space, time, and causality. We meet the situation by saying in effect that the old concepts shall still be one-half applicable to what happens on the elemental level and that the other shall be pure chance, that is, pure chaos." We fix the meanings of concepts such as position or momentum, particle or wave, potential energy or electric field, on the level of macroscopic experience, where they are operationally definite, but apply them on the microscopic level, where they have no operational meaning.

We account for the evolution of the ordinary concepts of space, time, and causality on the level of ordinary experience by postulating that the effect of the chaotic element gradually gets cancelled out as we proceed toward large-scale phenomena in virtue of the regularities of large numbers. *Is this honestly, from a perfectly detached point of view, a very impressive performance?* [Emphasis added] Is it not exactly the sort of compromise that we would have predicted in advance would be the only possible one if it should prove that we were incapable of inventing any vitally new way of thinking about small scale things? Do we really think that a dualism of this kind says nearly as much about the structure of external nature as it does about our minds?[19]

The unanalyzables of quantum mechanics, Bridgman thought, could not be described in any terms that tell us "the nature of the physical experience by which we recognize [their] presence." To him, it was "almost axiomatic" that unanalyzables should be *intuitively recognizable*, that is, self-evident.[20] Thus, quantum mechanics was a compromise, a half-baked theory illustrating once again the limitations of the human mind and reminding us that physical theory is an imperfect human invention.

Bridgman published these opinions in 1936 in *The Nature of Physical Theory*, the book that reaped so much unfavorable response. In the eyes of his critics, the "new" Bridgman—the "Hyde, not Jekyll" Bridgman—who made his appearance with this publication had become a stick-in-the-mud, a turtle who had withdrawn into its shell. He had repudiated the unifying, reconciliatory promise of early operationism and retreated into "mysticism." In all fairness, Bridgman had not, as he rightly claimed, ever made such promises. His goal was not synthetic or unitary. Operationism was a defensive stand, cautionary and conservative, purifying. As a practical advisory, it meant little more than "actions speak louder than words." As an evaluative prescription, it meant that in order to judge a knowledge claim, it is necessary to know the steps taken to arrive at the result.

On the other hand, as an instrument for interpreting scientific theory, especially the new twentieth-century developments in physics, operationism was handicapped by the classical empiricist preconception about the neutrality of measurement, the assumption that measurement does not affect the state of the object being measured, either actually or theoretically. This concept of noninteraction is grounded in the classical epistemology of ideal objects, which depends, in turn, on the metaphysics of permanent qualities, a metaphysics that does not recognize elapsed time, or process in action.

Operationism, despite appearances, was rooted in classicism. It was grounded in a "here and now" epistemology of measurement. Bridgman made this point explicit many times. The following quotation is illustrative.

But with regard to laboratory operations, one may ask why there has to be a universe of operations; why should there be any restrictions, or why should not our universe of laboratory operations be "all the operations that we can perform"? Such catholicity is certainly permissible, with the proviso that we should have said "all the operations which we can *now* perform" [Bridgman's emphasis].[21]

However, its classicism was masked by the implication, suggested by the word "operation," that it had captured the essence of modernity, process. Indeed, it had not. Operations were substitute permanent qualities. This ambiguity was one source of the misunderstanding between Bridgman and his interpreters, why so many readers thought operationism was synthetic. But the misunderstanding was not so simple as a conflict between two opposing viewpoints. Bridgman himself believed that he had found the solution to the challenge of action and knowledge. He thought that by making procedure the definitive reference for scientific reality, he had escaped from the metaphysical error of classicism. Moreover, as we shall see, operationism was based on an intuition that cannot be dismissed as trivial.

Nevertheless, operationism had just barely been able to accommodate special relativity and had accomplished that feat only through the creation of some unique definitions—definitions that did not acknowledge the real novelty of special relativity. General relativity (which Bridgman had simply given up, but which in some ways he could have rationalized as being more compatible with a classical epistemology—as Einstein himself admitted) and quantum mechanics were outside its explanatory range, each for a different reason, and each of these theories had modified the meaning of measurement in such a way as to make it uncongenial to Bridgman's particular operational assumptions.

When he presented his operational method, Bridgman offered no criteria for judging what qualifies as a valid measurement (operation). How is the physicist to know if there is any sense to what he is doing? Bridgman had taken for granted that this is intuitively obvious—self-evident—when in fact the meaning of measurement lies not in itself, but in theory and its underlying metaphysics. To be sure, operationism was not supposed to be a theory of measurement, but it was based on certain assumptions implicit in dimensional analysis, and these, in turn, originated in the static Newtonian universe of primary and secondary qualities, whose measured counterparts were roughly the fundamental and derived quantities of dimensional analysis. We have already noted that Bridgman's operations were substitutes for the categories of dimensional analysis, measurements taking the place of fundamental quantities that paper and pencil operations subsequently transformed into higher-level or derived quantities. The fact that operations were actions rather than properties did not alter the basic classical metaphysics.

Overall, the operational world was still a world of stasis or its equivalent, equilibrium. Fundamental measurement was nonintrusive, non-time-consuming (non-entropy-producing), carried out in principle by comparison of likes—a weight with a weight, a length with a length, time with a repeating cycle—with no intrinsic limitation to precision.

The definition was given by Aristotle: "The measure [a unit or standard] is always of the same kind as that which is measured; for the measure of magnitudes is a magnitude, and specifically, the measure of a length is a length, that of a width is a width, that of voice is a voice, that of weight is a weight, that of units is a unit."[22] This Aristotelian definition of measurement was not consciously overthrown until the twentieth century.

In addition, joined to the definition of measurement was an epistemology and a metaphysics, both also essentially Aristotelian, as well as a formal model of proof, whose cogency relied ultimately on the same epistemology—that of Aristotle for ideal objects. This package did not become irreparably unraveled until the twentieth century, although by then it was badly frayed (implicitly, severely damaged by Darwin). The dimensional analysis debates bear witness to its resiliency and illustrate the kind of problems it created for interpreting the categories of physics. We will return to this momentarily.

Classical physics was permitted, by virtue of the way Aristotelian concepts were fused with Newtonian principles, to assume that the properties selected as parameters for fundamental measurement had some absolute permanent meaning or ontological status, that exact measurement was possible in principle, and that error in determining measured values was due merely to human ineptitude.[23] Furthermore, Newton's principle of the uniformity of nature was the ruling assumption; according to this principle the laws of mechanics and thus the associated measurements were valid irrespective of scale. Reasoning and proof, as well as ontological organization, were governed by the principles of Euclidean geometry. However, when Bridgman repudiated the universality of Newton's assumptions, he denied only those principles dealing with the organization of nature—its rationality and its uniformity—and its implied material composition, not the basic principles that were assumed to make knowledge possible in the first place. He never noticed the submerged Aristotelian epistemology.

Nonetheless, Bridgman's physical operations took the place of primary qualities, and given their assigned role in operational analysis, they bear a striking resemblance to Aristotelian measures, which are "of the same kind as that which is measured." Accordingly, it makes sense for Bridgman to have stated that "meaning" is synonymous with the corresponding operation, an assertion implying that the operation is associated in some manner with physical quality—with what *kind* the object is. Yet the classical nature of operationism is more profound than a simple identification of operations with Aristotelian "measures" would suggest; the assertion that meaning is synonymous with an operation implies further that an operation is also a cognitive act.

Indeed, against the background of classical epistemology, an operation can be seen as an act that compresses or condenses into a single act of measurement a conceptual hierarchy or succession of cognitive acts for which the exemplar is Aristotle's epistemology of ideal objects. In other words, the idea that a measurement equals knowledge is an abbreviation of a chain of classical interconnections—measurement, proof, intuition of axioms—which, since each link is an unmediated cognition, culminates in certain knowledge, that is, immediate knowledge of a now-present truth. Let us isolate the pattern in this chain.

If the least elevated step in this sequence was physical measurement, a procedure that involved the juxtaposition of like magnitudes, above it was geometry, the science of formal measurement and, in particular, Euclidean geometry. Geometry linked on one side the physical and the formal by means of physical measurement and on the other side the formal and the ideal absolute through the direct intuition of first principles, or axioms. At the center was the process of linking formal objects, one to another, dependent in turn on being able to prove one formal geometric object equal to another.

All measurement, to repeat Bertrand Russell, depends on immediate judgments of equality. If we are to understand a theory of measurement, we must know how equality is determined. Euclid gave his criterion for equality in the first book of the *Elements*, as the fourth common notion: "Things which coincide with one another are equal to one another."[24] Euclid applied this notion in his proofs through the use of the method of superposition, a principle which has been subjected to much criticism by mathematicians. Russell, for example, charged that it was "contradictory to the law of identity" and based on "the supposition that a given point can be now one point and now another."[25] In other words, said Russell, it is the imaginary transport of one place in space to another and implies that "our triangles are not spatial, but material." Russell was suggesting that Euclid had not distinguished a material object from a formal spatial object. Yet its logical flaws notwithstanding, the principle of superposition, the idea of placing one object "on top" of another in order to prove equality, was the primary formal measuring operation of Euclidean geometry.

At the highest point of the hierarchy of cognitions was Aristotle's epistemology for ideal objects—the identity of thought and object. Aristotle tells us in his *Metaphysics* that knowledge of nonmaterial things results when mind is brought into coincidence with object, in which case knowledge and its object are the same: "Since the intellect and the object of thought are not distinct in things which have no matter, the two will be the same, and so both thought and the object of thought will be one."[26] Such identification of mind and object could happen only outside ordinary time, outside the world of matter and change, in the ideal world of thought. Similarly, the superposition of one mathematical object upon another did not entail any effects due to motion, nor did the juxtaposition of measured magnitudes involve any interaction between the entities being compared. In other words, none of these cognitive acts was subject to any conditions of empirical time.

Thus, despite the ambiguity Russell noted in Euclid's principle of superposition, all of the cognitive acts in the classical account of knowledge occur without the passage of time—in an ideal world. And if, as is probably the case, Euclid was influenced by Aristotle, his inclusion of the criterion for equality among the common notions (which were understood to be true in general) rather than the postulates (which held only for geometry),[27] this seems to indicate that the notion "things which coincide with one another are equal to one another" was the classically accepted criterion for all comparative acts involved in cognition. The pattern—juxtaposition, superposition, identity—forms a succession from the least to the most perfect embodiment of this notion. Or, in reverse order, each is an imitation of the higher perfection. Whatever is going on at the boundaries of each stage, a common principle is present, a principle that for the purpose of characterizing the inherited classical epistemological standpoint we may designate in a general way as Aristotelian.

In the broadest sense, then, the classical epistemology may be characterized as an epistemology of "intuition," or mental congruence. It is a nonphysical concept of gaining knowledge dependent neither on time or on subject-object interaction. These characteristics were the unstated underpinnings of the "classical" Newtonian concept of measurement (not to be confused with "scientific method"), as well as the unexamined assumptions underlying Bridgman's operational principles. Indeed, despite "modernist" appearances, operationism was a collapsed form of the classical epistemology of intuition.

However, these classical assumptions were precisely what modern science was rendering implausible. Modern science is the science of process, of action in time, of the temporal disjunction between subject and object. This disjunction is an intimate aspect of the twentieth-century "problem of knowledge." Moreover, the problem was not confined to science alone. The wide interest in operationism was stimulated by the hope that operationism, by virtue of its fusion of act with category, might provide a way to overcome this disjunction and restore the possibility of achieving unity of thought and object—the classical ideal. Nevertheless, while operationism had the appearance of modernity and gave the impression of being an epistemology of action, it was firmly committed to the epistemology of intuition. Yet it is not fair to suggest that Bridgman had missed the point altogether.

Bridgman's operationism was built on his realization that modern physics had shown that measurement is a *physical* (rather than ideal) process or action and, by extension, that the acquisition of scientific knowledge depends on physical activity. This he had learned from Einstein's 1905 paper on special relativity, which made the point emphatically clear. However, whereas Bridgman regarded the recognition of the active physical nature of measurement as the primary insight of special relativity, for Einstein it was merely the groundwork for his subsequent reasoning. Understanding the simple fact that measurement is a physical activity would not in itself have been an insight powerful enough to support the structure of relativity theory. Its underlying significance lies in Einstein's realization that the entire universe is not present to immediate perception or intuition and that physical knowledge is transmitted by a physical process, a process that takes time. In order to take advantage of this insight, Einstein had to introduce into physics a new concept of time, a modification of the universal, uniform, undifferentiated metaphysical time of Newtonian physics.

The Newtonian physical world, not without contradiction, was a mixture of the physical and the metaphysical. It was also a world without history. The time of Newtonian mechanics was classical timeless or eternal metaphysical time, time that did not go anywhere, spatial time. The primary properties (fundamental categories of measurement) that constituted physical reality were likewise eternal, always there to be measured. Therefore prediction was knowing and, in a sense, not prediction at all—not a forecast, but a statement of fact. Knowledge was not caused in any physical sense, but rather discovered or revealed.

Einstein created deformations in the Newtonian world fabric, but did not break or tear it. However, by showing that the outcome of measurement depends on the relative state of motion between the observed and the observer, he demonstrated that the classical Newtonian standards of measurement (the primary qualities)—mass, length, time— have no absolute or permanent universal status. The reason is that measurement involving distant or fast-moving objects depends on the time it takes for a signal (the fastest is light) to go from object to measurer. For such objects knowledge is mediated by a communicating signal, and measurement requires some stipulations about the behavior of the signal. This is what special relativity is about.

Thus, Einstein rendered the Aristotelian concept of measurement, based on the unmediated comparison of like qualities, physically untenable as a universal principle. Concomitantly, he showed that the Euclidean geometrical principle of establishing equality by superposition could not be applied rigorously in the physical world. We cannot transport material objects across the barriers of great distance or differences of velocity to make direct measurements; instead, we receive signals about which we reason.

On the other hand, Einstein did not give up altogether the idea of superposition. He retained it for local (where the measurer is) time. Simultaneity, still an intuitive judgment, is local temporal congruence, but it does not apply outside the immediate vicinity of the observer. The "measuring instrument" for distant time is the signal, in this case, light. The unchanging quality, or the property, that makes nonlocal measurement possible is its velocity. Thus, the constancy or invariance of the velocity of light is the fundamental postulate of special relativity; it is the primary stipulation about the behavior of the signal that allows the physicist to make statements about (or measurements on) distant or fast-moving objects.[28]

Bridgman's ambivalent interpretation of special relativity was a response to Einstein's treatment of time, from which he drew what was both a scientific and moral lesson. Having understood that simultaneity can be judged only locally, Bridgman concluded that only measurements made "here and now" have real physical validity or tell us something true about nature. Any other propositions, he claimed, are either arbitrary—for example, the assertion of the absolute constancy of the speed of light, which cannot be proven experimentally—or they are artificial projections by extrapolation, as in the case of determining the properties of nonlocal objects via the behavior of light. (In this division between the "natural" and the "artificial," we may recognize the source of Bridgman's operational dichotomy with respect to the meanings of measured physical categories.) Bridgman did not accept the second part of Einstein's reasoning—the idea that a postulate about the speed of light can be the instrument for discovering something physically true. Again and again, he complained that Einstein had not defined a "clock," by which we must assume he meant a "universal clock," when in fact, this is precisely what Einstein realized that the physicist cannot do. Bridgman surely must have been sensitive to this, for otherwise he would not have speculated so seriously on the possibility of making velocity a primary measured quantity.[29]

The moral lesson is a corollary to Bridgman's "here and now" principle. He expressed it in his refusal to accept the legitimacy of reasoning from nonempirical grounds: the physicist who overreaches the bounds of immediate measurements does so at the peril of mistaking knowledge of a paper and pencil variety for genuine knowledge derived from sensory experience. What is not sensibly intuited is merely an artifact of intellect. Carried to a logical extreme, Bridgman's "here and now" principle could result only in a self-centered theory of knowledge, for the empirical "here and now" ultimately resides in immediate—at the farthest extreme, even unreflective—individual consciousness. We have seen that this was, indeed, the direction that Bridgman's thought took, although he stopped short of the final destination. If we do not take account of time and communication, there are two extreme possibilities, both of which Bridgman rejected. Either there is no time because there is no memory, in which case there is no knowledge because any "here and now" event (if it can be said that there is an event at all) degenerates into a hypothetical mechanical reflex (behaviorism), or else the "here and now" blossoms into a metaphysical eternity encompassing all knowledge, throughout all time (mysticism). Both alternatives, as we have seen, were unacceptable to Bridgman.

Bridgman's understanding of relativity was based on only part of Einstein's insight. Special relativity granted the Aristotelian definition of measurement authority only for local simultaneity. To be sure, Einstein did restrict knowledge of the physical universe to what can be communicated to the observer, but he did not thereby demote physical knowledge to the local historical realm of human activity (individual operations) or subjective consciousness (incommunicable private awareness). Rather, by defining distant time in terms of the velocity of light, he was able to overcome the limitations of the Aristotelian definition, extend the validity of the geometric paradigm for theoretical physical knowledge, and affirm the ideal of truth as a rationally determinate structure.

It was in quantum mechanics where the classical epistemology and its associated measurement principle—equality by coincidence—came up against their critical limitations and had to make room for a modern physical, process-oriented theory of measurement, where the vehicle that communicated knowledge did not conform to the ideal conditions assumed to define causality. The new epistemology, however, did not simply replace the old, but submerged it, creating a situation that forced physicists into a fierce debate over which, if either, should take priority. The most famous of these debates was the one between Bohr and Einstein, Bohr arguing for modernism and duality, Einstein for tradition and unity. Here was the physicists' version of the twentieth-century crisis of meaning.

Bridgman claimed neutrality on this issue, stating that two epistemologies were certainly acceptable, so long as one could keep straight where one or the other applied, although he was skeptical about whether anyone could actually do this.[30] However, it appears that such a compromise was in fact distasteful to him, for as we recall, he made it clear that a half-and-half mixture was not a "very impressive performance." Despite all of his apparent modernism, or at least his attempts at agnosticism, Bridgman, even more than Einstein, was committed to a traditional epistemology.

A closer look at the measurement-epistemology problem will make this point clearer and also delineate in what respects Bridgman's operational outlook acknowledged, but did not come to terms with, the underlying difficulty. This examination will touch on the subtle issue of what a scientific theory of measurement has in common with judgments about human nature and the relationship of mankind with the divine and thus provide some insight into the nature of the cultural crisis of meaning.

The classical ideal world of thought, or theoretical knowledge, was assumed to be everywhere present in a "divine sensorium," to borrow Newton's phrase, where all knowledge is present to the instantaneous awareness of divine intelligence. Mind and knowledge—coexistent, coeternal, outside of time. This is an ancient idea. It was also expressed in the writings of Galileo and Newton, translated, of course, into Christian terms. The basic concept is the same. True knowledge is eternal and divine. To the extent, then, that the human thought process could come in touch with divine thought, theoretical knowledge was similarly instantaneous, perfect, known in the immanent coincidence of mind and object. (Here we might sense something of the motivation behind the semantic programs of Klyce and Korzybski.) Object and concept were equal, that is, self-same. There was no question of communication, at least in a physical sense. Knowledge was simply there, transparent to intuition or, alternatively, revealed by God (a nonphysical event). Of course, while mathematical knowledge, not being about material objects, might be included in this epistemology, the same principle could not apply to knowledge of material objects themselves. This difference gave rise to numerous difficulties when it came time to explain how the physical world could so neatly be assimilated to the mathematical world.

These problems have been well rehearsed in the philosophical literature and are exemplified above in the discussion of the behaviorist and logical positivist attempts to articulate a scientific methodology. Here the focus will be on the relationship between communication and measurement.

We have already seen how special relativity challenged the epistemology of intuition by demonstrating that knowledge of distant or fast-moving objects depends on the physically possible mode and range of communication between knower and object. Nevertheless, for relativity, this communication was unbroken, continuous, and, therefore in principle, perfect or complete. In other words, the message was completely decipherable.

In quantum mechanics, knowledge again depended on communication, but this time from an unseen microscopic world, where even in the furthest imagination, measurement by comparison or the establishment of equality by coincidence was impossible. Worse than that, the courier (an atomic entity) from the atomic world delivered a discontinuous message that could not be interpreted clearly because the courier would not reveal its identity or its route unambiguously. The message could not be completely deciphered because the behavior of the messenger could not be completely understood.

If the experimenter asked "Are you a particle?" the answer was yes. But if the experimenter asked "Are you a wave?" the answer was also yes. Moreover, if sometime during the journey, the experimenter tried to trick the messenger into providing enough information so he could reconstruct its path by asking "Where are you?" the messenger would not disclose how fast it was moving, and if he asked "How fast are you moving?" it would not tell where it was located. This is the Heisenberg uncertainty principle. It formally states that the path of the messenger cannot be precisely retraced. The messenger will not even appear on command. When it does appear, it merely announces "here I am" by allowing its presence to be registered as, for example, a dot on a photographic plate or a click of a geiger counter. The observer in the macroscopic world could do little more than ask a question by setting up his apparatus in a specified way and then wait and count—count the arrivals of individual messengers from the atomic world and see where they congregate.

What, then, can we as macroscopic observers say is "known" after this exercise? What is compared, or measured? We can compare the number of arrivals at one place to the number of arrivals at another. But what is thus measured? What knowledge has been communicated? Simply that over time a messenger is more likely to appear at a place of dense congregation than of sparse congregation. We can even give a numerical value—a probability—to this likelihood. Furthermore, the pattern of congregation, or probability, will be consistent with the numerical limits given by Heisenberg's principle and is theoretically predictable. But is this knowledge? From this, what do we know about the microscopic world? We know what theory tells us. That, as Bridgman made emphatically clear, was not good enough. It was merely the outcome of paper and pencil manipulation and had no connection with his physical intuition.[31]

It was evident to Bridgman that the classical epistemology could not serve as a model for quantum mechanical measurement. Measurement could no longer be predicated on the possibility of making a primary intuitive determination of coincidence, that is, an immediate judgment of perceptible elemental equality—temporal, spatial, or qualitative— equality sensibly present in the "now." It was clear that the classical concept of measurement was not appropriate as a theoretical instrument for capturing the modes of either microscopic or macroscopic experience. At the same time, however, a mechanical account of the communication process was being invoked. Thus, to "see" a microscopic object was to "touch" it with another microscopic object, although no one could say just exactly what touching was, except to portray it as a disturbance or interaction. On the macroscopic level, measurement was waiting for and recording the touch of a communicating microscopic object, the outcome of the microscopic happening. It meant waiting for a message from the microscopic level that would be indicated by the arrival at the detector of a discrete signal, a signal that could only be counted and recorded at an arrival point, but whose path could not be reconstructed, supposedly because of an uncontrollable or indeterminate microscopic disturbance.

That was the problem. Since there was no way to know exactly how the individual communicating object traveled from its origin to the observer, when it would arrive, or even whether it was a wave or a particle, a disjunction was created between the macroscopic and microscopic worlds. Therefore, whatever was believed to have happened on the microscopic level, on the macroscopic level knowledge of the signals could be treated only statistically. The consequence was an altered relationship on the macroscopic level between theory and experience, in fact, a new epistemology, where probability was the theoretical object of knowledge, and counting was its basic operation. The microscopic level was still described in terms of a vocabulary based on the old epistemology. It was this situation that Bridgman characterized as being "not very impressive."

The new epistemology of probability could not rely on continuously transmitted (geometrically represented in ideal time) mechanical motion (motion that was not really motion, but an ideal space-time trajectory) to ascertain the outcome of a given set of initial circumstances. What were theoretically continuous were the not-quite-real probability functions. Therefore, the "now-present" properties that could be "classically" compared (but only by means of paper and pencil operations) were probabilities.

Furthermore, theory and measurement involved a coupling of a "now" and a "then," in "waited" time (rather than topological geometric time), where probability represented a weighted count of "now" possibilities for the prospectively measured "then," and where the possibilities of the theoretical "now" drop off into the past when measurement precipitates one possibility into a newly actualized "now." This relationship provoked much argument about whether the probabilities are "real," or if the act of measurement actually summons reality into the present. If it does, must it not create a regress of successively nested potentialities that can end only in an act of human awareness calling the entire series into being? Such questions emphasized the inapplicability of classical epistemology as a model for the measurement process. There was no now-present "equality" in the Euclidean or intuitive sense of coincidence of object and concept—its selfsameness. The timeless or instantaneous epistemology of eternal being cannot be physically imitated by measurement in "waited" time—the time of "noncausal," probabilistic projection.

What this means is that the metaphysical problem of knowledge had become transformed into a problem of communication. The question was no longer "How is knowledge possible?" but rather "How is communication possible?" and "What counts as communication?" From this standpoint, measurement has a new function, now as part of the organization and interpretation of forms of communication; that is, as part of understanding patterns of interaction, itself included. However, although Bridgman was clearly sensitive to the general direction of change, his "here and now" epistemological criterion prevented him from reconsidering measurement in terms of a process of communication, just as it had kept him from accepting the postulates of special relativity. In fact, we have seen that Bridgman's reasoning led him to the conclusion that knowledge is private and that in the final analysis, communication is impossible.

Similarly, according to his physical "here and now" principle of measurable qualities, the quantum mechanical probabilities could have no status as knowledge. Probability, Bridgman asserted, is in the mind. It is a psychological posture, a state of mental preparation for future action. Furthermore, the method of probabilities is applicable only to collections and has no meaning with respect to an individual happening. The individual sensed event is what is real, and it either happens or does not happen. It has no "property" of probability that can be operationally checked after the event to see if it was what it was thought to be.[32] In Bridgman's opinion, the need to invoke probability theory was one more sign of the feebleness of the human intellect, "that the logical processes of the human intellect [he should have said 'intuitive,' for the logical processes were functioning quite admirably] are not capable of meeting all the situations with which they are confronted in practice."[33] Moreover, the new epistemology gave priority to the "reality" of group properties, ignoring the absolute reality of the concrete individual event.

Bridgman was stuck. He knew that there was a tension between the two epistemologies, but the classical assumptions buried in his operational method would not let him choose, or even let him acquiesce to a dualistic compromise. If he chose the old, he could formally save his assumptions, but then he would have to admit that in the unseen microscopic domain, operations, being macroscopic, were useless, meaningless, as primary standards for judging physical reality. If he chose the new, he could retain the priority of immediate macroscopic experience, but he would have to admit a set of operations (counting and recording) that were without qualitative significance and perhaps even the paper and pencil validity of theoretical probabilities. He could have changed his epistemology to one based on communication—on the logic of questions and answers—but this possibility did not even occur to him, and if it had, we do not know in what light he would have considered it. If he had accepted Bohr's compromise or principle of complementarity, operationism might have made sense, if it had meant that to be (become) real, a quantity had to be measured. But that would have violated the classical, Aristotelian definition of measurement— comparison of (already existing) magnitudes of the same kind—and, what is just as significant, the Newtonian assumption that the laws of nature apply exactly to individual events.

So Bridgman equivocated. First, he asserted that meanings are "ultimately to be sought in the realm of classical experience," but then he admitted that an analysis of the workings of the brain of the observer probably cannot "be exhaustively described in purely classical terms" and therefore "one must hesitate to maintain that the necessary physical processes never reach down far enough into the small-scale structure of the brain to encounter quantum phenomena again."[34] With this equivocation, he implicitly accepted an ambiguous mechanical explanation of how knowledge is possible.

To appreciate this, let us notice that the idea that the Heisenberg uncertainty relation is a consequence of a microscopic disturbance caused by the measuring probe is basically a mechanical concept extended into the microscopic domain; we are not necessarily confined to this interpretation. The uncertainty principle is a formal statement of the macroscopically exhibited wave-particle behavior of microscopic phenomena or, alternatively, the conjugate pairing of their wave and particle aspects. As it stands, it does not restrict us to a mechanical interpretation, and in fact, even today, its interpretation is a subject of vigorous debate. On the other hand, the Heisenberg principle is not an absolute barrier to the precision of measurement per se but tells us that if, with a single experimental arrangement, we succeed in obtaining a given amount of knowledge about one aspect of a conjugate pair, a certain amount of knowledge about the other becomes unavailable. It is not so much a prohibition on knowledge as it is an indication that a particular form of questioning does not make sense.

Bridgman understood this. He also understood that the question originates on the macroscopic level with an experimental setup. The answer depends on the question, and the question depends on how the experiment is arranged. The theory formalizes the questions and enumerates the possible answers. The experiment locates one of them. What better example of the constitutive aspect of human intervention in physical knowledge could Bridgman have asked for? The tool is macroscopic, the question is macroscopic, the answer is macroscopic. This answer is just not returned in terms consistent with the classical model of continuous qualities and mechanical cause. Yet operationally it is what it is. No more, no less. How can we disagree with Bridgman's critics when they accused him of abandoning his own operational precepts? Clearly, he had something else in mind.

That something else was, indeed, a philosophical bias favoring the metaphysics of quality, not so far removed, after all, from the assumptions underlying the dimensional analysis debates. Bridgman's operational interpretation of physical reality took for granted the exclusive legitimacy of the classical mode of scientific questioning. He did not actively explore the possibility that the form of the question might have to be changed, in this instance, from the type, "What is it [the object] or how much of property x is present?—that is, What is its nature (its properties)?" to "Will I encounter an event under specified conditions if I do the following?" The latter, even though it seems to be more operational in form and spirit, was not adopted by Bridgman.

The most consistent and plausible explanation for Bridgman's attitude is that he was unable to abandon the classical epistemology. Nevertheless, having decided (more or less) that the new epistemology was inevitable, a product of scientific progress and therefore the outcome of the increased power of science, Bridgman adopted the position that the new was a "more realistic" modification of the old and, at the same time, a lesson in humility. It was a chastisement to overambitious cognitive expectations, a recognition of the futility of aspirations to a divine perfection of knowledge, aspirations that were, after all, unfitting to the imperfect condition of human existence. The result of this attitude was a moral preoccupation with the finitude of human knowledge that overshadowed his concern with philosophical adequacy. Once we accept this, we can understand why Bentley labeled him Hyde and not Jekyll, why Bridgman replied in essence that it is too bad if "I give you not bread, but a stone," why he told Klyce that he was not bothered by logical inconsistencies, and why he always insisted that he was not doing philosophy.

CHAPTER 9

Thermodynamics: The Outer Bounds of Physical Knowledge

To the Socratic man the one noble and truly human occupation was that of laying bare the workings of nature, of separating true knowledge from illusion and error. . . . Whoever has tasted the delight of a Socratic perception, experienced how it moves to encompass the whole world of phenomena in ever widening circles, knows no sharper incentive to life than his desire to complete the conquest, to weave the net absolutely tight. . . . But science, spurred on by its energetic notions, approaches irresistibly those outer limits where the optimism implicit in logic must collapse. For the periphery of science has an infinite number of points. Every noble and gifted man has, before reaching the mid-point of his career, come up against some point of the periphery that defied his understanding, quite apart from the fact that we have no way of knowing how the area of the circle is ever to be fully charted. When the enquirer, having pushed to the circumference, realizes how logic in that place curls about itself and bites its own tail, he is struck with a new kind of perception: a tragic perception . . . . —Friedrich Nietzsche, *The Birth of Tragedy* (1870–71)

■ ■ ■

THE FIELD OF PHYSICS that best exemplified Bridgman's operational attitude, even if it did not exactly uphold operationism in its original sense, was classical thermodynamics. It was his home science, the place where he felt he could relax and be comfortable. Here the furniture was macroscopic, plain to ordinary observation. In Bridgman's opinion, "thermodynamics, with its large scale operations, stands as a sort of prototype of the best we shall ever be able to do."[1] Its principles are the most general in all of science, the necessary conditions to which all other branches of science must conform. However, thermodynamics possessed another important virtue. It affirmed Bridgman's moral position. Not only is the conceptual framework of thermodynamics macroscopic, but, equally significantly, its principles or laws are the scientific "can't do's." They are, in fact, admissions of our "universal failures." They "smell more of their human origin."[2] Thermodynamics, according to Bridgman, is built on a recognition of human limitations, on the admission that certain things simply cannot be done.

There are many alternative ways to state the laws of thermodynamics. Bridgman preferred (as most textbooks on the subject do) the forms that express these laws as statements of impossibilities—it is impossible to create a perpetual motion machine of either the first or the second kind. More precisely, the first law (the conservation of energy) says that perpetual repetition of mechanical motion cannot produce any useful work, and the second law (the natural dissipation of energy or increase of entropy) says that it is impossible to make a cyclically operating machine (one that returns repetitively to its original condition) that, without having outside work done on it, does nothing but transfer heat from a colder body to a hotter one. In other words, a 100 percent efficient heat engine cannot be made, or more popularly, heat does not flow spontaneously from cold to hot bodies. (The third law, which Bridgman did not analyze, states that the condition for maximum efficiency cannot be realized.)

Thus the laws of thermodynamics summarize the limits of human ability to exploit natural motion. Even so, Bridgman mused, it seems strange that science, which is supposed to formulate the "regularities of *nature*," should be able to do so on the basis of what *human beings* cannot do. The physicist must wonder why it works at all. Indeed, commented Bridgman, it strikes one as "a verbal *tour de force*, an attempt to take the citadel by surprise."[3]

With this observation Bridgman began his analysis of the concepts of thermodynamics, published in 1941 in a book entitled *The Nature of Thermodynamics*, further distilled and condensed in various subsequent articles. His goal, he said, was to achieve *understanding* rather than to attempt to throw his analysis into postulational form. The latter, he claimed, was not the kind of clarity he personally found satisfying. Understanding meant discovering to what degree thermodynamic concepts are "verbal" or artificial (the outcome of paper and pencil operations) as opposed to being "real" (corresponding to physical operations).[4] If Bridgman had ever expected his physical operational criterion to distinguish between concepts that are admissible in physics and those that are not, his analysis of thermodynamics impressed upon him how deeply "verbal" categories have infiltrated science, so much so that he felt obligated to defend the "verbal" simply on the basis of its survival value.

The concepts of thermodynamics—energy, heat, temperature, entropy—are, in fact, highly abstract, and their meanings depend on a subtle interplay between the controlling definitions and carefully organized phenomenal changes. Moreover, the macroscopic concepts suffer from this paper and pencil contamination, this deficiency of immediate operational "reality," as severely as their microscopic, statistical mechanical counterparts, whose legitimacy Bridgman dismissed not only because of their statistical nature but also because he sensed that behind them was an unjustified extrapolation of mechanical ideas to the microscopic level. If the phenomenon of heat had exposed the inadequacy of classical mechanics to represent the experienced world, Bridgman reasoned, by what logic does it make sense to transport the same imperfect concepts to the microscopic level and expect them to furnish an explanation for what they could not do in the first place?[5]

Heat was, indeed, a troublesome phenomenon, a prankster—perhaps the devil himself, dressed in his red costume—challenging the orderliness of Newton's perfect world machine, and the discovery of the mechanical equivalent of heat in the early nineteenth century did not suppress its disruptive activity. The persistence of heat was the reason that the "can't do's" of thermodynamics had to be formulated as fundamental principles of science. However, this capricious renegade from ideality not only upset the cosmic balance but actively participated in bringing about the downfall of the classical epistemology of intuition—the Aristotelian formula according to which measurement was comparison of existing objects or magnitudes of the same kind, and equality was a (lower-level) judgment imitating the self-sameness of ideal subject-object coincidence. Without fanfare, the science of heat gave impulse to the transition from an epistemology of permanent qualities to the epistemology of communication and thus to the dependence of measurement on communication. Thermodynamics was an early defector from Newtonian orthodoxy and a precursor of quantum mechanics. For heat, it turns out, signals its appearance or allows itself to be detected only by its effect on something else—for the purpose of physical measurement, the thermometer. Temperature, the parameter that measures this effect, is not the same kind of "thing," quality, or magnitude as heat, and the precise relationship between the two is not as obvious as might be assumed. The thermometer is a reporter that gives a secondhand account of what it experiences.

We have encountered this difficulty once before, in the dimensional analysis debates; there it appeared impossible to decide whether temperature was to be considered a primary quantity. We are now in a position to appreciate the problem. One temperature, of course, can be compared directly to another temperature, but equality of temperature is not the same thing as equality of heat. To give a simple example, ice and water may coexist at zero degrees centigrade, that is, have the same temperature, but it requires an input of energy (heat or work) to transform ice to water even when the temperature remains the same. Furthermore, how much heat or work is required to effect the change of state depends on the conditions under which the process is carried out. Ice and water may be equal in temperature, but they are not equal in heat, if the latter can be said to mean anything scientifically precise at all. Heat, in fact, is not a quantity existing in the "eternal now" that can be measured by comparing it immediately or sensibly to another of its own kind. It is only palpable as an aspect of change, or motion, motion not as represented by the ideal (static) geometric trajectory of a moving body, but motion as a real temporal process.

Nevertheless, classical thermodynamics makes use of a device to avoid or leap over the passage of time, preserving the "now-ness" of its parameters. It circumvents temporal considerations by referring its measurements to macroscopic equilibrium states—states that are stable under the selected boundary conditions and appear outwardly to be unchanging. These states can be imagined to be analogous to the spatial positions in classical mechanics that determine the potential energy function. Thus thermodynamics takes notice of and records only differences between such states, assuming that what happens in between happens in an ideal way. (Why this is permissible is a more technical matter, but the analogy is still useful. Just as potential energy, say in a gravitational field, is determined only by position and not by the route by which this position was reached, so thermodynamic parameters are independent of the path by which the change of state was brought about.) However, by adopting this approach, thermodynamics sacrifices the ability to make any statements about the amount of time involved in the process of change. To know when a process is completed, one is directed to wait until no more macroscopic change is detectable. This could be anything from a very short to a very long time, but as far as the equations of thermodynamics are concerned, elapsed time is not a consideration.

In addition, the measuring instrument—the thermometer—takes part in the same process. Therefore it is, strictly speaking, an observer that interacts with the system it is observing. Before a valid measurement can be recorded, the thermometer must have come to a state of equilibrium with the system whose temperature it indicates. The thermometer must assume the thermodynamic condition, or have communicated to it the thermodynamic state, of the system whose condition it reports or amplifies. Equality means being in the same thermodynamic state or condition. However, the report is communicated to the experimenter in a different language, the language of the property affected by the transfer of heat. The only reason that the thermometer does not itself affect the measurement or take an impractical amount of time to reach equilibrium is because it is made negligibly small in comparison to the system to which it is coupled. In sum, then, for practical purposes the measurement is nonintrusive, although theoretically it is not precisely so. Furthermore, the sensible measurement is not of heat, but a thermometric property—that is, one that externally responds in a detectable and reliable way to an internal physical change associated with the transfer of heat.

The discovery of a quantitative "mechanical equivalent" for heat, the accomplishment that made possible the formulation of the principle of conservation of energy, or the first law of thermodynamics, was the source of the dimensional analysis quandary. The idea of a mechanical equivalent of heat created a problem by reducing heat in principle to mechanical motion, or work, which is defined in terms of mechanical parameters—mass, length, and (ideal) time. (It is necessary to disregard the fact that the mechanical concept of "force" already insinuates an influence or communication.) Thus it implied that for the purposes of dimensional analysis heat is no different from work and can be treated dimensionally as a mechanical "compound," that is, that its measure is nothing more than a combination of fundamental mechanical parameters.

But the second law of thermodynamics contravened—imposed restrictions on the first law, according to Bridgman's way of speaking—by asserting that this is not an accurate representation of the motion associated with heat. Otherwise, perpetual motion of the second kind would be possible, at least in principle. But it is not possible, even in principle. Therefore heat must be something different, and temperature must be a unique, independent dimensional quantity. The dilemma was to decide whether heat is translatable dimensionally as a form of mechanical work or requires the introduction of an independent dimensionality, temperature. As noted earlier, the debate was never resolved on the basis of the terms that originally gave rise to it. The argument faded into the background with the ascent of statistical mechanics and quantum mechanics—with the increasing prominence of the physics of microscopic phenomena and the new metaphysics of process.

However, Bridgman was once again confronted with the lack of clarity in the distinction between work and heat when he undertook his examination of thermodynamic concepts. Beginning with the first law, he asked What is energy and what does it mean to say that it is conserved? He quickly discovered that there is no simple answer. In the first place, he noted both a physical and logical difficulty in the idea of the conservation of energy as based on the interconvertibility of work and heat. Minimally, it says that "we recognize energy in two forms, thermal energy and a generalized mechanical energy, and it is the two forms together which are conserved." That much, Bridgman said, was acceptable. However, to go further and postulate that "heat energy is mechanical energy of the small parts of which matter is composed" is to revert to the statement that the conservation of energy is "nothing but the conservation of mechanical energy, mechanical energy of the ultimate particles." Why, he queried, should the physicist be "so pleased with his idealization and so sure that he had the right answer?" There is no experimental evidence that microscopic systems do not have the same imperfections as macroscopic systems. And without evidence to the contrary, this means that the unsolved problems of the macroscopic level must eventually be reiterated *ad infinitum* on successive microscopic levels, until the solution is ultimately a metaphysical one—the postulation of discrete particles with no internal degrees of freedom.[6]

To Bridgman's mind, therefore, the logic was faulty and the order of explanation was backwards. It was tantamount to trying to snuff out the problem by burying it in the microscopic, while actually passing the buck to metaphysics. In addition, the fact that the properties of microscopic systems are probabilistic, or group properties, formally denies them the concreteness expected of single operationally distinct occurrences. All meaning, all explanation, Bridgman insisted, must originate on the macroscopic level of human experience. It will not do to explain the macroscopic in terms of the microscopic. "The reason is simply that we, for whom meanings exist, operate on the macroscopic level." One must recognize that "both quantum and classical statistical mechanics [are] paper and pencil devices, of which it is meaningless to ask whether the description they afford of microscopic phenomena is true or not."[7]

However, if the physics of the submacroscopic world inevitably led physicists on a lively paper and pencil chase toward the ultimate metaphysical atom, the more stately march toward the supermacroscopic, cosmic physics and its accompanying logical ascent toward more comprehensive, and ultimately universal, all-inclusive principles, Bridgman discovered, was also supported by a paper and pencil scaffolding that eventually disappeared into the insubstantial clouds of metaphysics. Energy, as a universal relationship and its categorical conservation as a condition framing the portrait of physical phenomena, seemed to be everywhere and nowhere, operationally undelineated, without specificity, and without a comparative equal. In short, there was no way to measure it. It was out of the question for Bridgman to consider energy a metaphysically prior and real subsisting universal relationship that creates the stage for phenomenal activity. He was too much a physicist for that. He knew that physics is of the world of phenomena, that without phenomena there would be no physics and, it might be added, no individuality. Yet he was puzzled, as he admitted, by the dependence of physics on its negation—the absence of phenomena, formalized by the stated universal impossibilities of thermodynamics. How, indeed, can a universal impossibility be seated at the foundation of a positive science?

Bridgman's search for physical meaning was predicated on the reality of the phenomenally immediate, the "here and now," and he made it clear that thermodynamic properties are "now" properties. However, "now-ness" was not sufficient for operational identifiability, for the "now" was not necessarily "here." Thus, Bridgman's attempts to find an instrumental operation corresponding to what he called "generalized" energy failed. He could find no instrumental counterpart to energy in the most universal sense. Bridgman assumed that for something to be real it must be somewhere, it must be localizable, it must have a place. But energy is not a substance; it is a function of a "state couple" (Bridgman's term for a pair of thermodynamic states, or states possessed of a potential difference). On this point, Bridgman laid great emphasis. Therefore energy must be "located" in a relationship. It must be "localized" in the difference between two distinct thermodynamic states or "positions," or within a region or isolatable system whose boundaries are defined with respect to an "outside." Only then can we talk about a difference. Nonetheless, Bridgman continued to search for the place where energy is, a place somewhere in space, similar to the position of matter. He was able to achieve a limited success, enough so that he could claim that for specific instances, energy is localizable. However, he could not do the same for the abstract concept energy. Here, he thought, to completely generalize or universalize energy and affirm its conservation would require in the last resort a summation over all possible state couples or relative regions taking account of all possible equivalent forms, a procedure so complicated that it defies instrumental possibility (and which would take, he neglected to add, an infinite amount of time).[8] It can only be done in thought or hypothetically on paper, asserted *a priori*.

Energy, Bridgman discovered to his obvious dismay, is as much an affair of metaphysics as of laboratory operations. In fact, in its most compact or efficient form, the principle of conservation of energy can be stated as the metaphysical principle that cause is equal to effect. Bridgman stumbled onto this relationship during his attempt to localize gravitational energy and did not quite recognize it. He was nevertheless made uncomfortable enough by it to frame a question, imagining it to have been raised by the classical mechanist Maxwell or Kelvin, who might "want to ask by what mechanism potential gravitational energy gets transformed *in situ* into kinetic energy (that is, how cause is transformed into effect)." Bridgman, who was himself vicariously none other than this classical mechanist, could offer no answer. No matter how he looked at it, energy turned out to be a "verbalization," a paper and pencil construction, an artifact of human thought to which he could not concede "'physical reality' in the fullest sense." The statement of the first law "in the grandiose form 'the total energy of the universe is constant,'" he complained, is "a statement which seems to me to be mostly a pure verbalism." Thus, after presenting in detail the ways physics measures, calculates, and talks about energy, he could advise only that one should properly speak of an "energy function" rather than "energy" to remind oneself of the conventional or man-made character of energy and to prevent oneself from thinking of energy as a primary universal substance. He himself frequently drifted into the language of substance (for which he apologized) as, for example, when he talked about the "amount of 'change-of-energy'" and asked where it is and how we can detect its movement or flux.[9] Bridgman knew perfectly well that he had not escaped the hold of the classical habits of thought he was trying to expose and replace.

The problem Bridgman had begun with at the outset of his analysis, the macroscopic distinction between heat and work, remained unresolved. The microscopic distinction, heat as incoherent motion, work as coherent motion, was not relevant to the solution of the difficulty as he perceived it. To him, the proof of the conservation of energy depended on being able to directly measure the independent value of each term in the canonical form of the conservation principle, the expression which states that the change in the energy of a system within a bounded surface is equal to the heat crossing the boundary plus the work done on the system. If the value of each term could not be independently measured, Bridgman felt, the equation would be nothing more than a definition or convention. However, Bridgman noted that there are situations where heat and work cannot be measured separately or independently. This, together with the fact that "no instrument responds to [a] stream of generalized energy," meant that there is no way to know for all cases that change cannot be effected by influences other than heat or work. Therefore, Bridgman asserted, "it is not known whether conservation is true or even whether energy has any meaning."[10]

Bridgman eventually decided to designate energy a "wastebasket" concept, a name used to talk about a collection of operations including both instrumental and paper and pencil operations that have "in common a common purpose and common use" in theoretical manipulations.[11] He was thus forced to conclude, to his utmost dissatisfaction, that "the paper and pencil element in the energy concept is inextricably interwoven with what we should like, if we could only make it mean anything, to call 'purely physical.'"[12] In other words, energy is not altogether physically real, even though we wish it were.

In a letter commenting on *The Nature of Thermodynamics*, Rudolf Carnap objected to Bridgman's idea of "reality" as being unclear. He sug-

gested further that perhaps "reality" is a concept that is entirely superfluous for the theory and practice of science and advised that it might
have been better for Bridgman to have avoided using the word altogether.[13] To this Bridgman replied that he imagined himself as the average physicist talking to himself and that, as a matter of fact, "the average physicist of my acquaintance does think in this way." However, he
added, it was his goal to make this average physicist see that "what he
means by the 'physical reality' of some process occurring at the boundary of a region to which he is applying the first law is merely that simple
physical instruments placed there give appropriate readings."[14]

Nevertheless and in sum, not only was Bridgman plagued by the
tendency to lapse into an Aristotelian language of substance and by the
inclination to expect mechanical causation, but his analysis of the first
law of thermodynamics exemplified the well-known dilemma of empiricist logic—the problem of induction: since all instances of a universal ideal category or relationship cannot be rounded up and presented
as proof of its general validity, we are consequently not permitted to
assert its absolute truth. To do so exceeds the limits of empirical evidence. This logical problem belongs to the metaphysics of a static world
picture, where the challenge to establish the truth of ideal categories is
based on the assumption that they are permanently subsisting entities,
"here and now" realities waiting to be discovered, inferred from and
substantiated by experience. The trouble is that there is not enough experience to do that. Experience is finite and thus inconclusive. Experience at best can only echo, but not establish with certainty, the truth of
the ideal, and an echo is not convincing evidence for empiricists, even
those inclined to accept the echo as a reflection of the truth being
sought. Bridgman's doubt about the conservation of energy is an illus-
tration of this dilemma.

The second law of thermodynamics inverted the challenge; it
changed the logical/epistemological question from "How do we prove/
know that the ideal is true when we cannot display all instances?" to
"How can we assert/know anything about the actual if no comparison
to the ideal can be formulated?" The question refers, of course, to irreversible processes, happenings in actual time, events that occur only
once in history. Strictly speaking, this includes all of experience. In this
case, however, the ideal to which experience is to be referred is no
longer an ideal category or relationship but rather the standard of perfect cyclical motion. Although there are repetitive processes that can

reasonably be assimilated to perfect cyclical motion, irreversible processes, which are by far more common, cannot be made to imitate this perfect motion. "The admission of general impotence in the presence of irreversible processes appears on reflection to be a surprising thing. Physics does not usually adopt such an attitude of defeatism," remarked Bridgman. "Of course this may be made a matter of words if one chooses, and one can say that thermodynamics by definition deals only with equilibrium states. But this verbalism gets nowhere; *physics* is not thereby absolved from dealing with irreversible processes, or we from trying to understand a little better this anomalous situation."

Bridgman attributed this impotence of physics to a complexity so great that there is a breakdown of the ability to formulate adequate descriptions of irreversible processes or to an origin so deeply set in "the structure of things that any detailed study by instruments of the thermodynamic scale is forever ruled out." But not all irreversible processes, he claimed, are scientifically inaccessible. Some irreversible processes, in his opinion, should be amenable to "treatment by methods thermodynamic in spirit," if not in name.[15] In this opinion he has been vindicated by the development of modern nonequilibrium thermodynamics. He need not have added the qualification.

On the other hand, Bridgman's judgment of the theoretical cornerstone of the second law of thermodynamics, namely, the reversible cycle, was not as embracing. In fact, he tried to brush it aside as a minor element, a "more or less accidental fact" deriving from the role of the ideal heat engine in constructing the proof of the second law or, secondarily, as an artifact of the mathematics involved.[16] This is not surprising, given Bridgman's empiricist viewpoint. The reversible cycle, the standard of perfect motion, infinitely slow, yet not time-consuming, a motion characterized by changeless change—motion completed without motion—is an ideal, impossible in actuality. However, it is the standard with respect to which the concept of entropy is defined. Entropy, the signature of irreversibility, strange as it might appear, is defined classically only for "reversible" processes. The entropy of irreversible processes is not defined. (Modern nonequilibrium thermodynamics defines a state of minimum entropy production to deal with a selected class of irreversible processes. It does not redefine entropy.) This seeming paradox is the consequence of the classical requirement that entropy should be a state function, independent of path and therefore of time. Thus, entropy is "precipitated" or "deposited" by the passage of time. It tells us that time has been used but does not itself participate in the movement of time.

Bridgman thought that the reversible cycle was a misrepresentation of empirical reality. He argued that if one were to insert into the hypothetical cycle a ratchet, a mechanical device preventing backward motion, nothing at all would be changed physically. The process would continue in the forward direction just the same, but now it could not be considered reversible and therefore the theoretical proof of the second law could not be carried out. The mathematics, he asserted, is impotent to recognize this physical constraint, and consequently reversibility is merely a formal requirement having nothing to do with the actual physical situation. The idea of reversibility should be replaced, Bridgman argued, by the concept of recoverability of initial conditions.[17] "Recoverability," unfortunately, was an ill-chosen word, for it would surely have been contrary to Bridgman's intentions to imply that the original condition was actually restorable, that time could be undone. He knew that, in time, nothing can be put back where it was. What he must have meant was reproducibility—imitation.

There is, as Bridgman argued, no such thing as a reversible cycle. It *is* quite impossible. But this is a matter of history, not of mechanics or ratchets. Reversibility is not a condition demanding that the possibility of backward movement is necessary for a process to occur (something Bridgman clearly perceived), but rather a requirement deriving from the fact that thermodynamic measurements have meaning only for equilibrium states. The concept of reversibility is a way to extend the reach of thermodynamics to active processes by introducing a hypothetical or ideal process so slow that it is always neutrally poised in a state of equilibrium. It is a formal way to represent change that approximates rest, change so slow that it creates no disturbance in the surroundings. This must be change that does not communicate outside itself, change, indeed, that we would not see. The idea of change that is not really change, change without concomitant or "running" effect, even as an ideal, was surely too paradoxical a concept for Bridgman to seriously consider, although it was perfectly consistent with the representation of time as it is treated in Newtonian mechanics.

However, although the forward direction of history, as Bridgman so graphically illustrated with his example, does not require mechanical enforcement, knowledge, on the other hand, does require an organizing standard, or ideal reference principle. Otherwise measurement would establish nothing. Here is where Bridgman turned off his operational spotlight. He did not use it to sweep through the full scope of the subject. Bridgman was not of a mind to entertain the concept of a reversible cycle as a necessary though impossible ideal without which scientific knowledge cannot be formulated (as, for example, in the case of irreversible processes) or as an ideal that makes the science of thermodynamics itself possible. To him, the idea of a reversible cycle was an accident of technique and, ultimately, an artificiality imposed by the formal inflexibility of mathematical language.

Nevertheless, the reversible cycle is far more than an accident of historical and logical circumstances. It is a subtle transformation of a very old idea and the result of genius that was able to translate a purely metaphysical principle into a productive scientific principle and, moreover, to use it and negate it at the same time. Here is a clue to answering Bridgman's opening question: How can a statement about what is impossible form the basis for knowledge about nature? A more global view of the history of the philosophical problem of natural motion will facilitate discussion and help us to follow Bridgman's subsequent reasoning.

Aristotle had divided the cosmos into two major dominions—the divine celestial sphere where motion was eternal and unchanging and the sublunary world of corruption and decay. Each realm had its own principle of natural motion; that is, motion that proceeded on its own or was unenforced and that did not require explanation. Natural motion in the celestial sphere was cyclical, whereas natural motion in the sublunary sphere was rectilinear and either away from or toward the earth, which was at the bottom of the hierarchy of celestial perfections. Whatever was contrary to natural motion was violent or forced motion. Aristotle went to great lengths to prove that there is only one perfect motion, and that is circular motion.

Newton erased the Aristotelian distinction with his law of universal gravitation and showed that the celestial and sublunary motions were governed by the same principle. He did this by introducing an "imperfect" motion—uniform rectilinear motion—as the universal standard of natural motion. This is Newton's first law. As a consequence, celestial cyclical motion became a form of enforced or violent motion, a gravitationally forced deviation from the newly postulated "perfect" rectilinear inertial motion that was now, according to Newton's first law, the universal natural motion.

However, in casting a slight haze over the purity of the heavens, indeed, by establishing a common mode of influence or communication between the heavens and the earth, Newton did not abandon the ideal of eternal celestial perfection. Gravitational influence acted instantaneously, light took no time to carry its message, motion was not in principle constrained to a one-way path in time. That was because there was as yet no scientific concept of real elapsed time. Ideal time had no actual direction. Newton's world was a divinely created clockwork that still operated in Aristotle's celestial sphere, outside of time, untouched by the sublunary motions that cause corruption and decay. Of course, this is not exactly correct. Newton knew that the clockwork mechanism could run down, but he trusted that God, in His providential wisdom, would see to it that this would not happen. (It is also worth noting that Newton's universe was infinite. The principle of conservation of energy was not yet a part of physics.)

It was thermodynamics, the science of heat—heat, the companion of corruption and decay—that eventually discovered time and imperfection and that closed the universe again. Thermodynamics gave science a concept of "actual" natural motion, and it did this by defining what it is not. Thermodynamics, a nineteenth-century invention that grew out of the investigation of the efficiency of the steam engine, is the physics of nonideal physical processes, of temporal phenomena where imperfection—heat—leaks from around the edges of ideal mechanical motion, from the cracks in God's not-so-perfect clockwork. The frictionless motions of Newtonian ideality do not occur in the world of temporal experience. The eternal cycles of timeless time do not represent the progression of natural processes. If there were such motion, says the first law of thermodynamics, we would not know it, for it would not communicate with us; it would not transfer any of its motion to us. Therefore, it is impossible. There can be no perpetual motion machine. Natural motion, says the second law, is expansive, indiscriminately communicative, directed away from itself, outreaching. It spends itself in a profligate manner, automatically spreading out to where it is not, using (or perhaps, creating) time. We may exploit (order) some of this natural motion, but we cannot divert all of it to our own purposes. This is the second impossibility. Energy, the fuel for time, the ideal difference that makes action possible, is not completely renewable in the world of phenomena.

Thus, thermodynamics discovered that there is a price to pay for activity. Its currency is time and energy, and Newton's God was not around to turn the clock back to zero or to restore the energy. Nor was Aristotle's divine abstract final cause available to lift the motions of the world of activity toward ideal perfection. The vision of the universe as a divinely contrived machine had no place for final cause. A machine is not motivated by a sense of purpose. It receives its motion from an original cause, not a final cause. Nevertheless, a machine with a finite source of energy (one subject to a conservation condition) will eventually grind to a halt because it leaks heat. The potential for change will be used up as history inexorably follows its course. The cyclic return to the ideal pristine revitalizing origin is impossible. Time is a one-way street into the future, and as time ticks on, energy flattens out, becomes impotent, useless. The universe will suffer a "heat death," come to a dead level, suffocate from uniformity—that is, if the universe is nothing more subtle than a steam engine, or if it has no infinite, constantly renewable source of motion (a possibility that Bridgman did not categorically rule out).

To the degree, then, that Newtonian mechanics was the physics of the celestial sphere, classical thermodynamics was the physics of the sublunary domain, the world of corruption and decay. The principles of impossibility mark the limits of the "sublunary" realm, the outer bounds of phenomenal experience, and thus the outside limits to physical knowledge. They tell us where phenomena cease, where physics stops. To cross over them is to enter the world of pure metaphysics. They delineate the boundaries of the knowable universe from the perspective of human experience. They tell us how much we can do by telling us what is not nature—what we cannot observe, and by extension, what we cannot make happen. (In a similar manner, Einstein's special relativity tells us how far we can "see.") That is why Bridgman could justifiably say that they "smell more of their human origin." He meant that these principles originate in an explicit acknowledgment of what human beings have realized they can and cannot do and know. They are not metaphysical assertions. Stated as principles of impossibility, the laws of thermodynamics do not assert the possibility of knowledge of a metaphysical or superphysical existence.

Bridgman wanted to take this reasoning one step further. He wanted to impress on his readers that the laws of thermodynamics themselves (like all scientific laws) are expressions of a human perspective and whatever universal significance is attributed to them is the result of paper and pencil extrapolation. The operations that give them validity are macroscopic, local, human operations. Therefore, there is no warrant for imputing even to them a generality of universal scope. That judgment is not one a human being can make, for he is not in a position to know this. Man cannot, as Bridgman concluded in his *Harper's* article, know as a god knows. He is constrained to know and act within the boundaries defined by the reach of human operations. He cannot lift himself above or place himself outside these limitations. His thought is not equipped to do this. Man, Bridgman asserted, is "isolated on an island of phenomena between the very large and very small, which he cannot transcend because of the nature of thought and meaning themselves." [18] He is, Bridgman might have said, confined to the Aristotelian "sublunary" world.

Bridgman's position was called into dispute by the eminent Catholic philosopher Jacques Maritain, who was, fittingly enough, a neo-Aristotelian (more accurately, a neo-Thomist; the circumstances of the confrontation are described in Chapter 11). The issue was not dissimilar to the one that dogged Bridgman in science, namely, the dependency of knowledge on being able to formulate a relationship to the ideal. Bridgman was so intent on showing that the ideal is manufactured by the human mind that he did not make the effort to consider how the ideal might at least have a necessary function, if not ontological presence (reality, in other words). Thus, he lost sight of the lesson of irreversibility and relinquished the opportunity to answer his original question—How can the thermodynamic principles of impossibility work at all? Perhaps he did not want to admit the answer—they work because they have definable implications—or, perhaps this is what he meant by calling them a verbal *tour de force*.

The example of irreversible processes illustrates that science cannot know what it cannot define. The advance of science depends at least as much on the formulation of new or more refined definitions as it does on turning up new empirical data. Thermodynamics can define a reversible process, and by so doing, it can state what an irreversible process is not. What is imperfect, unique, and individual is not definable. The ideal is what is definable. Nature, in its manifold of individual expressions, cannot be defined. It is impossible to define everything. Thus, for thermodynamics—the science of nonideal processes, the science of all of nature, of all phenomena—the most general principles, if they are to apply to any phenomenon whatever, must be expressed in terms of what nature is not, in terms of what can be defined. The lesson of irreversibility is that to know the actual, we need to reflect it against an ideal, and in this respect, the ideal cannot be nothing, it cannot be less "real" or less "true" than the knowledge it makes possible. Yet this was precisely the conclusion that Bridgman wished to deny.

Therefore, on his own assumptions, the alternative was to locate the dividing line between what is real and what is merely verbal or "ideal," something he found that he could not do. There is no such clean dividing line. Our analysis has thus brought us full circle back to Bridgman's stated goal in the introduction to *The Nature of Thermodynamics*, and Bridgman's analysis left him with nothing more precise than a moral pronouncement about the limits of human knowledge.

However, let us not underestimate the difficulty of the problem facing Bridgman—how to think about microscopic molecular motion, motion with no Aristotelian or Newtonian archetype, motion in a realm where there is no rest, no equilibrium, just a chaos of random fluctuations. What operations might make knowledge from molecular chaos? This situation, emphasized by the challenge posed by the Maxwell demon (the hypothetical being who can cheat the second law of thermodynamics by trapping the most energetic fluctuations, the "hotter" molecules), Bridgman said, was the "crux of the matter."

The mere existence of fluctuation phenomena shows peremptorily that the cardinal assumption of thermodynamics can be only an approximation. Temperature, the fundamental concept that distinguishes thermodynamics from other disciplines, is defined through the equilibrium which any two bodies reach when allowed to remain in contact with each other indefinitely. But two bodies do not come to eventual equilibrium; the large-scale bodies of ordinary experience come to apparent equilibrium, but if the bodies are small enough we can see their failure to come to equilibrium in their Brownian motion under a microscope. Bodies sufficiently small do not come to equilibrium and hence do not have a temperature. In the same way, if we penetrate far enough toward the atomic domain we find that the steady pressure recorded by our pressure gauges on the walls of an enclosure containing a gas gets resolved into a rain of microscopic impulses. Hence the concepts of thermodynamics have a meaning only in the realm of not too small instruments and systems, and in this realm their meaning cannot be logically sharp because they must always be infected with some of the fuzziness which they obviously acquire when made small enough.[19]

In other words, if we make our instruments delicate enough, we find that equilibrium never occurs. Therefore, Bridgman concluded, "the validity of thermodynamics is in some way connected with the scale of our measuring instruments." The universe of thermodynamics is the "plateau" where fluctuations smooth out, become insensible, where our instruments do not detect microscopic motions. "The great laws of thermodynamics, however, are formulated with no recognition of the existence of this plateau [they were formulated 'too soon,' as he phrased it elsewhere], or any attempt to instruct us as to how near we may get to the edge of the plateau without danger, or what sorts of effects we might expect if the inhabitants of the outer regions should invade this plateau."[20] The conclusion, then, must be that the laws of thermodynamics are not exact, not universal.

What about the Maxwell demon? It is, after all, a hypothetical microscopic being who must see the molecules by some method, who must operate a trapdoor whose motion must not be subject to the restrictions placed upon all mechanical motion, who must possess some kind of intelligence. It seems pretty clear that this is a paper and pencil invention with little physical validity, Bridgman reasoned, except for two empirical facts: first, the fluctuation phenomena can be observed in the form of Brownian motion, and second, small-scale phenomena can be amplified and thereby used to trigger large-scale events. The microscopic fluctuations are real, whether or not the Maxwell demon can act as hypothesized, and therefore the idea that the second law of thermodynamics can be violated is not in principle out of the question. Perhaps some kind of ratchet device could be employed for this purpose. Bridgman did not accept the argument that a ratchet would also be subject to fluctuations because, he said, the equations of motion that have been extrapolated to the microscopic domain simply do not have anything to say about ratchets. Furthermore, he thought it possible that his ratchet "could be constructed out of phenomena which had not yet left the plateau," that is, phenomena of a different order or kind than the phenomena exhibiting the fluctuation.[21]

Whatever the merit of his position, these are mechanical ideas, and Bridgman was applying them to the domain where he himself argued that mechanical ideas do not necessarily pertain. Nevertheless there was a purpose to his line of reasoning, and that was to set the stage for an assault on statistical mechanics (and quantum mechanics) and what he believed to be the error of probabilistic logic. This may appear to be a non sequitur, but behind it is his objection that probability does not tell us anything about concrete individual situations, which is what we really want to know about. In fact, Bridgman was trying to retain the validity of the classical macroscopic epistemology and find a reason to discredit the newly emerging epistemology, first, because its objects are not directly measurable and, second, because it did not recognize the individual event as its primary object. Introducing the ratchet was a way to attack the mathematics as being inadequate to the actual situation. This, we recall, was the strategy he used to dismiss the importance of the reversible cycle. Bridgman's logic may appear to be a bit obscure, but it should become more understandable shortly. The point to remember is his motivation to make mathematics, which is just a paper and pencil activity, the culprit.

Two epistemologies, Bridgman had claimed, were not unacceptable, but what he feared (and what he thought he saw happening) was that the "statistical epistemology" might be taken to be "more correct" than the epistemology of common sense,[22] when both are equally correct with respect to the corresponding context or equally incorrect with respect to any absolute claims. Nevertheless, there can be no doubt of Bridgman's allegiance to the classical epistemology, despite his reluctant acknowledgment of the reality of microscopic phenomena. Otherwise he would not have made such a point of arguing that man is isolated on the "island of phenomena" and that all else belongs to the paper and pencil domain.

Yet to assert that upon this island of phenomena, which is none other than the "plateau" of classical thermodynamics, rests the entirety of possible scientific knowledge, to restrict the scope of scientific knowledge to the categories of macroscopic perception, is surely too severe, even granting that scientific knowledge must eventually be validated by some form of macroscopic experience. Furthermore, to demote the principles of thermodynamics to the status of mere approximations because of the discovery of phenomena unknown at the time they were formulated is to fall prey to the inductivist fallacy and to ignore the potential for extending their applicability through new definitions, new interpretations. The problem is difficult, to be sure, and the question of the possible incommensurability between the two modes of representation—macroscopic and microscopic, or deterministic and probabilistic—obtrudes an element of irrationality offensive to the classical sensibility. But the solution cannot be to withdraw into a fortress and deny legitimacy to the new growth of reality, to refuse to behold its luxuriance because its form is alien to cultivated sensibilities.

Furthermore, Bridgman knew better. He knew well enough from

his experience with measuring instruments in the laboratory that one instrument does not necessarily have the capability to sense the entire range of the property under consideration and that it often needs to be replaced by another when its limits have been reached. He never argued on that account that the results obtained by the second instrument are less true or that it does not measure the same property measured by the first, provided the proper calibration has been carried out. It would have gotten him scientifically nowhere to claim that pressure measured by the primary piston gauge is not the same property, pressure, measured by the resistance in the manganin wire. The experimentalist is always looking for new ways to extend the range of measurements. Why not allow the same freedom to the theorist? Part of the answer is that Bridgman did not accept the possibility that ideas—principles and categories, even languages—can be measuring instruments, too.

Now there are, indeed, good reasons for wanting to believe that the second law of thermodynamics is not final or absolute. For one thing, it would be comforting to be assured that the heat death of the universe is not inevitable. For another, more practical reason, it would be useful to find relief from the restrictions on the efficiency of energy conversion. The statistical picture of heat as random molecular motion suggested to many physicists a possible way to escape from the second law, to circumvent entropic decay. Thus was invented the Maxwell demon, which can sort out preferred motions from a random mixture and create the potential for work from a source declared out of bounds by classical principles. This, however, is not a task that can be entrusted to a mere demon, especially a demon who has not been given the power to undo and reconstitute time. The whole project has too much the aura of cosmogony.

Perhaps here is the root of Bridgman's argumentative strategy, why he sought to link his critique to the inadequacies of mathematical language. We may recall that during the dimensional analysis debates, the cosmogonical implications of Tolman's principle of similitude were what motivated Bridgman to publish his opposition. Whether similar considerations played a role in this case cannot be determined. However, it is a fact that Bridgman's objections to the statistical epistemology were framed in terms of the inadequacy of abstract categories, of the inexactitude of language, in particular, the language of mathematics, instead of the activity of the Maxwell demon. In this way, he could, consciously or unconsciously, avoid direct entanglement in metaphysical issues. Accordingly, then, if the second law of thermodynamics can be circumvented, it is because the law is not exact in the first place, and this has to do with the arbitrary nature of mathematics and the misfit between mathematical categories and actual phenomenal experience. It is not a problem of physics. Physics must admit whatever happens. The problem is that the language of statistics and probability simply does not represent a true record of events. And since it says nothing about ratchets, the question of whether they can be employed to exploit the fluctuations of molecular motion is an open one.

Yet all this seems beside the point. It does not confront the issue of the character of the new epistemology or what it has to do with the second law. The inductivist logical objection is the same whether the members of a class are macroscopic objects or microscopic sets, whether they are individual elements or relative states of collective organization. Thus, for example, Bridgman's complaint about the idea (basic to enumerating probabilistic outcomes) that certain processes, such as water freezing on a fire, which have never been observed and which no one realistically ever expects to observe should be assigned a positive probability, even if it is extremely small, only reiterates the inductivist position. It does not identify what is disturbing about the new epistemology.

Nor does the operational method successfully deal with the issue of what will be the paradigm for measurement or knowledge in the new epistemology, that is, what "perfection" measurement will imitate. It cannot be an operation (as Bridgman evidently intended), at least in Bridgman's sense, for the simple reason that his operation is modeled after Aristotelian principles—measurement as comparison of objects of identical quality, knowledge as the ideal Aristotelian coincidence or self-sameness of thought and object. The examples of special relativity and quantum mechanics have already made that abundantly evident. These examples have indicated, each in a different manner, that communication is a vital feature of the new epistemology.

Thus the basic failure of operationism was the failure to recognize the role of communication, the transmission of a signal, in creating knowledge. Operationism did not succeed in making the distinction between the classical logic of "now present" categories and the logic of time-dependent processes. It was still based on the epistemology of intuition, on the expectation of knowledge unmediated by the process of communication. However, the new physical and epistemological problem had to do with distance or separation, be it spatial/temporal, or psychological/social, and how this distance is accommodated in what we understand to be measurement. Replacing a quality with an operation did not take account of the distance between subject and object. It did not confront the new questions How do we make and interpret measurements across the separation? and What form of communication results in measurement? Operational analysis was equipped to deal with only the "here and now," not the distant.

Nevertheless, although Bridgman was not clear about the explicit character of the difficulty with which he was struggling, he was very close. Although he was not quite able to bring his intuition into full consciousness, his operational criticism was steadily circling its objective. Consider the following statements.

Even if the time at which the operation is performed need not be specified, the temporal aspect cannot be entirely absent from the specification of the operation, for the operation is an activity, performed in time, and must consist of parts performed in a specified sequence in time.[23]

Or:

Perhaps the most important conclusion of that discussion [of probability] was that there is no logical method of bridging the chasm between the verbal, paper and pencil structure and our application to concrete cases. It is not that a similar chasm is not always present whenever we attempt to justify the application of any theoretical argument, whether involving probability or not, to a concrete situation; it is only that the chasm is a little more yawning and obvious in the situations presented by probability than in classical situations.[24]

Clearly, Bridgman had sensed that time and distance were involved in the new epistemology, but at the same time we can plainly see that he was still thinking in terms of a logical separation rather than a physical or temporal separation. But he was ever so close, and how close he actually came to isolating the difficulty can be seen in his treatment of the Gibbs paradox.* It is here, in Bridgman's analysis of the Gibbs para-

---

*The Gibbs paradox is a thought experiment based on the following model: Imagine a closed box with a partition in the center (We may arbitrarily choose the center for greatest simplicity). The partition separates two ideal gases, A and B. A is on one side of the partition; B is on the other side. If the partition is removed, the gases will spontaneously diffuse into each other, creating a mixture. (The process cannot be undone. For example, I cannot remove the cream from my coffee once it has been poured in. In fact, I do not have to stir my coffee to get a uniform mixture if I am willing to wait for a while.) This

dox, that we will discover the key to understanding his philosophical attitude, where we will discover the underlying insight that informed the judgments of the "new" Bridgman—the Bridgman who offended philosophers with his plunge into subjectivity, who asserted the privacy of scientific knowledge, and who urged the duality of language and the isolation of the individual. Even more, we will see an example of the pattern of thought that brought Bridgman to the conclusion that man is isolated on an island of phenomena.[25]

Bridgman considered the Gibbs paradox an indication that "entropy has no absolute or universal significance, but is relative to the

---

change is accompanied by an entropy increase which, according to the mathematical expression characterizing it, does not depend on what A and B actually are. They could be any two ideal gases. The increase of entropy is a consequence of the fact that when the partition is removed, the particles of each gas can spread themselves throughout twice the volume previously available to them.

But what if we start with the same gas on both sides of the partition? Will it not be the case that when the partition is removed the particles of each portion will similarly spread themselves into twice the original volume, thus resulting in the same entropy increase? While this reasoning seems to be consistent with the theoretical model, the answer is no. There is no entropy increase when two portions of the same gas are mixed. Why? Simply because we cannot tell the difference between what we started with and what we ended with. It looks the same to us, although from the point of view of an individual particle, the space available to it has doubled. The paradox arises from the fact that we cannot "tag" individual particles, and therefore, from our point of view, nothing has changed; or more precisely, we cannot *perceive* that anything has changed. We cannot make a measurement that will "document" the change. From our point of view, there has been no "motion." Or, from our point of view, no time has passed.

The Gibbs paradox raises difficult questions about the relationship between macroscopic events and microscopic events, as well as calling into question the scientific understanding of the nature of time. It also indicates the close connection between phenomenal change, time, and the way we do our counting. From the phenomenal (macroscopic) point of view, we say, if we do not see any change, there is no change, and therefore, in our corresponding statistics, we may not count the interchange of two indistinguishable particles as a different arrangement (as we do, for example, when we count the possible combinations which come up in throwing dice—5 and 2 is different from 2 and 5 because we know which die comes up 5 and which comes up 2). When we are dealing with gas particles all of the same kind, the arrangement A(1)A(2) is the same as A(2)A(1). We must count the two as one because we cannot tell the difference, even though, from the hypothetical point of view of the individual particles, the two arrangements are different. We must choose the point of view that is consistent with what we can observe, even when our theory indicates otherwise. The demand for consistency can be accommodated by changing our counting procedure.

It is not unreasonable to surmise that for Bridgman, the Gibbs paradox represented a scientific example of the discrepancy between intimate knowledge and objective knowledge, that is, self-knowledge and knowledge of the other, including subjective time and objective time. (See Yu. B. Rumer and M. Sh. Ryvkin, *Thermodynamics, Statistical Physics, and Kinetics*, trans. S. Semyonov [Moscow: Mir Publications, 1980].)

universe of operations." The paradox may be resolved, he asserted, by "observing that we have discontinuously changed the universe of operations in the limit when we allowed the two different gases to become identical." In other words, the universe of permissible operations cannot (because it introduces a discontinuity) include actually carrying out this particular limiting process. We must stop short of this limit. Otherwise we would have to "ascribe a change of entropy to the process of self-diffusion of a gas after the removal of a partition [which divides the gas into two portions]."[26] Therefore, in Bridgman's view, the concept of entropy is not absolute or universal because it cannot be extended continuously without contradiction to the extreme case of self-interaction. At this limit, the entropy concept loses its meaning. We cannot take this extra step; we cannot pass over this boundary into an impossible state of paradox.

The Gibbs paradox is a thought experiment that translates the paradox present in the idea of the impossible reversible cycle into concrete terms; indeed, it highlights in a most compact manner the implications of the definition of entropy. It displays the consequence of defining temperature, and thus entropy, in terms of a fictitious equilibrium that does not exist on the microscopic level and, for that matter, does not even exist on the macroscopic level because of the impossibility of the reversible cycle or, what is equivalent, the passage of time. However, Bridgman did not examine the paradox in terms of the definition of entropy and the principles structuring the definition. Instead, he attributed the paradox to an inadmissible formal procedure that does not recognize the discontinuity between "self-relatedness" and "other-relatedness" (taking self-relatedness as a condition of non-communication where no entropy change occurs, and other-relatedness as the situation in which interactive difference generates an entropy change).

The elements of Bridgman's thought are now beginning to converge. We have already seen that Bridgman had a distaste for paradox. Paradox stood in the way of his quest for absolute clarity. Therefore he was bound by his explicit intentions to seek a resolution that would dissolve the ambiguity inherent in paradox; that is, render the paradox specious. In the case of the Gibbs paradox, the flaw in thinking responsible for the apparent paradox (for Bridgman, all paradoxes are merely apparent) is the failure to recognize the self-other discontinuity. The Gibbs paradox is thus due to a procedural error, the error of performing an illegitimate operation, an operation, it might be noted, that is illegitimate because it results in paradox and, furthermore, just in this one instance. The operation of taking a limit is not one that Bridgman considered illegitimate in general (otherwise he would have had to renounce all of calculus).

Thus, what could have been the occasion for an in-depth examination of the foundations of thermodynamics was sacrificed by attributing the Gibbs paradox to the application of an inadmissible formal device. Bridgman chose to keep a safe distance from metaphysical questions that could tumble down on him if he were to get too close. On the other hand, the fact that he singled out the self-other disjunction as the central operational incompatibility is at the heart of the meaning of operational reasoning and provides the key to understanding its connection to the new epistemology. We can see now that Bridgman did not actually oppose all categorical knowledge, or even necessarily statistics and probability, but rather the epistemology of communication.

Bridgman had the "uncomfortable feeling that we have lost something." Since he was not able to pinpoint the epistemological reason for that discomfort, however, he attributed it to a loss of explanatory capability due to the replacement of the traditional physical models by abstract mathematical models, never considering that the physical (mechanical) models were just as abstract. Bridgman held out the hope that it might someday be "possible to glimpse through our present mathematics something much more fundamental beyond it, which we may ultimately be able to reach more directly." [27] However, what was lost was not explanatory power but the applicability of this classical epistemology to the problems of modern physics. The mechanical model of physical reality simply followed. The loss or sense of distance Bridgman experienced was not a result of the abstract nature of mathematical language but an outcome of the new physics itself.

The classical "here and now" epistemology supposed that knowledge was immediately present to the knower. There was no attenuating gap between subject and object that would diminish the quality of the knowledge. In the case of special relativity, there was a loss of immediacy, but it was bridged by a rational assertion about the communicating vehicle. For Bridgman, the rational assertion was not a legitimate cognitive instrument, but by dividing cognitive tools into the "natural" and the "artificial," he found a way to come to terms with special relativity. However, the new epistemology of unknown distance and discontinuous communication entailed the consequence that knowledge is necessarily imperfect. For as long as measurement, and thus knowledge, depends on communication, and as long as communication takes time, there will be some loss, some increase of entropy. This is the situation to which mankind must become resigned. Only in the ideal world is communication perfect, non-entropy-producing, and as the Gibbs paradox illustrated, that is self-communication. The crucial difficulty, then, the difficulty that Bridgman intuited but could not express, was not, after all, the inductive or logical relationship between individual and class, but rather understanding process, which means taking account of communication and elapsed time. Operations, being actions, were supposed to take care of that, but in fact they did so only superficially because they were still based on assumptions belonging to the "here and now" classical epistemology. The result was an unintended and unfortunate confusion about class definitions in science and language in general.

The real problem was the relationship between the individual and the other—the problem of communication. However, it was obscured by a haze of classical preconceptions, and it was thus misconstrued logically, scientifically, and morally by everyone involved—Bridgman's critics, his supporters, and even by Bridgman himself. No one, not even Korzybski, who preached with such zeal about the need for a "non-Aristotelian" system of values, conceived the crisis of meaning in terms of the process of communication—language, yes, but not communication. And no one saw that Bridgman's radical and demoralizing conclusions about the isolation of the individual, the self-other dichotomy of language, or the privacy of knowledge might somehow be associated with the new epistemology of communication, indeed engendered by its conflict with the classical epistemology.

Furthermore, despite the fact that his analyses kept on turning up the same constant—the self-other disjunction—Bridgman, instead of recognizing that the problem was about communication, imposed a universal moral interpretation on his discovery, an interpretation filled with exhortations for humility in the face of human imperfection and warnings against overestimating the capabilities of human intelligence. Just as his analysis of the Gibbs paradox had shown Bridgman that the operation connecting the world of imperfect intercommunication with the world of perfect self-communication is not legitimate, so man cannot leap over the chasm separating the actual and the ideal. That, too, is an impermissible operation. Man is isolated on an island of phenomena, bound by his very nature to a state of imperfection.

Perhaps surprisingly, the person who came closest to recognizing that communication was the focal issue was Edwin Boring, who judged Bridgman's public-private dichotomy from the viewpoint of scientific publication.

You get quite stirred up and vehement about your dichotomy at the end. Perhaps we ought to discuss some of these things sometime. . . . As to public and private, it would seem to me that science is something that could be put in a book, that it is either something that eventually gets in a book or, as you made the point when we discussed the matter in November in the Faculty Club, could get in a book even if it never does (the science of the scientist marooned on the desert island who never even mutters to himself). . . . I do not see a dichotomy here.

. . . I should say that privacy was lack of relationship, that publication is the formation of other relations, the fitting of a thing into a larger system. . . . And since even the most private thing is the effect of something else, you won't find that you can actually start with privacy and move out into publication but that you are always entering in media res a continuum. . . . In other words, that is private which is not published. And publication is a continuous process of enlarging effects. In raising the question as to whether any event is private, you are asking how isolated it is, how much it has in the way of effects. Thus there would be all sorts of degrees of publication for an event within a given organism and most of those can not be got so that the organism can respond verbally to them. . . . The behavior of the bones in the inner ear get[s] published a little by the electrical behavior of the receptor cells of the inner ear. . . . If the central auditory tracts are blocked by lesion, publication can occur through electrodes placed on the inner ear. Privacy is relative, not absolute.[28]

It is clear that by "publication" Boring meant communication. It is also obvious that he regarded the natural state of the experienced world— the world, both physical and social, into which an individual is born— as being communicative, interactive.

While Bridgman was advocating in principle that humanity should give up the ideal of perfection, it was Boring, not Bridgman, who accepted the world as an imperfect place, a place where relationships exist in mutual imbeddedness, shadings, and alternatives. Yet he could not follow Bridgman's reasoning. We are thus compelled to ask to what kind of imperfection Bridgman thought we should become resigned? And if only an ironic answer makes sense—pure human imperfection, uncontaminated by the ideal or the divine—then we must wonder further whether Bridgman did not confuse mental clarity, a disinterested state of awareness or consciousness, with self-consciousness and mental discipline, and the latter with an experiential state of phenomenalistic purity.

Although Bridgman's operational approach had evolved into a fairly consistent pattern of reasoning, whose application to nonscientific situations, as we shall see in the following chapters, was bound to produce similar conclusions, still he did not rest content that his quest had been fulfilled. He continued to search and probe, looking for ways to improve his understanding and at the same time to uncover, as Einstein had, unjustified assumptions that might invalidate scientific results. Therefore, Bridgman maintained an active interest in new attempts to expand the sphere of scientific authority. His correspondence with Leon Brillouin, Walter Elsasser, and Jerome Rothstein, for example, provides evidence of a keen interest in information theory. He was especially impressed with Rothstein's *Communication, Organization, and Science*, of which he wrote that he was "fascinated by the brilliant exposition of the potentialities implied by recent progress in information theory and its related disciplines, not only for technological advance, but also for improved social control and philosophic understanding."[29]

It is not at all surprising that information theory should attract Bridgman's attention, since it has a close affinity to thermodynamics (in fact, it borrows directly from thermodynamics). It also makes formally explicit some of the problems to which Bridgman responded only intuitively. Furthermore it introduces a new kind of measurement problem. In the words of Leon Brillouin, the problem is "How is it possible to define the quantity of information contained in a message or telegram to be transmitted?"[30] The solution, which could not have been pleasing to Bridgman, is based on utilizing the probabilistic interpretation of entropy. Here was yet another challenge to Bridgman's understanding of measurement that defied the classical definition. A detailed discussion of his reaction is beyond the scope of this work, but the subject has been mentioned in order to suggest that the complexity of the problem of measurement has been generally underestimated. We have come a long way from dimensional analysis and S. S. Stevens's classification of measurement scales.

However, if curiosity about information theory was a logical step in the evolution of Bridgman's thought, there are no early hints that Bridgman would take an interest in the Duke University experiments on extrasensory perception (a supposedly nonphysical mode of communication) and that he could do so unhindered by the hostility displayed by so many of his colleagues. Indeed, to his utter consternation, he became involved in a minor controversy connected with an article that he published on the subject.[31] Again the problem of measurement was involved, and again, Bridgman's intuition pointed him in the direction of a sticky scientific-philosophical problem that was not recognized for what it was—the problem of specifying the meaning of "randomness" in such a way as to make the measurement of "nonrandomness" both philosophically and experimentally unequivocal and, furthermore, knowing what kind of proofs can be built upon arguments that certain events are nonrandom. A full treatment of this provocative episode is also beyond the scope of the present work, but it underscores once more the difficulty of understanding what makes a measurement valid.

The ESP experiments, among other things, brought this difficulty into full relief and challenged commonly accepted ideas about both practical and formal measurement procedures, if simply for the reason that the methodology of the ESP researchers looked respectable enough, but most scientists did not want to believe the outcome of the research. This is not to say that there is or is not such a thing as ESP, but only that measurement is a more complicated affair than its advocates or opponents realized. Reference to a methodology, however "rigorous," is not sufficient to validate experimental results. On the other hand, anyone who wishes to dispute the legitimacy of the experimental findings must show, without prejudice, why the results are not valid, why the measurements do not measure what they are thought to measure. Bridgman attempted to do this, but his attempt was misunderstood, not because his thinking was clouded by prejudice, or because he thought that ESP was a subject that science could not judge, but because he was unable to formulate his objection in a manner sufficiently cogent or formally precise. Whether it was correct or not is not a simple matter.

In any case, operational analysis was of no help. Bridgman was left to depend on his intuition alone, an intuition that selected as problematic an important and philosophically sensitive issue but could not focus sharply enough to bring it into full clarity. As it had earlier, operationism, his methodological hope for achieving clarity of vision, failed him again, and he was left to blame the shortcomings of probability theory. The lesson that Bridgman had learned from Einstein was simply not adequate.

PART V

Science as a Vocation

*Today the spirit of religious asceticism—whether finally, who knows?—
has escaped from the cage. But victorious capitalism, since it rests on
mechanical foundations, needs its support no longer. The rosy blush of its
laughing heir, the Enlightenment, seems also to be irretrievably fading,
and the idea of duty in one's calling prowls about in our lives like the
ghost of dead religious beliefs.*
     —Max Weber,
       The Protestant Ethic and the Spirit of Capitalism (1904–5)

■ ■ ■

IN 1918 MAX WEBER delivered a speech at the University of Munich that
was later published as the well known essay "Science as a Vocation." In
it he asked why one should devote one's life to science when science is
"apt to make the belief that there is such a thing as the 'meaning' of the
universe die out at its very roots," when by its very nature it cannot
give an answer to the question, "the only question important for us:
'What shall we do and how shall we live?'" In a world disenchanted by
intellectualization and rationalization, he continued, has the "progress" to which science is linked as a motive force a meaning that reaches
beyond the merely technical? "To raise this question," Weber declared,
"is to ask for the vocation of science within the total life of humanity,
What is the value of science?" [1]

In America, where modern science was yet at the beginning of its
rise to cultural dominance, before it had irreversibly usurped the Christian cosmogony, and when it was still enveloped in an aura of progressivist utopian ideology, doubt such as Weber's was an undercurrent,
making its appearance in episodes of antimodernist reaction. This reaction, it should be noted, was directed toward the naturalistic scientific
worldview, the theoretical expression of science, in contrast to science

as technique or practical know-how, against science as Truth, but not against science as the instrument of social and material progress. However, doubt was soon to swell to proportions that would affect the mainstream of American consciousness, calling into question the assumptions behind the "American civil religion," that philosophically unstable alliance between Enlightenment rationalism and Puritan morality that forms the underpinning of the American democratic consensus.[2]

Both strains of thought are equally fundamental to American tradition. Although the formal political framework is founded on principles derived from Enlightenment rationalism, strongly influenced by Newtonianism and eighteenth-century neo-classicism, the moral impulse is deeply religious and practical, imbued with the spirit of ascetic Protestantism. Metaphysically, these two traditions represent antithetical worldviews, united nevertheless by a utopian animus. Historically, the relationship is more complex and subtle, but they find common ground in the assumption of natural theology that the Word of God and the book of nature, which is God's creation, cannot be in contradiction. Thus, a commonly shared ideological vocabulary acts to obscure the original philosophical differences. The stress of twentieth-century scientific, political, and economic developments worked to undermine this implicit compromise. We will see that Bridgman's response to the Weberian dilemma brought this antithesis into sharp relief.

Before the twentieth century, when America was basically rural and self-occupied, concerned with the practical matters of maturing nationhood, science was not yet a pervasive organizing social force or a significant moral threat to religious authority. However, after Darwin, who effectively canceled the assumptions of natural theology, a secularized science began to exert its naturalistic challenge. At the same time, the increasing urbanization and industrialization of American life and a growing emphasis on professionalization in education enormously expanded the social utility and prestige of science. World War I accelerated the trend toward secularization by further loosening the hold of traditional religious authority, and by the 1920's, despite resistance by religious fundamentalists, science was well established as a social authority. Ever since Darwin liberal Protestant theologians had been adjusting their doctrine to the "modernist" trend,* and the Scopes

*Indeed, in the 1920's, "modernism" came to refer to liberal, scientifically sympathetic Protestantism, as opposed to "Fundamentalism," the anti-modernist Protestant traditionalism.

trial severely diminished the image of the fundamentalist conservative religious camp.[3] Science was now confidently asserting its cognitive hegemony and optimistically rejoiced in its ability to advance social and material well-being.

The mood of optimism, however, was short-lived. We have already seen how developments in modern science refuted the assumptions of natural theology. In addition, the shock of the Depression and the rise of totalitarianism presented a cruel repudiation of the confident Progressivist idealism. They challenged the American system of values in the immediate and concrete world of human activity, rather than in the realm of abstract thought. Americans had to reassess their ideological resources. What reasons could be given for the moral failure that these humanly devastating events represented? On what cognitive basis could moral standards be affirmed? On what principles could the superiority of the American way of life be rationally argued? Democracy, freedom, and individualism needed a theoretical, as well as a practical, defense. This was not the time to invoke the Enlightenment faith in the natural goodness of humanity or the utilitarian belief in free enterprise. The catastrophic events of the 1930's belied such naive assumptions.

Scientific naturalism, with its relativist implications, was in a philosophically even weaker position. Not only was science unable to contribute to the resolution of the moral crisis, but its very claim to be value free made it vulnerable to attack as a culpable agent. In the words of historian Edward Purcell, "The disillusionment of some intellectuals with naturalism, the hostility of church groups, and the frustrations created by the depression combined with the rise of Nazism to galvanize many Americans into a renewed attack on scientific naturalism as a destructive and inadequate world view."[4] Scientific naturalism, it was charged, by fostering skepticism and moral cynicism, had allowed Western man to lose sight of his values. Furthermore, it was obvious that the miracles of technology alone were insufficient to create a sound and just social order. The suspicion arose that science had overplayed its hand. Perhaps science was not the benefactor of humanity, after all. Weber's question was now America's question.

As critics of the 1930's sought explanations and prescriptions for this crisis of values, their thought veered toward conservatism. Intellectually, this conservatism was expressed in one of two general ways—either as a call for a return to some form of philosophical absolutism or a reassertion of religious transcendentalism. Among scholars in jurisprudence, social science, political science, and theology, the cry went up that a belief in absolute ethical principles was necessary for a humane social order. Catholic intellectuals took part in a revival of neo-Aristotelianism, brilliantly represented by the neo-Thomist philosopher Jacques Maritain. Liberal Protestantism produced a school of neo-orthodoxy, whose American spokesman was Reinhold Niebuhr. Neo-orthodox Protestant theology retracted the scientific liberalism of mainstream Protestantism, and harking back to the Reformation, it re-asserted the fundamental sinfulness of man, the imperfection of knowledge, and the need for humility. "The collapse of liberal ideas was as rapid as the social order on which they rested."[5]

On the other hand, the most vociferous anti-modernist Protestants of the previous decade, the Fundamentalists, had surprisingly little to say. After their defeat at the Scopes trial, they retreated from visible political involvement, though they did remain an active subculture, collecting their strength, to re-emerge as a political force again in the late 1970's. The philosophical profession had similarly little to offer. Increasingly positivistic in orientation and academically specialized, American philosophers were committed to disassociating philosophy from metaphysics, science from values, truth from justice, and they defined their task in an ever-narrowing technical framework. Liberal Protestant thinkers had the most difficult problem. Science had let them down. Yet their choice appeared to be between the moral relativism implied in scientific naturalism or the political authoritarianism they associated with metaphysical absolutism. Without ever achieving a clear philosophical resolution of this conflict, many turned toward cultivating a social conscience. Philosophically compromised and thus morally weakened, they welcomed and encouraged the intervention of government as the overseer and guarantor of social welfare.

Bridgman would have nothing to do with any of the prevailing strategies. To him, they all reeked of absolutism of one kind or another. To turn to religion in any form was in his view culturally retrograde, philosophically indefensible, and morally irresponsible. As far as he was concerned, religion (or equivalently, metaphysical absolutism) should be discarded as an obstacle to human progress. He did not even consider it worthy of discussion. The defect of religion, he believed, was "in the idea that [the] springs of conduct can be found outside of oneself (or the race)."[6] To be freed from religion was to escape from an oppressive and false authority. Bridgman felt similarly about social activism. A commitment to the collective welfare was in effect an acquiescence to a surrogate god, Society. And to further invite government to be the social and economic patron of the good life was contrary to everything he thought America was about.

Yet despite his disapproval of the emerging attitudes, his own thinking did not remain unaffected. The shift in the meaning of Bridgman's operational method coincided with the shift in the mood of American thought. In fact, the "new" Bridgman of the 1930's adopted an epistemological outlook in science formally analogous to the pessimistic stand of the Protestant neo-orthodox theologians, but his was even more severe, for he rejected their social mandate. Nevertheless, even though from the point of view of scientific detractors his new restrictions might have confirmed judgments about the moral bankruptcy of science, Bridgman did not believe that science was at the root of contemporary problems.

In Bridgman's opinion, the critical response to science was based on a misconceived notion of its nature. "It has become the fashion in some quarters during the last few years," he observed,

to depreciate the entire scientific outlook, and we hear much of the "bankruptcy" of science. This scepticism has doubtless arisen from simple misconceptions about the nature of the scientific approach, which has often been thought to have applications to such irrelevant questions as those of value. More particularly, however, it has arisen from plain disappointment because the easy Utopia which brilliant technological advance encouraged the uncritical to think was just around the corner has proved to demand for its actual realization the exercise of intelligent cooperation by all parts of the community.

Nevertheless, Bridgman asserted, "In spite of this popular reaction, I believe that science was never less bankrupt than at the present time."[7]

Indeed, Bridgman regarded science as the emancipator of humanity and the guardian of individual freedom. In his eyes it was not only the exemplar of intellectual honesty but also the instrument by which the human race may achieve intellectual integrity. Intellectual integrity for Bridgman was not the relativistic Weberian inner consistency that functions as a substitute for cognitive certainty. It was not a logical condition but a state of virtue where the lower drives and passions have been subdued and one no longer serves more than one master. Furthermore, intellectual integrity can be reached and maintained only with the most intense effort and the most determined will, since so many personal and cultural forces work against it. Intellectual honesty is at once a moral condition and a state of higher consciousness that permits one to distinguish and resist the temptations of false gods, to freely and voluntarily view the truth without prejudice.

Science, in Bridgman's view, embodies a commitment to intellectual honesty that at the same time augments itself. The growth of science causes an increase in intellectual honesty. The reason is that science is at the mercy of no other authority but fact. The scientist, Bridgman declared, has subjected himself "voluntarily to a single supreme control, the control of agreement with the facts." He accepts the challenge to adapt to the external world as he finds it, rather than constructing a mystical world that might be more comfortable for him. The scientist takes pride in the

consciousness of integrity which is the gift of the intellectual honesty that dares to discover the correct answer irrespective of personal discomfort. This consciousness of integrity abides with the scientist and is, I think, one of his most precious compensations.[8]

Bridgman introduced this idea in 1933 in an article written for *Harper's* entitled "The Struggle for Intellectual Integrity." His intention was to rebut two previous articles whose authors had lamented the decline of religious belief among the younger generation. For Bridgman, this was nothing to mourn about. Instead, he argued, it was a manifestation of an increase in intellectual honesty brought about by a greater educational emphasis on science, more popularization of science, and a greater proportion of scientists in the workforce. The hold of religion has weakened, he asserted, because many of its teachings have come to be seen as simply "not 'true,' to express it very crudely."[9]

Bridgman argued that intellectual honesty is a product of intellectual power and an advanced state of civilization. It is "the last flowering of the genius of humanity, the culmination of a long cultural history, and the one thing that differentiates man most notably from his biological companions."[10] Bridgman distinguished "intellectual honesty" from "intelligence" or the "method of intelligence," the faculty by which an individual adapts to the environment, that is, by which he discovers knowledge. Intellectual honesty is a moral attitude, a courageous and disinterested disposition of the mind, and thus a capability for accepting truth. Neither is to be confused with the "intellect," which is the rational or logical (verbal) function of the mind. It is the "intellect" that is responsible for verbal maladaptations and disingenuous communication. Intelligence is a problem-solving faculty superior to the intellect:

It is, perhaps, easy to misunderstand the role of intelligence in dealing with emotional situations involving values. Depreciation of the attempt to run one's life on an exclusively logical and rational basis is popular and common, and lends itself to easy caricature. The mistake is to assume that such an ordering of one's life is intelligent. The role of intelligence in questions of value is primarily a neutral one—that of a tool by which values may be effectively realized. . . . By its neutrality, intelligence acquires universality; it is the one common denominator of mankind, independent of creed or culture, spanning the hemispheres and the centuries.[11]

By this assertion, Bridgman did not mean to subscribe to any form of transcendental or collective idealism. He meant only to say that intelligence, which is epistemologically superior to intellect, is a faculty common to all, independent of culture or education. Intellectual honesty, however, is prior to both, since it is the condition for their proper functioning. Bridgman did not think of it as a radical skepticism but rather as a purified state of mind.

It is not possible to mistake Bridgman's idea of intelligence for the innate faculty of reason of Enlightenment rationalism. It has too Darwinian a cast. However, the relativism of evolutionary epistemology and morality has been mitigated by an overlay of Puritan values. Indeed, once quantum mechanics had dealt out a world with no intrinsic order, Bridgman leaned more and more on Puritan logic to rescue him from the embarrassment of the Enlightenment metaphysical error. And, as the next two chapters illustrate, a harsh Puritanism, one worthy of Calvin himself, overshadowed whatever Enlightenment sympathies that may have been present. We will see that Bridgman projected a Puritan mission onto science and impressed the scientist into the role of the Puritan saint. His vindication of the role of science in the crisis of values was undertaken on Puritan ground.

And, if Puritanism was, as has been forcefully argued, historically an agent of modernization, intellectually and politically an ideology of transition particularly suited to accommodate and adjust to disorder in society,[12] what more appropriate time to invoke its strategy than at this moment of cognitive uncertainty and moral confusion? What ideology could better serve Bridgman's conservative emotional requirements as well as his revolutionary rejection of tradition, his belief in the absolute value of scientific truth and his commitment to scientific progress? Formally, he had already applied this logic to cope with cognitive distress in science. After all, what was operational analysis if not an instrument for purification of scientific truth and a principle of watchfulness against further scientific error?

Intellectual honesty was the moral counterpart to operational analysis. It would be a defense against what Bridgman regarded as a growing laxity in individual moral purpose and a collective tendency toward the establishment of a corporate state mysticism. Indeed, as we trace the evolution of Bridgman's thought, we become aware of the Enlightenment cloak dropping away to reveal the steel undergirding of a Puritan temperament. The logic is unassailable, for the Puritan worldview was invulnerable to the naturalistic erosion of the classical tradition. Man might have killed Nietzsche's God, but the Puritan Jehovah, willful and free in His absolute transcendence, was inscrutable and untouchable.

The scientist, Bridgman wrote, "must be willing to follow any lead that he can see, undeterred by any inhibition, whether it arises from laziness or other unfortunate personal characteristics, or intellectual tradition or the social conventions of his epoch." [13] Thus, science, clothed in Puritan garb, the embodiment of intellectual integrity, would be an agent of purification, and the scientist, an unflinching soldier of truth. It might even appear, for the time being, that Bridgman had outsmarted Weber.

CHAPTER 10

Puritan Logic and the Politics
of Science:
Neither Shalt Thou Serve Their Gods

*What Calvinists said of the saint, other men would later say of the citizen:
the same sense of civic virtue, of discipline and duty, lies behind the two
names. Saint and citizen together suggest a new integration of private
men (or, rather, of chosen groups of private men, of proven holiness and
virtue) into the political order, an integration based upon a novel view of
politics as a kind of conscientious and continuous labor. This is surely the
most significant outcome of the Calvinist theory of worldly activity, preceding in time any infusion of religious worldliness into the economic
order. The diligent activism of the saints—Genevan, Huguenot, Dutch,
Scottish, and Puritan—marked the transformation of politics into work
and revealed for the first time the extraordinary conscience that directed
the work.* —Michael Walzer, *The Revolution of the Saints* (1965)

■ ■ ■

THE OUTBREAK OF World War II changed the nature of the question
What is the value of science? from a philosophical to a practical one.
The crisis of meaning was shunted aside by the exigencies of war and
was replaced by a crisis of action. For the scientist the question became
one of choosing where his loyalties lay, whom to serve, what authority
to accept. For many of Bridgman's European colleagues, this was an
extremely painful decision.

The subtitle derives from Deuteronomy 7.16: "And thou shalt consume all the
people which the LORD thy GOD shall deliver thee: thine eye shall have no pity upon
them; neither shalt thou serve their gods; for that *will be a snare unto thee.*"

In Bridgman's thinking this cultural change of focus was reflected in his shift of interest from purifying science (its cognitive categories) to treating science itself as cultural purifier. To him, the solution to the problem of loyalty was self-evident. The interest of science transcends all else. It serves no end but its own. The scientist is engaged in a mission of absolute worth that is not to be diverted to serve less dignified purposes.

In some ways Bridgman spoke in concert with a number of his colleagues who also argued that science is done for its own sake, but at times the harshness of his position was unintelligible to them. They did not recognize the Puritan cunning behind his reasoning. In fact, the coherency of thought underlying Bridgman's political and social judgments is most evident when held up against the model of Puritan ideology.

At the same time, the problem of meaning reappeared, now in an ideological rather than a cognitive framework. And as Bridgman pressed the consequences of his presuppositions to their logical extreme, the operational principle was still at work, dissolving the internal compromises contained within ideological symbols. In effect, Bridgman was decomposing the Enlightenment/Puritan synthesis by selecting the Puritan meaning, discarding what he regarded as the metaphysically tainted Enlightenment contamination, casting out the elements of rationalist heresy in his drive to recover and defend his democratic ideal.

In contrast to Enlightenment thought, which is predicated on the concept of a rationally ordered universe, transparent to human reason, the universe of the Puritan is ruled by an inscrutable, unpredictable God whose ways are forbidden to human knowledge. In the words of the historian Michael Walzer:

In Calvinist thought nature ceased altogether to be a realm of secondary causation, a world whose laws were anciently established and subject to God's will only in the extraordinary case of a miracle. Providence no longer consisted in law or in foresight: "Providence consists in action." The eternal order of nature became an order of circumstantial and particular events, the cause of each being the immediate, active (but inscrutable) will of God. . . . God's commands did, of course, create something of a pattern, and it was this pattern that Calvin sometimes called natural. Yet he always insisted that it arose from no "perpetual concatenation and intricate series of causes, contained in nature," but from "God the Arbiter and Governor of all things."[1]

Thus, the Reason to which the universal, self-evident truths of Enlightenment thought are made known is not the intellect of the Puritan, nor are the general principles constituting these truths the kind of knowledge to which the Puritan is privy. The Puritan intellect is constrained to activity in a concrete, practical, this-worldly theater. It is entitled, indeed obligated, to seek into the meaning of the manifest Word of God, but it does not inquire into His actual working. Beyond that it is a moral watchdog and disciplinary agent, a keeper of order in a world of uncertainty, rather than a diviner of universal truth or natural social harmonizing principle.

Moreover, according to Calvin, man is a fallen, depraved being. He is not instinctively social in the classical sense; he has no natural social nature that needs only to be freed from oppression, as Enlightenment thinkers would argue. The Fall of Adam had turned man into an asocial being, forever trying to dominate his neighbor. Therefore, social order is realistically achieved by repression; for the Puritan, by a double repression—a self-repression and a mutually enforced social repression. Paradoxically, this internal repression was also a freedom, for it provided a means to social mobility and an escape from imprisonment within the rigid formalities of a corporate Church. The price for this alternative was self-discipline.

The Puritan entered into a voluntary covenant with God and, in so doing, bought individual insurance for salvation for which the premium was acceptance of his vocation; more accurately, the Puritan placed a bet on salvation for which the ante was dedication to his calling. The covenant was between the individual and God, and agreement was the result of painstaking inner search, but his hard work was a public declaration, testimony to his piety and his acceptance of the terms of the covenant. Those who refused God's offer, those who would not work—the beggars and indigents—did not qualify to receive Puritan charity, since charity was reserved for the spiritually educable, those with some discernible potential for salvation. Of course, salvation was preordained because God could not be limited by human solicitation, but nevertheless it was not automatic, for such an admission might be taken as sanction for resignation or laziness. And while ignorance of the ways of God gave all equal hope for salvation, worldly success was taken as a sign of favor, a positive forecast of chosenness, but never a certainty.

In one other important way, Puritan society was not entirely egalitarian. There were God's preordained elect, to be sure, but their identity would not be made known until the Millennium. In the here and now, there were the ministers of God's Word, the saints who by virtue of their talent and ability could convey God's Word to those with lesser insight. Puritan faith was not blind or instinctual; it was not a mere gamble, but a voluntary act mediated by the Word of God, and that required an intellectual effort, an educated judgment dependent on intelligence and ability as well as on a willingness to enter into the covenant with God. Thus Puritan society was ruled by an aristocracy of the intellect, by "advanced" individuals who were "experts" in the Word, and who formed a voluntary association, a brotherhood of saints.

The delicate balance of understanding that must be maintained to comprehend how repression can simultaneously be freedom is also necessary to encompass the meaning of progress. For even though the Puritan saints were engaged in revolutionary activity leading to the establishment of a new social and political order, and even though they and their followers were embarked on a spiritual journey whose end was deliverance to a promised land, nevertheless an overriding concern with maintaining social control subordinated next-worldly salvation to this-worldly improvement. More to the point, the fear of social chaos concentrated attention on the means rather than the end, since from a practical standpoint disciplined activity is a social harness of energy to which utopian and eschatological concerns have only secondary relevance. The work was more important than its product or its reason, except insofar as the product was social control and its reason provided motivation.[2]

Bridgman was a scientific puritan. When he defended science, he defended it as a true vocation, a voluntary dedication to a higher calling. Bridgman asserted that the scientist does not recognize the jurisdiction of any authority other than that dictated by science itself. The duty of science transcends mundane social or political interests. Therefore he felt impatient with contemporary debates about the responsibility of science to society. To his view they were, at best, irrelevant, at worst, simply wrong.

The most publicized instance of Bridgman's commitment to his principles was his decision, early in 1939, to close his laboratory to citizens of totalitarian states. At the same time, he issued a manifesto presenting his reasons for doing so. His statement, which he handed to visitors, was also published in *Science* on February 24, 1939. It read:

I have decided from now on not to show my apparatus or discuss my experiments with the citizens of any totalitarian state. A citizen of such a state is no longer a free individual, but he may be compelled to engage in any activity whatever to advance the purposes of that state. The purposes of the totalitarian states have shown themselves to be in irreconcilable conflict with the purposes of free states. In particular, the totalitarian states do not recognize that the free cultivation of scientific knowledge for its own sake is a worthy end of human endeavor, but have commandeered the scientific activities of their citizens to serve their own purposes. These states have thus annulled the grounds which formerly justified and made a pleasure of the free sharing of scientific knowledge between individuals of different countries. A self-respecting recognition of this altered situation demands that this practice be stopped. Cessation of scientific intercourse with the totalitarian states serves the double purpose of making more difficult the misuse of scientific information by these states, and of giving the individual opportunity to express his abhorrence of their practices.

This statement is made entirely in my individual capacity and has no connection whatever with any policy of the University.[3]

The reaction of Bridgman's colleagues was not entirely favorable. While they respected his courage, they said, still they felt that a certain sensitivity to humanitarian ideals was wanting. Max Born, for example, wrote, "Your Manifesto against the Totalitarians pleases the feelings of my heart, but not quite as much my reason. Would you exclude also men like von Laue, [one name illegible], who are permanently risking freedom and life by opposing the Nazis in Germany?"[4]

Harlow Shapley was not as diplomatic. He told Bridgman that he disagreed with him "so violently that only the fear of too messy publicity" prevented him from speaking out.[5] Shapley felt that Bridgman's position was equivalent to the "stand that we should fight intolerance with intolerance, and that we should, so to speak, bomb the women and children behind the lines in order to avenge ourselves on the generals and war-makers and front line fighters." From his acquaintance, he wrote, of the "half a dozen 'citizens of totalitarian states' in the [Harvard] Observatory," he knew that these people bitterly resented the "degradation of intellectualism in their countries" and were "highly appreciative of the sympathy that is afforded them by free intercommunication with other men of science." Bridgman, charged Shapley, was "quite sympathetic with the idea that we should humiliate these victims still more."[6]

If there was reason for the discomfort that these men felt with Bridgman's stand, it was not that Bridgman was being vindictive. A higher principle was involved.

On the surface, Bridgman's refusal to share scientific knowledge with citizens of totalitarian states might appear to be an acknowledgment that there can be exceptions to the vocational priority of scientific interest, that Bridgman was softening his stand and acting in the belief that he was facing one of the few instances in which a scientist could legitimately be regarded as having a social responsibility above and beyond the pursuit of scientific knowledge. In fact, however, he was denying access to the scientific brotherhood to those who could not free themselves from a competing authority. The threat of totalitarianism canceled the right to free exchange of scientific information because totalitarianism had abrogated the voluntaristic premises upon which this right is founded. When the individual is in a position where he must bow to the interests of the state, he cannot meet the demand of intellectual integrity. The value of truth is debased, and that individual no longer truly serves the scientist's vocation.

Bridgman characterized the totalitarian threat to democracy as an intellectual crisis, a loss of confidence in the efficacy of rational thought, a failure of intellectual morale. He called on physicists to show the way to its recovery. The physicist, Bridgman believed, by virtue of his superior intelligence and his training in straight dealing, is especially qualified to lead. He is accustomed to having to face facts. He cannot succeed otherwise. But the mandate to responsibility is self-willed, not imposed by external authority. And he leads not by acting on behalf of society, but by his individual example.

Thus, the responsibility is not in the first place to society. It is to science. Bridgman believed that in serving science, one automatically serves society. But the service is not utilitarian. It is moral. And the science of which Bridgman spoke was not applied science, but pure science—the ideal search for understanding, a good in itself. (Pure science is not necessarily theoretical science.) Science, Bridgman argued, presents to society the fruit of its activity, an activity that in "more idealistic phraseology" is "sometimes described as the pursuit of truth."[7] This has to take place on a purely voluntaristic basis. And although Bridgman contributed willingly to national efforts in both world wars, he did not believe that science should be pressed permanently into economic or political service. It did not make any sense, he asserted, to fight for freedom and then proceed to curb the freedom whose preservation was the ultimate justification for entering into war.

Bridgman was consistent in his belief that science should not serve mundane utilitarian interests and expressed his view by vigorously opposing efforts by the American government, stimulated by the experience of World War II, to institute a formal science policy. He saw no difference between these attempts to control science and those of totalitarian governments. Either way, science would be in the situation of having to serve two gods.

Thus, when Senator Harley M. Kilgore introduced a bill in 1943 to set up a permanent Office of Scientific and Technical Mobilization in the federal government, Bridgman lodged an impassioned protest. He wrote to Kilgore, with copies to the congressmen from Massachusetts, that he wished to enter "the most vigorous protest in my power against Bill S.702 entitled, 'A Bill to mobilize the scientific and technical resources of the Nation, to establish an Office of Scientific and Technical Mobilization, and for other purposes.'" He emphasized his opinion that it would "lead to complete control of all scientific activity by the Central Government, and would, in principle, deprive every individual scientist of scientific freedom." Furthermore, he argued, it is "profoundly undemocratic and constitutes virtually the acceptance of the totalitarian philosophy that the individual exists for the benefit of the State." He continued, "If this bill were enacted, I would feel that I had been made a member of a slave class exploited for the benefit of the majority, and my attitude and actions would become those which are always elicited by the consciousness of exploitation. My brains are my own. I may give them to the State, but the State cannot take them from me."[8]

He also criticized as shortsighted the idea, expressed in the Steelman Report to President Truman, that science should be made the backbone for the expansion of the national economy. "Do we," he asked, "want to be committed to such an economy that we are compelled to be always ahead of the whole world?" Such a commitment requires that we produce more than we can use. The consequence is that we must export the surplus to other nations who "find themselves in the position of being continually less and less able to repay us for what we force them to take. The situation is self-defeating, and furthermore is a potent source of feelings of bitterness toward us by the other nations."[9]

Equally illogical in Bridgman's opinion was the notion that pure science should be exploited in the technological race with the Soviet Union. It was not Sputnik that disturbed him (that was nothing to get excited about), he claimed, but the misunderstanding about the nature of pure science implied in the response to the Russian challenge. Pure science is done for its own sake. Therefore, Bridgman argued, "the man who does pure science cannot help being a little bewildered by the clamor of the public that he should get busy and turn out pure science in order that the United States may stay ahead of Russia, nor can he help feeling that if he should yield to the clamor he would lose something in personal integrity."[10]

A clever turn of logic, but how stubbornly incognizant of the irrevocability of the radical transformation World War II brought to the political face of science. Bridgman's unwillingness to acknowledge that science had been inextricably drawn into the web of world affairs was made starkly evident after Hiroshima by his response to Oppenheimer's anguished remark that science now knew a profound guilt. Bridgman would have nothing to do with such depreciation of science. He was reported to have retorted, "If anybody should feel guilty, it's God. He put the facts there."[11] Responding to an inquiry whether the quotation was correct, Bridgman wrote:

The quotation was verbally substantially correct, but taken out of context so as to give a diametrically wrong impression. It was not God that I was against but the sense of sin. I told the Time reporter that if Oppenheimer felt a sense of sin he could speak for himself, but I resented his speaking for the rest of us. I have never had any use for the sense of sin. If a man does his damnedest after due consideration, as did all the physicists who worked on the atomic bomb, and then still feels a sense of sin, it simply means that he hasn't grown up yet and is pathologically unwilling to accept the construction of the world about him.[12]

Bridgman's statement underscores his attitude that the nature of truth is such that mankind cannot be held responsible for what God has made possible. Man does his best, but cannot arrogate himself to the position of God. In Bridgman's eyes, this must have seemed to be the implication of the responsibility Oppenheimer so pathetically assumed. Man is not obligated to take responsibility for what is God's prerogative. No matter how unpleasant the truth turns out to be, man cannot judge the ways of God.

The role of the scientist, then, is moral and exemplary: to discover the truth and accept it no matter how disagreeable it might be. This is the meaning of intellectual honesty. Yet, already in Bridgman's highly principled action against totalitarianism lurked a contradiction. For surely it was not truth as a moral good that he wished to deny his enemy, but the practical advantage it might confer upon him. And it is at this point that the constricted viewpoint of the Puritan ethic seems to break down. Scientific activity itself might be evidence of intellectual integrity or virtue, but how is one to deal ethically with the practical by-products of this endeavor? The problem is only exacerbated when the outcome of this highly virtuous activity is to place at the disposal of mankind an enormously enhanced destructive power. A wealthy man may give away his fortune, but should science do likewise with dangerous knowledge? The Puritans had no problem comparable to that facing the twentieth century. Theirs was a moral and political challenge that did not include the problems created by modern theoretical and technological prowess.

One might argue that a true Puritan would not restrict the dissemination of knowledge and leave the decision for the outcome in the hands of God. But that would be to overlook the role that the covenant with God had in securing individual freedom from corporate organization. The Puritan, through the self-discipline pledged in the covenant, had won freedom from the medieval ecclesiastical corporation and had gone on to transform this freedom into a political individualism. It was this freedom that Bridgman believed was at stake, a freedom that in his mind paradoxically took precedence over the survival of the human race itself (we will consider this issue in the next chapter).

By now it could not have escaped notice that the values Bridgman was professing were not entirely a matter of scientific principle. They were very much a part of a defense of a scientific lifestyle that was rapidly approaching extinction. Bridgman was not a modernist. His values had been shaped by the sensibilities of the premodern era, an era of isolationism and scientific innocence, of scientific leisure and genteel freedom. The advance of the twentieth century, quickened by the demands of two great wars, transformed science from a gentleman's vocation into a crucial element in the overall economic and political well-being of the nation, a nation increasingly urbanized and industrialized. As a contemporary witness, Bridgman was acutely sensitive to the direction of change, and he by no means welcomed it.

Early on, in 1931, he had taken notice of the change that had taken place over the past 25 years in the institutional form of science and called attention to the growing commercial, industrial, and military utility of basic research. He pointed out that physics was becoming more and more specialized and theory was growing in importance. He cited the development of the motion picture industry, especially the introduction of the talking film, and the art of television as examples of the commercial applications of physics that depended heavily on basic research. He noted that the use of X-rays in materials analysis and the treatment of cancer illustrated the enormous potential for artificially produced radiation. He placed great emphasis on the development of the vacuum tube, which had been "greatly stimulated by military demands" and which was responsible for "the recent spectacular increase in radio broadcasting." Its application in airplane navigation, he thought, was an area in which the theoretical physicist could effect "great improvement." At the same time, he noted that the cost of research in some areas was escalating. Furthermore, World War I had underscored the value of cooperation among scientists, as well as having sensitized "some persons [to] the responsibility of the scientist to the entire social life of the community." [13]

Not only had all of these practical functions of science been inconceivable 25 years earlier, but even more significantly for Bridgman, physics had undergone an enormous conceptual recasting. The new ideas that had replaced previously held notions of space, time, and causality were increasingly abstract and mathematical, requiring revised understanding of what was meant by physical reality. At that time, it was the latter that primarily concerned Bridgman and motivated the enunciation of his operational interpretation of physical concepts.

However, sixteen years later in 1947, his metaphysical preoccupation had yielded to an intensified interest in the ideological involvement of science and its institutional setting. World War II and the events surrounding it had accelerated and magnified the organizational and economic trends Bridgman had recognized earlier. Physics had become big science, politically, financially, and organizationally. To Bridgman, this meant the loss of some very precious values.

At a dinner given at the Harvard Club in Boston on January 11, 1947, in honor of his receipt of the Nobel Prize, Bridgman outlined the contrast between the old and the new with reference to his own work. The most important condition for carrying out his work, he said, was its freedom from external constraint. Second in importance was its small scale, which permitted close contact with all aspects of the research. He had few assistants and could do everything by himself. In addition, he had the leisure to think, to work with his hands, to try out partially formed ideas, and, if necessary, to abandon them with impunity since his obligation was only to himself.

On the other hand, the modern trend in science was toward cooperative large-scale enterprises in which engineering took such a large fraction of the physicist's time that he had less time to spend on "the calculation of results and rumination on their significance." Competition was so intense that there was "little opportunity for leisure or scholarly digestion of results before publication." The great expense involved also demanded continual operation of the apparatus, and this situation was "not conducive to a feeling of leisure." Furthermore, "each of the teams which is the slave of one of these instruments has to be driven by some one at the head who has the ideas." The danger was that the creative scientist would be swamped with administrative duties to the extent that "his purely scientific activity [was] destroyed." In other words, the scientist would be diverted from his vocation.

Worst of all, in Bridgman's eyes, was that teamwork organized for maximum efficiency had the effect of submerging the individual, and the younger men who had never known independence would never know what it was like. "The result is that a generation of physicists is growing up who have never exercised any particular degree of individual initiative, who have had no opportunity to experience its satisfactions or its possibilities, and who regard cooperative work in large teams as the normal thing. It is a natural corollary for them to feel that the objectives of these large teams must be something of large social significance."[14] These younger men did not understand that scientific freedom is not a question of self-indulgence but is fundamental to the existence of science itself.

It was not just bigness with its diminution and absorption of the individual or the stepped-up pace of scientific activity that Bridgman deplored. It was as much the subordination of science to a larger purpose outside itself. Twenty years before, Bridgman had acknowledged in *The Logic of Modern Physics* that his metaphysical footing had been knocked loose by the impact of relativity. His reaction had been to retreat to particularism as the sanctuary for truth. Only concrete individual events could qualify for the status of truth. General principles were mere approximations.

Now his ideological beliefs were being assaulted in another scientific revolution, this time social and political, whose outcome was a radical change in the relationship between pure physics and government. Again, he denied the legitimacy of the larger claim and defended the prerogative of the individual against the corporate interest. Bridgman was consistent in his bias. The collectivity, whether conceptual or social, had no absolute validity.

Whereas practical or applied science involving the development of the natural resources of the nation had an established history of government sponsorship, basic academic physics had been insulated, if not from industrial, at least from government involvement. But now that pure science had shown itself to be a national resource, its development had become the focus of national concern. Hiroshima, especially, had turned the spotlight onto the terrible efficacy of pure science. Government could not afford to remain at a distance. However, Hiroshima had also awakened an ethical response. Who should be held responsible for creating and controlling this fearsome power?

Bridgman's opposition to harnessing the power of science to national goals, making it a "servant to the state," has already been made abundantly evident. He believed that pure science could not be done unless it was left on its own, without restriction, without direction. Any government support, he thought, no matter how innocent, can only foster a dependence that broaches the possibility of eventual control. Furthermore, generally speaking, when the resources of the country drift into the hands of government, an irreversible process sets in, "with everything eventually getting down to the dead level of government management—a sort of second law of social dynamics." [15]

Bridgman was even more adamant in his disagreement with those who proposed that the responsibility for the consequences of scientific discovery should rest with science itself. He interpreted the idea as an attack on the independent status of the voluntaristic commitment. In his eyes, this meant that the individual scientist—for him there is no abstract entity, science, any more than for the Puritan there is a corporate Church—is being asked to do more than his share. It is a form of discrimination against men of high ability and is based on a social philosophy that espouses the "right of the stupid people to exploit the bright ones." In an economy based on specialization and division of labor, he argued, it is unrealistic to expect any individual to take responsibility for the extension of the products of his labor. No one would expect a miner of iron ore "to see to it that none of the scrap iron which may eventually result from his labors is sold to the Japanese to be used against his country."[16]

Furthermore, Bridgman asserted, it is not necessary to restrict scientific activity to prevent misuse of scientific knowledge. Mechanisms already exist to accomplish this. After all, Bridgman asserted, if society "had not wanted to construct the atomic bomb, it need not have signed the check for the two billion dollars which alone made it possible. Without this essential contribution from society the atomic bomb would have remained an interesting blueprint in the laboratory." The problem, as Bridgman saw it, was that lazy people were trying to pass off to someone else, and in a tremendously clumsy and backhanded way, what they ought to do for themselves. In other words, the unregenerate were trying to blame the saints for their own sins. "It is obvious," declared Bridgman, "that if society would only abolish war, 99 percent of the need for the control of scientific discoveries would vanish."[17]

There is an unresolved difficulty behind Bridgman's disclaimer. Who, we may wonder, is society? According to Bridgman, "Society is composed of you and me; society does not have an individuality of its own, but is the aggregate of what concerns you and me." Society is a collection, but not a collectivity. It has no corporate identity. However, if everyone minds his own business, cultivating his own little plot, taking responsibility for only that much, who is left as the society that makes the decision to pay for the bomb or to abolish war? Bridgman's answer is based on the expectation that all members of society would discipline themselves uniformly. There would be no such problems in an ideal society in which people make no claims on each other and where the scientific life is held up as the ideal of the good life; where the average man has been educated to appreciate the necessity for scientific freedom, where he has been made to "see that the need for this freedom is born with us, and that we will practice it in the inmost recesses of our thoughts no matter what the external constraints"; where the average man has been "made to see that the imposition of restraints on the freedom to be intelligent betrays fear of the unknown and himself, and . . . that this fear is an ignoble thing." "In a society so constituted I venture to think the problems created by scientific discoveries will pretty much solve themselves."[18]

For a man who eschewed absolute principles, Bridgman's utopia was predicated on a rather uncompromising position. In a tone almost defiant, he posited scientific freedom as a principle of unconditional

importance. For those who regarded it as less than absolute, Bridgman
had few kind words. In his opinion, a scientist who accepts responsibility is demeaning himself and yielding to social pressure. To all scientists, Bridgman issued this exhortation: "And let the scientists, for their
part, take a long-range point of view and not accept the careless imposition of responsibility, an acceptance which to my mind smacks too
much of appeasement and lack of self-respect." More specifically, he
had in mind the younger generation of scientists. Their philosophy he
characterized as youthful, "enthusiastic, idealistic, and colored by
eagerness for self-sacrifice," one that "glories in accepting the responsibility of science to society and refuses to countenance any concern of
the scientist with his own interests, even if it can be demonstrated that
these interests are also the interests of everyone." [19]

The first generational estrangement Bridgman suffered had been
associated with a theoretical advance in science. Although it left him
with a feeling of extreme discomfort, he had at least found a reason to
be optimistic insofar as he could interpret this as a general increase in
intelligence and mental adaptability among the younger generation—
an increase in genius. His operational reinterpretation of physical concepts had seen him through the crisis. Now, however, the crisis was
of another kind. It was ideological. He felt that his entire way of life
and the values that supported it were being betrayed. To his mind, a
collectivist mentality was overshadowing the libertarian outlook that
nourished pride in freedom and individuality, and this mentality was
threatening the free pursuit of scientific truth. This he resisted with a
vehement sense of righteousness.

However, if the younger generation was not upholding the dignity
of free science, this was merely a symptom of a wider degeneration of
democratic values. In a dinner speech given at the annual meeting of
the American Physical Society in January 1949, entitled "Sentimental
Democracy and the Forgotten Physicist," Bridgman poured out his resentment. Citing the diminished relative earning power of the physicist
as an indication of withheld social recognition, he alleged that "one of
the most obvious and sinister of the social changes of the last fifteen
years is just such a deterioration of the admiration and respect with
which the community holds the man of unusual intellectual ability." [20]

The social changes Bridgman found so reprehensible (among them
the graduated income tax, the establishment of a minimum wage, the
increased power of the labor unions) were Marxist in substance, he

charged, if not in name. Such social equalizers, in Bridgman's opinion, did not represent progress in social justice but the triumph of mediocrity, the exploitation of the superior man by his inferior, theft hiding behind the mask of justice, democracy gone soft. As far as he was concerned, no man is entitled to reap more than he sows.

It is easy, he conceded, to see "how the mediocre man can fight for his own exaltation at the expense of his superior fellow." But the only way to understand how the superior man can acquiesce in his own abasement was to see him as a victim of a shortsighted sentimentalism, the product of a "doctrinaire attitude toward the fundamental philosophy of democracy which springs directly from its roots in the French Sentimentalists who so influenced our founding fathers. This doctrine glorifies the common man simply because he is a common man." Bridgman called on the man of exceptional ability, the physicist as a special case, to "discard his sentimentality and his false modesty, and be willing to urge more articulately the worth of his contribution to society." [21]

The triumph of mediocrity, the object of so much of Bridgman's scorn, however, was much more than the discrediting of excellence or even the subjugation of the scientist. It was for him a moral betrayal and a philosophical error. Bridgman, the moral absolutist, saw his stern, ascetic ideal being overridden by laxity and mundane interests. The life of the mind was losing out to the life of the body. "The mediocre man prefers security to freedom—security at the expense of freedom is what we are getting. The exceptional man finds his highest values in the satisfaction of the creative impulse, while the highest good of the mediocre is creature comfort—we are getting a society in which the supreme good, to which the government should bend all its efforts, is raising the standard of living." [22] Bridgman, the philosophical particularist, argued that the degeneration of old-fashioned democracy to the democracy of the welfare state could take place only behind the screen of the universal absolutes Society and Government. Only when "the forest obscures the trees" is it thinkable that an "individual would have the effrontery to steal from his neighbor on the plea that he needs his neighbors' goods more than does his neighbor." [23]

A good deal of ground has been covered—from the threat of totalitarianism to the question of the moral responsibility of science, from the degrading specter of the welfare state to the stifling anti-intellectualism of the 1950's—but in Bridgman's mind all these situations have in common their origin or justification in an absolute principle, a "thesis of some kind." Human progress, he believed, was achieved through the insights of individuals of genius, whose accomplishments have been to expose the untenability, indeed the falsity, of universal absolutes. It is but a short step, then, to his conclusion that when the intellectual activity of the intelligent individual is inhibited, progress is arrested and degeneration sets in.

Furthermore, his own experience had taught him that for the task of unmasking metaphysical imposters, there is no one as qualified as the physicist. He had learned this from Einstein (the Einstein of special relativity). The burden for the progress of the human race falls to the physicist, to him who is of the "elect," who is privileged to possess the intelligence and moral stamina to resist the temptations of metaphysical delusion, social dependency, and political collectivism.

The physicist leads by his example, by demonstrating his dedication to science, which has no master outside itself and serves no end but that of truth. Yet, the question remains—what is that truth? As Scudder Klyce had observed many years earlier, Bridgman had decided that the truth is that there is no Truth. Science is *about* itself: meaning is no more than method. In a similar fashion, Bridgman had now concluded that science is *for* itself, that science is pure activity with no ties to its social surroundings. But where does that leave the scientist? Has he been condemned to merely running in place, carrying on a dialogue with himself? Surely the scientist is doing something more than virtuous busywork while he waits for Judgment Day. Surely this could not be what Bridgman meant to say.

CHAPTER 11

The Broken Covenant: Science, Society, and Individual Freedom

A too sophisticated society . . . like ours has outgrown not merely the simple optimism of the child but also that vigorous, one might almost say adolescent, faith in the nobility of man which marks a Sophocles or a Shakespeare. . . . When its heroes are struck down it is not, like Oedipus, by the gods that they are struck but only, like Oswald Alving, by syphilis, for they know that the gods, even if they existed, would not trouble with them. —Joseph Wood Krutch, *The Modern Temper* (1929)

In a society that has lost its simple faith in progress . . . [the] tragic spirit can promote a saving irony. . . . Modern tragedy is particularly deficient in Spirit; it seldom exhibits or promises Redemption. At most it may help to redeem us from fear or despair, or from the vanity of cheap hopes. —Herbert J. Muller, *The Spirit of Tragedy* (1956)

■ ■ ■

WHEN MAX WEBER ASKED what contribution science can make even though it cannot tell us how to live, he answered that besides offering technology and methodology, science can serve as an instrument for achieving clarity of thought, confronting one with the rational alternatives involved in making decisions. It can, he said, go no further, however, for ultimately a practical position must rest on inner consistency,

hence integrity, [derived] from this or that ultimate *weltanschauliche* [worldview] position. . . . Figuratively speaking, you serve this god and you offend the other god when you decide to adhere to this position. And if you remain faithful to yourself, you will necessarily come to certain final conclusions that subjectively make sense. This much, in principle at least, can be accomplished.[1]

It remains to be judged, however, whether all gods are equally be-
nevolent, and in Bridgman's life, we witness the outcome of one man's
choice. Nevertheless, Bridgman did not believe that he was so constrained. He would pay homage to *no* god. He would simply know the
truth, the pure unembellished truth. He would be deceived by no human artifact. He would permit no intellectual abstraction or metaphysical invention to parade as absolute and deprive him of freedom. By
adopting the scientific attitude and employing the operational method
of analysis, he would locate the source of truth and thereby secure his
individual freedom. Bridgman would follow the example of Einstein,
who in overthrowing the false Newtonian metaphysics, had awakened
him to the revolutionary efficacy, the purifying power, of the operational approach. Thus, Bridgman not only denied contemporary com-
plaints that science had become bankrupt but asserted that the operational method of science, as taught by Einstein, was applicable to
understanding social truth as well.

Bridgman did not advocate the development of a new or improved
social science. He thought that approach had been tried to the point of
evoking "revulsion" at the attempt. Nor did he entertain any proposals
for social engineering. Rather, he held up the scientific life as the ideal
toward which a rational man would be inclined to gravitate and, by direct implication, as the norm against which the whole of social life
should be judged. However, behind this seemingly modest and genteel proposal ran a deeply subversive intention, which was, in fact, a
challenge to the cognitive foundations of traditional social authority.
For science, Bridgman believed, is an activity organized to discover the
truth, and the members of the scientific community are human beings
who value rational thought and intellectual honesty. That means they
do not accept the imposition of supernatural sanctions by arbitrary
authority.

In Bridgman's opinion, social truth should be established in the
same self-critical way in which Einstein had laid the groundwork for
special relativity, with the same concern for nonmystical, naturalistic
foundations. If this were done, he thought, the irrational beliefs upon
which social authority is founded would be exposed and the intelligent
individual would be freed from the "stultification" that he presently
suffers.

Convinced that the insight he had gained from physics was applicable to understanding social reality, Bridgman believed that an exact

parallel could be drawn between the cause of contemporary social problems and the troubles that had plagued pre-relativistic physics. In particular, he thought that man's traditional conceptual equipment was inadequate to deal with the expanded range of modern social experience. Physics, he declared,

discovered that its traditional concepts were not capable of dealing with the situation . . . [and] . . . had to devise new ways of thinking about the situation, and in doing this it has had to examine to a certain extent the nature of human thinking itself. It has come to see that thinking is merely a form of human activity, performed with the brain, subject to the limitations of its evolution and its organ of production, and with no assurance whatever that an intellectual process has validity outside the range in which its validity has already been checked by experience.[2]

It was time to recognize that social concepts adequate in the past had simply become outmoded because of new social experience created by modern technological advances and the impact of contemporary political events.

Given the obvious parallelism, Bridgman argued, the solution to social problems is to duplicate what has been successful for physics; namely, old mysticisms must be identified and replaced with a new, more realistic (i.e., naturalistic) attitude. This was what Bridgman believed that Einstein had achieved in physics: "Probably the greatest contribution made by Einstein in his theory of relativity was his insistence on the realistic nature of the concepts of physics." And, we have seen, Bridgman regarded Einstein as having brought this improvement into the body of physical theory by means of the operational method. Contemporary social thinkers, Bridgman observed, were now experiencing

a growing appreciation of the realistic nature of social conventions and institutions as opposed to the traditional idealistic point of view, which is perhaps best exemplified in the disappearing notions of abstract justice and abstract rights, and of the nature of law, with which legal practice and theory are still encumbered.[3]

To Bridgman this meant that the time was ripe for an operational analysis of social concepts.

If philosophers of science had grounds for misunderstanding the meaning of operational analysis and its relationship to Einstein's special relativity when it was originally enunciated in *The Logic of Modern*

Physics, no ambiguity obscured Bridgman's intentions when he turned to social analysis. Operational reasoning was unequivocally an instrument for demystification, a defense against error and deceit. It was a vehicle that would free the individual from external or received authority, an authority whose legitimacy, he thought, rested on the perpetuation of falsehood.

Bridgman felt that operational analysis was his guide to a continued intellectual awakening which had been forced upon him in science by Einstein, and, one may reasonably hypothesize, which he had experienced in the first instance with his discovery of the superiority of scientific truth over religious dogma. Now the revolution was to be extended to social thought. Henceforth, no inherited idea could be considered trustworthy unless it passed the test of operational analysis. Furthermore, since one was certain to experience considerable disillusionment in the process, it was a measure of a man's intellectual honesty that he could face up to and accept the unmasked truth.

Thus, in a 1933 article entitled "The Struggle for Intellectual Integrity," Bridgman asked,

What will the man do in whom has been suddenly born an appreciation and capacity for intellectual honesty, with its disregard of ulterior consequences, when confronted with our social institutions and the demand to accept them and to live with them?

To this he answered,

The first and inevitable reaction will obviously be a complete repudiation in his own mind of the bunk that he is asked to accept. So much he must do, though it slay him. But he must also continue to live in society as he finds it, and he must try to work out for himself some code of conduct which he can pursue without intellectual stultification.[4]

Bridgman was no more a social activist than he was a scientific theorist. He was not a creator of new realities. He did not imagine that he was issuing a manifesto that would inspire a social movement. His was a private revolution, a revolution in consciousness. It could perhaps more accurately be thought of as an intellectual inquisition. The individual whose freedom he sought to establish was not the universal human being of the Enlightenment, but rather a concrete existential person, himself to be more precise. The intelligence that was to be the agent of liberation was not the natural reason of the Enlightenment, which participates in the harmony of an orderly universe. It was intellect as inquisitor and tyrannical disciplinarian.

And as Bridgman's inquisitor pressed the inquiry with ever more rational earnestness and operational righteousness, the ties that create social meaning were destroyed, leaving Bridgman stranded at the limits of the meaning of human freedom and value. Truth, he found, is loneliness, virtue is endurance, sin is seeking escape from self-sufficiency.

Moreover, in this bleak, rationally purified universe, there seemed to be no place for life itself. Bridgman noted immediately that the practice of intellectual integrity was an unsocial virtue insofar as it encompassed no impulse to human life. He could not see how rationality could serve as the foundation for survival of the human race. And in a universe where life has no meaning, death has no meaning either. "Death," he wrote in his 1933 article, "from the point of view of any individual free from mysticism, can be no calamity; for death itself is not experienced." If, for example, the future held the prospect of more pain than pleasure, "a completely rational conduct of one's affairs would then demand that . . . one should immediately find the way out by suicide." A society made up of this kind of rational individual might never survive, he admitted. On the other hand, he speculated, it is conceivable that intelligence in society would move in "Sisyphean" waves, growing to a certain level and annihilating itself, leaving behind the possibility that the process would repeat itself.[5]

This was not a momentary insight or passing phase in Bridgman's thought. Once it had occurred to him, he found that he could not undo or revoke what he had discovered. The only protection against a new idea, he decided, is stupidity. He wrote in 1938, "we cannot just put back the vision we have had and pretend we never saw it."[6] And in 1960, a year before his death, he wrote to Raymond Seeger at the National Science Foundation, "What *scientific* justification can you give for the universal assumption of all ethical systems that the preservation of the human race is desirable?"[7] Despite his best intellectual efforts, he was never able to overcome his nihilistic vision, philosophically or psychologically.

The fullest development of Bridgman's social analysis is to be found in a remarkable book entitled *The Intelligent Individual and Society*, published in 1938.[8] The book did not receive especially favorable reviews. It was characterized as having little originality, and Bridgman's ideas were compared to those of the eighteenth-century philosophes and the British utilitarians.[9] To be sure, there were superficial resemblances, but Bridgman's meanings were based on quite different assumptions. The Darwinian revolution had intervened, canceling the deistic metaphysics that supported eighteenth- and early nineteenth-century philosophical speculations. Indeed, this book may be read as a clear indicator of Bridgman's growing existential sensibilities. It is of significance that, as he remarked in a letter to G.E. physicist Willis R. Whitney in 1951, it was his favorite, and he regretted that it had not received more attention.[10]

Written in a form styled after *The Logic of Modern Physics, The Intelligent Individual and Society* is an inquiry into the nature of social reality and the condition of the thinking individual within society, that individual being, naturally, Bridgman himself. Analysis was the key, he believed, to the intellectual clarity that would enable him to distinguish the elements he must consider so that he could order his life according to rational principles, a need, he confessed, that was growing "to an almost physical intensity."[11]

For Bridgman this was no disinterested or abstract intellectual exercise. His earnestness did not leave room for philosophical romance. He was playing for keeps. Making it emphatically clear that the book was not written for philosophers, he represented himself as the *enfant terrible*. In his own mind, he was like the child in "The Emperor's New Clothes" who innocently announces what everyone knows but dares not say. Yet his line of reasoning was far from innocent.

As a preliminary to his analysis, Bridgman felt it was necessary to spell out the limitations of the cognitive instruments constraining human understanding. In doing this, not only was he acting in accordance with his commitment to empiricism, but he was also applying Einstein's operational methodology as he understood it. At the same time, he was affirming an assumption now characteristically associated with existentialism. For Bridgman the imperfection of human reason was not a formal *a priori* truth or logical judgment. It was basically a biological and/or psychological condition. His stand was certainly compatible with both religious transcendentalism and philosophical empiricism. Nevertheless, it was not Einstein (who had sounded the alarm), or even Calvin (who supplied the defensive strategy), but Darwin who had set down the ultimate limitations.

There are no universal "intellectual tools," wrote Bridgman, and there is no escape from self-illusion or error. All events are unique. The categories of thought—words and objects—are approximations, mnemonic devices, or pragmatic inventions, but not realities in and of themselves. Any certainty attributed to the processes of reasoning— logic and mathematics—is the consequence of induction from experience and is, in fact, only a frame of mind. Traditional philosophy and religion, in Bridgman's opinion the two most egregious examples of man's "bad faith" (to borrow a phrase from Sartre), are "permeated with intellectual cancer" and should be scrapped and started over. As an example of a similar situation in science, Bridgman declared, in a stab at Einstein, to assert the constancy of the velocity of light is to claim more than we can know.[12]

Having presented the general claim that the instruments of thought are intrinsically imperfect, Bridgman went on to consider social beliefs as a particular class of these flawed human devices. His procedure illustrates his broadened conception of operational analysis, which by this time no longer carried the ontological suggestions that it once did. He did not look for an operational equivalent to a social concept; he asked how it was that he came to have the idea and in what situations the idea is applied, expecting, thus, to discover a historical and naturalistic origin.

He began, "Let us now examine by way of introduction and explanation the meaning and implications of a few of the concepts that hold society together."[13] Selecting "duty" as the first target for operational analysis, he asked himself how he decided "whether a certain line of conduct is my duty or not." What guidelines did he follow? Where did they come from? After reflecting on the origin of his concept of dutiful behavior, Bridgman concluded that duty was an idea imposed on him by someone else. In a wider social setting, he suggested, it appears to be merely sublimated public opinion. And from a utilitarian standpoint, it is a device to induce conformity of behavior. But beyond that, he asked, what is the source of the felt compulsion associated with the apprehension of duty? It lies, he answered, in the sense that one is somehow serving higher purposes, absolute obligations outside one's own interests. Analysis, Bridgman asserted, shows that this presupposition has no objective basis, and therefore the rational individual will reject the conventional concept of duty because he does not surrender to outside authority the prerogative to dictate what his conduct should be. Historically, Bridgman speculated, the idea of duty could have appeared only by accident since no human being could have had the "diabolical intelligence to deliberately invent such a concept and sell it to his fellows." [14]

Next, Bridgman took up the question of the meaning of "responsibility" and immediately encountered its concomitant, "free will." Here, operational analysis was insufficient to its task, or rather, it led to conclusions whose implications offended Bridgman's distinctly Puritan conscience. Operationally, the sense of free will was merely a psychological accompaniment to decision making, since there can be no operation by which one can actually know if he could have chosen differently. This was, he asserted, in no way incompatible with his own belief in a deterministic universe. After all, one can think he is exercising free will when as a matter of fact he is not. Modern psychoanalysis, he added, confirms this possibility when it teaches that there are cerebral processes of which we are unaware.

However, such a conclusion, even though the product of operational reasoning, seemed to Bridgman at variance with both his moral sense and common experience. Does it mean, he asked, that all effort is therefore futile? No, he said, empirical evidence suggests a correlation between trying and achieving. The problem is that the concepts of determinism and free will are themselves imperfect; it does not follow therefore that one should give up trying. Indeed, had Bridgman admitted the absolute validity of his operational conclusion, it would have deprived the motivation for writing his book of all significance.

Therefore, undeterred by the necessity to supplement, or attenuate, operational analysis with a bit of Puritan casuistry (one cannot know God's purpose, but one can proceed on the basis of encouraging signs), Bridgman continued his vendetta against social abstractions. The idea of a right, such as the right to property, he argued, was entangled with absolutist assumptions, when in fact operationally it is only an approximate concept, often denoting merely something intensely desired. Similarly, morality cannot be referred to absolute principles; it is something that "may change with time and circumstance and opinion." Even if it might still be possible to agree on a unique standard of morality, he ventured (and hoped), it nevertheless would retain some quality of mysticism. And justice, too, has no absolute status. It is only a program and an aspiration. In sum, except for a certain coercive utility, Bridgman was unable to attribute any operational meaning to the social concepts he selected for consideration.

Finally, the state, as the enforcing agency for social ideals, was operationally demoted to a mere collection of individuals. It is not, Bridgman emphasized, a "superperson." Its only authority derives from force, not from absolute or transcendental principles. Thus, to Bridgman physical force was the true or "realistic" foundation of social authority. He had concluded, in tones reminiscent of Nietzsche, that authority has no divine sanction but is the manifestation of human will, enforced by superior physical power.

So far, Bridgman has enunciated nothing more than a crude pragmatism or utilitarianism. He has offered no original insights. Nor has he introduced the subjective aspect of operational analysis that made his epistemological investigations into the nature of physical reality so unacceptable to his colleagues. This occurred when he took up consideration of the absolute limitations of analysis itself, what he called the "great limitations." It was in the finality of these great limitations where Bridgman again encountered his subjectivity.

According to Bridgman, the first limitation, one which guarantees that certainty is forever an impossibility, is the need to assume the validity of our mental processes. Of this, there can be no independent proof. "Certainty" therefore can be only an approximate concept and is, in fact, intrinsically self-contradictory. (Self-contradictory only if regarded as approximate, Bridgman forgot to add.)

The second limitation, which made by far the greatest impression on Bridgman, was the essential and unconquerable isolation of the individual within the confines of his own mind. "Never have I escaped by a hair's breadth the fate that has decreed that I shall lead my life alone in my own 'consciousness,'" he wrote.

So absolute is the ring that encircles me that it is meaningless to ask whether I might not escape my own "consciousness" and share the "consciousness" of another. So far as "consciousness" is capable of definition at all it is that which comprises my experience. . . . The concept of consciousness is perhaps the ultimate unanalyzable, and when we attempt to express the possibility of getting away from it we attempt the impossible.[15]

This "inexorable isolation of the individual," Bridgman complained, "is a bitter fact for the human animal, instinctively so social, and much of his verbalizing reflects his obstinate refusal to squarely face so unwelcome a realization."[16]

Operational analysis revealed to Bridgman that consciousness can never be other than self-referential. Operationally, there is only one consciousness, and that is one's own. There is no operation by which the consciousness of another can be verified. It is social utility which justifies the "trick" by which one projects onto another a consciousness just like his own. It "is the device by which one endeavors as far as possible to anticipate the probable future actions of his fellows and so to put himself in a position to make the necessary preparations." One of the greatest social fictions invented is the idea that "the best and deepest interests of the individual are coincident with those of society." [17]

Bridgman believed that this essential isolation and loneliness was his most significant discovery. He spent a great deal of effort trying to communicate the ultimacy of this operational truth. It was, he wrote, "central to the entire point of view of this essay." [18] The reason, we will see, is that in his isolation and subjectivity he found the key to individual freedom, a lonely and pointless freedom, to be sure, but one that was unconditional. Just as significantly, it must have seemed to Bridgman that here he had found operational *proof* that the idea of universal truth is an artifact.

The most difficult argument Bridgman had to convey was that he was not advocating solipsism. He acknowledged that he was frequently accused of this, but his understanding of the solipsist position, "that nothing exists except my own thoughts," was "so contrary to common sense as to be its own refutation." [19] While he agreed that the solipsist has "caught sight of a fundamental problem," he thought that the difficulty of conventional solipsism was one of language, that language does not possess terms refined enough to deal with the problem the solipsist has discovered. Evidently this was a problem for Bridgman also, for he had little success in distinguishing his position from what others took to be solipsism. In fact, Bridgman was not trying to make an ontological generalization. This would have betrayed his existential epistemology. His was a psychological criterion that made subjective experience the condition for the possibility of truth itself. Anything outside of immediate experience was an intellectual construction and therefore suffered from the inadequacies of all idealization. But even more than that, knowledge is subjective, it belongs to the individual, and this is all that true knowledge can ever be. It is locked into the individual consciousness, or locked out, according to the ability of the participating mind to judge and accept what is presented to it.

Here, in the inviolability of subjective consciousness, Bridgman finally found the freedom from authority that he set out to discover. He wrote with great passion:

I think examination will show that *all attempts to secure certainty involve somewhere a transcendental element, something over and above the individual in the face of which the individual is supposed to be powerless.* But if our analysis is right, this transcendence can never be achieved, for at the very beginning is the hurdle of this act of acceptance on my part. No one can force his authority upon me until I have recognized his right to authority; *without acceptance it is not authority,* by the very meaning of authority. . . . Failure to recognize this I think permeates the most diverse fields of thought and activity. I think the professional logician fails to recognize it when he represents symbolic logic as something with absolute validity. The concept of validity is meaningless until I have made the logic live by my vision and my acceptance, and it is only a living logic that has significance. I think Einstein fails to recognize it when he tries to get rid of all taint of the individual observer in his general theory of relativity. I think the professional theologian or moralist fails to recognize it when he claims an absolute authority for some code of conduct or some dogma. The voice of God which I am told speaks to me directly through my conscience with an authority that cannot be evaded is no more than a babbling brook and not the voice of God until I have accepted it to be the voice of God. I think Stalin and Mussolini and Hitler fail to recognize it when they exact unquestioning allegiance. [Emphases added][20]

In other words, as is the case with knowledge, authority does not exist without the individual's consent.

There was one further "great limitation" Bridgman presented to round out his argument that social idealizations are illusions perpetrated to mask the realities of social authority. At rock bottom, he believed, no mere idea can prevail against physical force.

No superiority of ethical or logical or esthetic position avails against the actual exercise of superior physical force by another. . . . This it seems to me is another of those things that need only be said. . . . There have been attempts to blink this fundamental human limitation, so derogatory to the dignity of the human reason, as there have been attempts to blink every other unwelcome limitation.[21]

This meant, among other things, that to invoke codes of conduct in dealing with hostile parties is simply naive and unrealistic. An example, Bridgman wrote, is the "childishly naive eagerness" of the attempt "to get Hitler to enter into another contract with regard to militarization of the Rhine in order to prevent him from violating still another contract."[22] Similarly, there is no point in exacting an oath from someone, since one can never know if the swearer is lying or not.

At first glance, by thus acknowledging the finality of physical force, Bridgman appears to have compromised the absolute freedom that he sequestered in his seemingly impregnable and indomitable subjectivity. Must not the individual ultimately yield to overwhelming force and accept the authority of the physically superior agent? No, said Bridgman. A completely rational and intelligent person realizes that death "annuls all sanctions for conduct," and in the case of suicide, society is impotent to impose its sanctions. Society has tried to hide this by putting up a "smoke screen" and inventing the mysticism of the sacredness of life. But for the individual, death is not a calamity; it is just the cessation of experience.[23] And, for Bridgman, the perspective of the individual is the only one that matters. No "cosmic purpose" makes the survival of the race or society of transcendental importance. There is no grand universal scheme in which it makes a difference whether the human race perishes. And in the last instance, the ultimate act of individual freedom is self-destruction. This conclusion descended upon Bridgman with the inexorability of scientific reason. And it was not to be contemplated with a sense of tragedy. It was, when the veil of mysticism is lifted, simply what is true.

Among the clouds of irony that settled over Bridgman's life, none was heavier than that which grew out of his quest for intellectual security and clarity. Bridgman's operational method was ultimately self-defeating. The skepticism of his operational approach acted as a powerful corrosive to the foundations of knowledge, and as the meaningful universe was operationally dissolved, his own consciousness became for him the last refuge of order and the only source of value.

The problem he then faced was one created by his own operational analysis, and what he found was not easy to accept. "Operational analysis really does make a difference; to properly appreciate this with my backbone," he wrote, "I find a more formidable task than the analysis itself." The emotional readjustment required to come to grips with the egocentric position to which he was reduced was one of resisting its demoralizing force, of resisting "the feeling that with our new intellectual vision our lives have become more selfish and the whole enterprise of living more sordid." It is not easy, he confessed, to maintain "an attitude of proper scepticism toward the traditional intellectual machinery of mankind without developing cynicism."[24]

Here was the challenge to this earnest and sensitive individual who had been unwillingly awakened to an unwelcome existential self-awareness, to this dedicated scientist who set out to purify his understanding of science only to find himself as the final term. He believed that science is discovery, not invention. Science is not art or philosophy, it is not constrained by aesthetic standards; nor is it engaged in building a unified picture of the world. It simply discovers the truth, the bare and unadorned truth.

As a consequence of his epistemological interpretation of Einstein's special relativity, Bridgman had concluded that truth was not an affirmation but a disillusionment in a very literal sense—a sweeping away of historically accumulated errors, both individual and cultural. Truth is what is left over when everything that is not operationally concrete has been eliminated. Furthermore, one does not embrace truth triumphantly; rather, one faces up to it alone, with fortitude and perseverance. And it is in this private stoicism and the strength required to sustain it that the source of human dignity is to be found.

No matter how disagreeable the truth might be, then, no matter how nihilistic its implications, Bridgman did not on that account conclude that there was nothing that could or should be done. That would have been admitting that he did not possess the intelligence to discover the truth or the moral strength to endure its loneliness.

However, if endurance is a measure of virtue, sinfulness is the tendency to escape from the lonely pain of truth to the comfort of the universal or communal. In this spirit Bridgman undertook an inquiry into what he called human traits, in particular those traits that encourage the creation of the abstractions he so disdainfully designated mysticisms. Bridgman's analysis took the form of a critical exposé of what he regarded as human tendencies to surrender independent judgment and place faith in abstractions. These tendencies were either social (such as the "herd instinct") or intellectual (for example the impulse to generalize). However, in contradistinction to the "great limitations," the human traits did not constitute absolute limitations. They can, he believed, to some extent be purged and thus overcome.

Of the traits that Bridgman found especially worthy of contempt was the common tendency of human beings to rationalize. It was to him equivalent to lying. Pragmatically, he argued, it is not safe to rationalize, since there is always the chance that "the consequences of rationalizing will come home to roost before one dies." But more than that, it is an indication of stupidity or mental laziness, that is, of un-willingness to take the trouble to "recover primitive complexity" (at-tend to details). The desire to make things easier is also at the root of the impulse to generalize, but so is immaturity. "Youth rushes to facile generalization." In a way, Bridgman noted, it cannot be helped because youth has less experience and must rely on the trustworthiness of "programs for the future." However, "at the end of life the program aspect is gone, and we have left merely a bundle of particulars, tied together with analogies which we will not have time to check."[25]

Judging the "drive to consistency" was more difficult. Bridgman could not decide whether it was a virtue or a vice. On the one hand, it appeared to derive from the same mental limitations as the generaliz-ing tendency, namely, the inability to cope with complexity; on the other hand, it was a sign of the "stability of drives," and this was in-dicative of greater intelligence.[26] Still, from a logical standpoint, self-consistency could be seen as an excuse to extend meanings beyond their domain of applicability.

Turning to the "herd instincts," the desire for approbation or the tendency to follow the crowd, for example, Bridgman remarked that they also "impress themselves in one way or another on our intellec-tual behavior."[27] This is because the crowd does not value using the mind. But there are also psychological drives that, when expressed, re-veal the urge to dependence. Among these, the desire for confession and forgiveness are related to guilt. With intellectual maturity, how-ever, comes the realization that the mysticism giving them meaning is a rationalization of the desire to find something external to provide relief from fear and anxiety. That something external does not exist.

The facts being thus exposed, the moral is clear. The individual is on his own. And since "the individual" is a concept one step removed from direct experience, Bridgman emphasized that his conclusions re-ferred in the first instance to himself. "I" am alone.

I think the one result of my analysis which is going to make the most difference is that I have seen no way of getting away from my own central position. I find that I have inherited a verbal machinery which treats my fellow exactly like my-self, and gives the same kind of meaning to his "consciousness" that I give to my own; there has been invented for me a super-person, the State, in whose identity I am told it is my highest duty to merge my own; and Nature has been populated with beneficent forces. I find very congenial features in all these things, and I would like to be able to take advantage of them. But nevertheless, *eventually I have got to stand on my own two feet, with no adventitious help from any-*

thing outside. The things that are to make me go I am to find inside me. *The reason is that nothing else makes sense. What I am trying to express when I say "There is something bigger back of it all" simply won't carry through; I am groping for something that isn't there, because it can't be made to mean what I want.* Furthermore, I ascribe no "objective" significance to the fact that I have groped for it, because it is easy to see how very pleasant I should find it if it were only there; this is motive enough for whatever search I may have made, and for the age-long search that the human race has made. [Emphases added][28]

Yet for all of the freedom from external coercion Bridgman has achieved, at least in principle, nevertheless he has not escaped the tyrannical taskmaster within his own consciousness. For in the underlying struggle between intuition and rationality, he has willingly submitted to the stern discipline of the intellect. The consequences of his choice made him distinctly uneasy, and the result of his reasoning, Bridgman admitted, was not easy to accept. It "runs counter to deepseated urges," but the intellect demands it. "If one wants to follow his brain to the limit I think this is where it lands him. After one has seen this he cannot put it back and just pretend he never saw it." Realizing that it was impossible to defy the authority of reason, Bridgman decided that "fortitude is the name of the virtue required to meet the situation."[29]

A good measure of fortitude was indeed called for to bear the sacrifice that Bridgman thought necessary to earn the respect of reason, for subsequently he declared that happiness was not one of his goals. In 1945 in a letter to anthropologist M. F. Ashley-Montagu he explained why.

The problem as I see it is how is a man who wants his life to be capable of some degree of rational scrutiny, and who also is a social creature who needs the sympathy and understanding and cooperation of his fellows, going to live a happy life in a society in which most people not only do not desire to live a rational life but are actively hostile to any endeavor to introduce rationality into social institutions many of which are beneath intellectual contempt. I think the thing simply cannot be done, and that any one who demands of himself that he lead a life which he can contemplate with intellectual self respect must renounce the expectation of attaining happiness in society as at present constituted. The most that such a person can expect is that he shall find a pursuit which shall be so absorbing that he can lose himself. I think that most of us do not ask more, and I think that many of us have been fortunate enough to have found as much as this. Perhaps you will want to call this happiness, but if it is, it is a drab and neutral thing, which comes only incidentally and not by conscious effort.[30]

The inquisition is nearly complete. Bridgman has been forced to admit that there is no Truth, no universal meaning, no Society; there is only self, the lonely, courageous, and all-important self. But the inquisitor is not satisfied. There is one last admission to be wrought, and Bridgman does not fail to yield. The fact that "there is nothing bigger in back of it all" meant that society must exist for the individual rather than the individual for society. "I very much hope that the social ideals of the future may be in the direction of rationality; that the individual will be recognized as the only thing that can give any justification for the whole." But more ominously he also warned that if society cannot be organized so that the group serves the individual, "then *society must be allowed to perish, and the race also, if that is involved*" (emphasis added).[31] Even the vindictive God of the Hebrews did not pronounce destruction upon *all* of mankind when they fell into sin.

Bridgman's social views remained essentially unchanged in later years, and in and of themselves they had little impact. Reiterated and published in his 1959 book, *The Way Things Are*, his social philosophy was characterized by the publisher's referee as being highly idiosyncratic and "one that might conceivably have been written by a gifted mind a century ago."[32]

Perhaps so in some respects, but what was not evident to this reviewer, and to a good many other readers as well, was the persistent shadow of existential despair that darkened Bridgman's philosophizing. He was constantly aware of a harsh disillusionment that mocked traditional values and made it impossible to turn back in history, impossible to look to the past for solutions to contemporary problems. To a large extent, Bridgman noted, science is responsible for this predicament, historically and intellectually, because it has changed the way we perceive the world. In a paper published in the *Proceedings of the American Academy of Arts and Sciences* in 1954, entitled "The Task Before Us," Bridgman made this especially clear. After reviewing recent developments in logic, quantum mechanics, and psychology, he wrote:

These considerations are to some extent all matters of detail. Out of all the detail emerges the one stark fact that *we can never get away from ourselves.* Yet apparently this is the one thing that the human race feels it has to do. The philosopher with his eternal principles and truths, the man in the street with his real external world demonstrated with a kick, the mystic with his transcendental visions, and the scientist Einstein with his general theory of relativity, are all equally engaged in the search for an absolute by which they may get away from themselves. Yet the kick of the man in the street is a kick activated and apprehended by his nervous system, and the vision of the mystic would be no vision without his nervous system. *Wherever we go we find ourselves; an observation which has the profundity of tautology, a tautology which reduces the age-long quest of the human race for standards and springs of action outside itself to the ultimate futility, the futility of meaninglessness.* What sort of peace we shall eventually make with this insight I do not know, nor do I know how the founding fathers would have acted if they had been confronted, as are we, with the necessity of finding acceptable terms for a temporary armistice until the terms of a final peace can be hammered out. But *it seems inconceivable that the precise form of such of their concepts as human rights, or the dignity of the individual, or freedom or justice, so confidently enumerated in our Declaration of Independence, would have come through unscathed, for they are all tainted with the odor of the absolute.* [Emphases added][33]

If the substance of Bridgman's social philosophy did not gain widespread consideration (*Harper's* rejected "Prospects for Intelligence" twice in 1944),* Bridgman did receive local recognition as a commentator on the relationship between science and human values. After the war, the problem of the crisis of meaning reappeared as a subject of public discussion, and Bridgman received many invitations to take part in such discussions. As a matter of principle, he refused to participate in those that included the subject of religion as part of the agenda.† Otherwise, he joined in the spirit of debate.

Thus at a symposium organized under the auspices of the Harvard Law School Forum in late February 1949 entitled "Values for Modern Man," Bridgman was invited to participate not as might be expected in the session on the physical sciences (which was represented by geologist Kirtley Mather, physicist Phillippe Le Corbeillier, physicist-philosopher Philipp Frank, and astronomer Cecilia Payne-Gaposchkin) but as a speaker in the concluding discussion, also entitled "Values for Modern Man," together with philosopher William Hocking, theologian Walter Muelder, and jurist Lon Fuller. The thrust of Bridgman's remarks was to express his hope that scientific method, the method of intelligence, could contribute to the understanding of social problems

---

*Upon receiving a rejection slip from *Harper's*, Bridgman responded in a letter on Oct. 9, 1944, that he did "not write this sort of thing easily and spent much time on it." His article was subsequently reconsidered but rejected a second time because "the merits of the general argument are obscured by their apparent dependence on observations which would strike many readers as doubtful."

†See, for example, Bridgman to Finkelstein, May 17, 1948. Bridgman refused to take part in the Conference on Science, Philosophy, and Religion because of his antireligious feelings. "So long as the name Religion appears in the title of the Conference, I could not with self-respect associate myself with it."

by providing an instrument to clarify the meaning of socially important concepts and thereby establish the groundwork for social consensus and harmony of purpose. However, as one might now expect, he was not especially sanguine about this possibility, and he ended on a pessimistic, indeed nihilistic, note.

If in the fullness of time it should appear that the human race is not so constituted [as to respond to the educative function of intelligence], I suspect that the majority would want to patch it up some how by suppressing the free use of intelligence. But there are others [among them himself?] whose system of values is such that they would esteem such a solution unclean and would prefer that the human race itself should perish.[34]

Nevertheless, Bridgman was willing to try an experiment to see if such an approach might work. He would test the efficacy of the operational method to induce consensus on the meaning of fundamental social concepts. The following year he proposed to his Harvard colleague Talcott Parsons that he (Bridgman) be allowed to offer a course to graduate students whose goal would be to establish operational definitions for social abstractions, formulate the properties of these abstractions, and arrive at results upon which all members of the group could agree.[35] Parsons consented, and the seminar, called "The Logic of Agreement" at the suggestion of Harvard president James B. Conant, was organized. Students working for doctoral degrees in widely varying fields were selected to participate. (Among the participants was the young Ph.D. candidate Henry Kissinger, who contributed a dimension of historical and philosophical knowledge absent from the perspective of the more technically minded participants.) However, the results were disappointing. Operational analysis did not render agreement automatic, as Bridgman had hoped. It seemed that the problems involved were too complex.* He did not admit, however, that this experience necessarily invalidated the principle of the operational method. It just meant that the details would take longer and more work to reach.

Following the Harvard Law School Forum symposium by less than a month was the Mid-Century Convocation of the Massachusetts Institute of Technology at which Bridgman again spoke on the subject of science and human values. The format was a panel discussion of the "part played by science in producing [the crisis of Western values] and

*Bridgman to Dingle, Jan. 28, 1952. In this letter Bridgman also thanked Dingle for "not lumping me in with the logical positivists." In addition, see The Way Things Are, p. 245.

the contribution it can make to a solution." [36] There were four discussants: Bridgman; Julius Bixler, president of Colby College; Walter Stace, Stuart Professor of Philosophy at Princeton University; and Jacques Maritain, also a professor of philosophy at Princeton and formerly the French ambassador to the Holy See. Stace was known as "an outspoken expositor of a practical philosophy for today," and Bixler, as a liberal Protestant theologian. The extremes of opinion, however, were represented by Bridgman and Maritain. Bridgman was introduced as a "Nobel-prizeman from Harvard University," known to have "distinct and not always orthodox views on the subject under discussion," and Maritain, as the "greatest Roman Catholic philosopher of the day." The stage was set for a confrontation between two opposing worldviews. Maritain was the spokesman for metaphysical absolutism, and Bridgman, for scientific empiricism.

In his paper, Bridgman returned to the theme that "by voluntarily making himself the complete slave of the fact, the scientist [wins] complete freedom for himself in all other respects." In other words, it is only in accepting his fundamental limitations and discovering his isolation, which together "make meaningless his absolutes," that man will ultimately find freedom. This was, Bridgman asserted, the lesson of science and also the meaning (decidedly not the traditional one) of the biblical words "The truth shall make you free." [37]

Nonetheless, the truth offered by science was a truncated one, as Bridgman testified, since the "discovery of unsuspected physical structures in the direction of the very small, and in the direction of the very large," discloses the inadequacy of the human intellect to comprehend them directly. The concepts we use to deal with the world of experience are simply not applicable in these regions. "Thought demands its permanent objects and its consciousness of recurrent situations; how shall we think about a world that has not these intellectual necessities?" It is evident, Bridgman concluded, that man is "isolated on an island of phenomena between the very large and the very small, which he cannot transcend because of the nature of thought and meaning themselves." Science, according to Bridgman, tells us that we are prisoners, trapped within the phenomenal world, and that any hope of escape is illusory. And freedom means accepting this fate, so that one can stop "attempting to do with his mind things which cannot be done because of the nature of thought itself." [38]

Maritain, on the other hand, held the view that science operates on its own plane of knowledge, and that spurious extrapolations of science, in repudiating metaphysics, undermine "human reason's myth-making suggestibility" and cut off the possibility of approaching ontological truth. Furthermore, "the human person is threatened today with all-pervading slavery, not through the fault of science, but through that of the enlarged power granted by science and technology (that is, by reason mastering natural phenomena) to human foolishness." If wisdom is to be restored and human liberation is to be achieved, man must reach beyond the scope of the merely scientific to the contemplative love inspired by the Gospel. "Since science's competence extends to observable and measurable phenomena, not to the inner being of things, and to the means, not to the ends of human life, it would be nonsense," he argued, "to expect that the progress of science will provide men with a new type of metaphysics, ethics, or religion." [39]

It is hardly possible to conceive terms in which Bridgman's operational reasoning could have received a more direct repudiation. The problem was that he did not know how to respond. This was a universe of discourse in which he was a stranger, and he candidly admitted it. Thus, in the discussion following the formal presentation of papers Bridgman complained, "I knew I would be sorry when I accepted the invitation to appear at this panel. I can't speak the language." [40]

Nevertheless, he did not back away from confrontation, although he did move toward ground where he felt that he could gain some advantage. He wanted to know, he said, what Professor Maritain could mean by a metaphysical fact. The charge that the scientist neglects the facts of metaphysics, Bridgman answered, is not valid. There are many eminent physicists who have spent much time analyzing the evidence for telepathy. But this is apparently not what Professor Maritain means.

I think the fundamental reason the scientist does not accept the facts of metaphysics in what I imagine is the sense of Professor Maritain is that he has not been able to find in the realm of metaphysics anything to which he can apply the term "fact" in the sense he is accustomed to use it. The scientist simply does not understand what is going on in Professor Maritain's head when he speaks of the "facts" of metaphysics. [41]

A scientist, Bridgman went on, would not use the word "fact" or "truth" to refer to a condition that could not be checked. The same must be required of a metaphysical fact. The problem as he saw it was that no one can say "what the nature of the checks is to which you may subject these metaphysical facts or truths after you have got them, to find whether you have got what you think you have."[42]

Maritain's reply could hardly have been satisfactory to Bridgman. "I would affirm," he stated,

that metaphysics has its own criterion of truth, which is intelligible necessity brought out from experiential data. Metaphysics does not depend on any practical confirmation; our conduct can judge our moral philosophy to a certain extent, it cannot judge our metaphysics. Truth in this sense is above human behavior.

It is certainly factually true, Maritain admitted, that the consequences of treachery, murder, and perfidy can be observed in their effects upon society. But, he continued,

if this observation is to become an obligation, binding me in conscience, I think that the mere scientific or experimental observation is not sufficient and this fact must be related to more profound things. . . . If I only think that cooperation and goodness are improving the general state of affairs, this is not sufficient to make me bound in conscience. Something more is necessary. I mean an absolute, unconditional value which cannot be taken from merely empirical facts.[43]

The papers and ensuing discussion were published as a chapter in the volume *Mid-Century: The Social Implications of Scientific Progress*, edited and annotated by the dean of humanities at MIT, John Burchard. As a preliminary to publication, all participants were encouraged to submit additional comments. The most significant of these was the critique Maritain addressed initially to Bridgman and Stace but directed overwhelmingly to Bridgman. Maritain's remarks stand in sharp contrast to the judgment of Bridgman's operational approach by the analytical or positivistically inclined philosophers. Whereas they viewed Bridgman's operational thought within the narrow scope of technical— logical and methodological—criteria, Maritain judged it from the standpoint of traditional and scholastic philosophy. If Bridgman felt out of his depth with technical philosophers, as he had admitted on many an occasion, he must have felt even more the distance between himself and Maritain. Maritain did not believe that it was necessary to refer his metaphysical arguments to empirically established findings. For him metaphysics was grounded in a higher order of truth, beyond the fallibility of sensory experience.[44]

The magnitude of intellectual disparity between the positivistic concept of philosophy and traditional metaphysics will become ever more apparent as we follow Maritain's comments.

Maritain quickly dismissed Bridgman's suggestion that the facts of metaphysics had anything to do with extrasensory perception. He then considered Bridgman's observation that in the domains of the very large and very small the concepts basic to human thought, for example, existence, identity, causality, no longer apply. This, Maritain admitted, was true for their scientific application, but it was a consequence of the fact that

in using those concepts in its own noetic [cognitive] vocabulary science has recast them according to the requirements of empiriological or empirio-mathematical knowledge, by defining them only with respect to verification by sense observation and methods of measurement.[45]

But, he went on, this does not affect the philosophical meaning of these concepts, because

in the noetic vocabulary of philosophy, the concepts in question are defined with respect to the intelligible being which is perceived by the mind through experiential data, and which does not depend on our methods of sense verification. For instance, "existence" is philosophically defined, or rather described, as the act by which something "stands outside nothingness." And the meaning of this concept subsists even there where the nature of things makes it impossible to verify its application, that is, to know whether these things exist or not; whereas a scientific concept loses its meaning in a field which excludes by nature any possibility of verifying its application.[46]

This means that philosophical absolutes "are not made meaningless by the fact that the corresponding concepts, recast by science, lose meaning when science crosses the threshold of the very small or the very large." Their meaning is drawn from ontological knowledge, not empirical science. Similarly, according to Maritain, it is philosophy that inquires into the possible bearing of physical indeterminism on the question of free will, as well as on the ultimate significance of the estrangement Bridgman indicated.

Furthermore, Maritain added,

If we were to admit the essential limitation and isolation of the human mind such as Professor Bridgman describes it, we would have to give up both the coextensivity of the intellect with being [Bridgman's basic conclusion], and the analogical variety of the degrees of knowledge; we would have to assert that knowledge is a completely univocal notion, only realized in sciences of phenomena [this, of course, though Maritain did not say so directly, is the premise upon which Bridgman's argument rests]. Finally we would have to conceive the human mind as a single-track mind, just as a bee's or an ant's mind is.[47]

Turning to the implications of Bridgman's conclusion for the possibility of self-knowledge (the problem that the behaviorists had so clumsily glossed over, and for which Bridgman had called them to task), Maritain asked:

If man were isolated on an island of phenomena, how could he discover the springs of action within himself? A self is not a phenomenon. [It is being.] My isolation on an island of phenomena includes the same condition for my own knowledge of myself and makes it impossible for me to go out from my island to enter myself. [If the self is merely a phenomenon, self-knowledge is impossible.][48]

Finally, addressing the general problem of human values and science, Maritain wrote:

I wonder, moreover, what a naturalistic, non-metaphysical foundation of morality could be, if not science? Now we all agree that science as such "is not concerned with values." But morality is essentially concerned with values.

Moral values have only to do with human life. Yet they ask to govern human life unconditionally, and therefore have their ultimate foundation in the absolute Intelligence which is the author both of human nature and the world.

Morality lives on absolute values; that is why it is finally appendant to the Absolute. A man is bound to die for justice. That means that he stakes his all on the moral value of justice. To stake my all, to give my life, I need to know that the intrinsic value of justice and the obligation to justice are unconditional or absolute—a thing which no statement of fact, such as all statements in sciences of phenomena, or in a "naturalistic" philosophy, can ever establish.[49]

.Maritain had appealed to the Absolute, the very Absolute Bridgman so vehemently rejected time and again, to resolve the agonizing dilemmas that Bridgman found so painful and that he stoically accepted as his own, to be borne in the dignity of solitude. The enfeebled intelligence to which Bridgman entrusted the search for truth was not Maritain's intelligence, a transcendental value coextensive with being and expressed analogically in various orders of existence and knowledge. To Bridgman, claims to higher knowledge were disguised sublimations of social authority. Therefore to his ears Maritain's words were devoid of meaning. They were representative only of the verbalisms operational reasoning had shown to be without foundation. His reply, then, is no surprise.

I am afraid that in his added discussion Professor Maritain has not removed what for me is the fundamental difficulty, namely to see that there is any essential gulf between the intellectual activities which are fruitful in science and those of any valid philosophy. It seems to me that our intellectual activities are all of a piece, and that any intellectual activity that claims to have validity is subject to the necessity that there be some method by which the validity may be checked. Any such method of checking validity must be a method that can be applied by me and cannot be accepted from an outside source of authority. Neither is it clear to me how the meaning of a term can be divorced from the question of its application. I cannot sense what it means to say, for example, as does Professor Maritain, that the concept of existence continues to have meaning even when there is no method by which it may be determined whether things exist or not. I cannot see what one would do to remove a concept of existence like this from the verbal level.[50]

Bridgman's method of intelligence clearly did not embrace the rational grasp of being Maritain placed at the core of his epistemology. Not surprisingly, this idea was unintelligible to him. Bridgman was at a distinct disadvantage in many respects. First, since he was not a theologian or metaphysician, he could not formulate his rebuttal in terms commensurate with Maritain's erudition. His education was not of that scope. Nor could he perceive to what extent the argument was not purely one of empirical science versus religion, but included elements of the opposition between ascetic Protestantism and Catholicism. Moreover, Maritain's scientific training in biology (he had studied under Hans Driesch, in whose vitalism he was interested) only added to the disparity between their perspectives.

In addition, Bridgman was not an intellectual historian. Had he been, he might have recognized that Maritain's neo-Aristotelianism was grounded in a premodern worldview that assumed an entirely different ontological organization, one that was organic and hierarchical rather than sequentially causal and whose organizing principle was eternal—microcosm reflecting macrocosm—rather than linear and temporally successive, as is assumed in modern metaphysics. The analogical reasoning accepted as valid within the micro-macrocosm system of organization has no legitimate place in the modern logical worldview. This is another reason Bridgman found Maritain's arguments incomprehensible. They were not talking about the same reasoning process.

However, the gulf was cultural as well. The intellectual and moral values to which Bridgman had been acculturated were those of pragmatism, ascetic Protestantism, and political and economic self-determination. The ideologies based on these values do not recognize the authority of claims—cognitive, moral, or sociopolitical—that refer to metaphysical absolutes. To the degree science is associated with these ideologies, it is also antimetaphysical. And insofar as these ideologies are part of American sensibility, American science is antimetaphysical.

However, while science may be antimetaphysical when it is functioning as a revolutionary force, or for ideological reasons the claim may be made that it is antimetaphysical, it cannot be said categorically that science is antimetaphysical. That would simply be naive. However, Americans, being generally unmetaphysical, a pragmatic, rather than a philosophically minded people, do not usually concern themselves with such distinctions. They tend to opt for practical compromise before intellectual purity.

Bridgman, on the other hand, set conceptual purity as his goal and believed further that its cleansing, revolutionary power enables science to pierce through convention—to discover truth and set the human race free. He did not distinguish the revolutionary aspect of science from its cognitive ordering capacity or its capacity to focus on new realities. In this role, science walks with metaphysics, and the realities to which it gives birth not only are structured by metaphysical principles but also share in their content. By overemphasizing the revolutionary and destructive impact of science, Bridgman neglected to give proper acknowledgment to the positive, creative mandate of science. He failed to appreciate that it is order which gives meaning to truth and freedom.

Bridgman's confrontation with Maritain, like his confrontation with Einstein, once again brought the empiricist scientific program into critical light, this time, moral, by calling attention to its most egregious deficiency—its inability to command unconditional commitment to its truth claims. Ironically, it was just the moral purity of this program that Bridgman thought he had rescued from the idealistic error of the logical positivists and the social authoritarianism of the behaviorists. However, once operationally purified of the philosophical idealism and social deference he judged to be so objectionable, he found science to be radically subjectivist—epistemologically self-limiting and socially isolating. He did not abandon it, though. His personal commitment was too strong. He refused to admit that we can transcend our immediate experience, or that we even have the right to try. This intransigence was the basis of his quarrel with Einstein. To Bridgman, Einstein's Platonist affinities were symptomatic of an arrogance unfitting to man's humble status. At the same time, however, even though relativity theory represented a glaring contradiction to Bridgman's antimetaphysical scientific expectations, he could not find any scientifically convincing counterevidence to undermine Einstein's position. Similarly, Bridgman's ethical nihilism was, from a strictly practical standpoint, disregarding any philosophical considerations, hardly a viable alternative to Maritain's philosophical absolutism. Bridgman struggled in vain against these elusive idealistic forces, so insubstantial and aery, yet so very persistent.

Bridgman had promised that in his search for truth, the scientist is undeterred by "unfortunate personal characteristics, or intellectual tradition, or the social conventions of his epoch."[51] Nonetheless, the exchange with Maritain, whose neo-Thomism contrasted so sharply with Bridgman's modern scientific naturalism, demonstrates that he fell short of his ideal. He had not, as he believed, escaped either the intellectual prejudices of his era or his own prejudices. The antimetaphysical attitude he took to be an indication of an enlightened mind was itself a product of an age and a culture. Maritain's arguments, because they derive from a premodern ontology, call attention to this in a way that Bridgman's disagreements with the logical positivists and the behaviorists, for example, could not, since the latter all shared the same basic positivistic attitudes.

Bridgman's antimetaphysical ideology prevented him from seeing that he had taken for granted certain assumptions about the nature of human reasoning that depend on prior metaphysical judgments. He did not see that what is taken as being within the purview of human reasoning power is itself a function of an assumed ontological order. He thought that he was not making any assumptions. On the other hand, Bridgman made no secret of his motives. He thought he could free himself from the oppressive yoke of Christianity. Nonetheless, despite his resolve to think things through independently, to purify knowledge, he had bought into an ideology of an era that had blinded itself to the interdependency of humanity and God.

Yet if Maritain and Bridgman, perched on their high academic posts, were pronouncing upon the general conditions for knowledge and morality, they were talking at and beyond each other, to no one and everyone. It was from a less exalted position that a voice more intimate, more personally directed, spoke out. It was a voice guided by a moral certainty—a faith—that knowledge is born of love, and validated in love, that this love is in the first place the personal love we give to one another, but which nonetheless has its origin in a higher source, a source beyond the reach of reason, yet whose infinite reassurance is still accessible to the human spirit.

Polly Campbell was a young woman with a mind of her own, a Randolph summer resident, and an admiring neighbor of Bridgman—almost family. To her, Bridgman was Uncle Peter. Polly had enormous respect for Bridgman, and it was the strength of her loving respect, as well as her confidence that she had something of value to offer Bridgman, that permitted her to speak as a moral equal to this man, many years her senior, socially and educationally her superior, and to suggest an alternative to the imprisoning conclusions of his operational reasoning.

I conceive of life as a pyramidal shape with the large body of the pyramid in the air, the whole weight resting on the small end. This is not a perfect simile because the boundaries are fluid and, at the same time all encompassing. Also this pyramid is open-ended, its wide end extending indefinitely on into infinity, really, though infinity is an ungraspable conception. . . .

. . . It is literally impossible for one person to exactly understand another. Not only must he make guesses from out of his own experiences and habitual attitudes about the other's ideas and feelings which are of course based on a different set of experiences, but, with each instant of time, the whole for each is altered. Therefore each person can only approximate another's intent and the success of the approximation depends upon the intelligence, sensitiveness, and love of the guesser.

The one common desire which man shares, has always shared, and always will share with his fellow man is his desperate need for faith so that he can accept the pyramidal structure which looms above, beyond, and out. . . .

There's the floor.

The only bridge to the beginning is FAITH.

Willing it, will not bring it.

Desperately desiring it leaves one open to receive it but does not give it. It can only come from some indefinable source. That source has been called innumerable things and its actual title is unimportant. I can sum it up adequately enough when I say it is "divine." [52]

Shortly afterwards, Polly sent an addendum. "Dear Uncle Peter," she wrote, "More details keep coming to me and I shall send them on quickly before I find out whether you think this all poppycock of an elementary nature."

About this idea of faith: I don't want you to think I mean bells and incense and living incarnations etc. etc. though these can be genuine paraphernalia of a faith. I meant that man shares with his fellows the need for a faith which will make all else (above the floor) acceptable to him. Thus, a baby is taught the rudiments of faith when his mother says to him (in effect): "you may know that this is a spoon, this is a tree, that your Father and I are 'we,' because I love you." Each time his hunger is satisfied, his hurt remedied, his affection fulfilled he adds to this faith. Later he is taught to incorporate these into some church form. All his life he needs to reassure himself as to this faith whether he is doing it thinkingly and in awareness or not. He tries to find this reassurance in his family, work, church, all, all of his contacts with everything outside of his person.

The important thing about this handed-down faith, this system we have developed over the centuries to stabilize ourselves and our progeny is that *its* source must lie in what I have called "divinity," or in the unknown.

For some people who are too reasoning to accept the dogma and pattern of comforts available to them, the only course is to strike out on a lonely path to discover, if possible, the answers for themselves. Each to his own path! If these people are fortunate and have enough sweet humility—for isn't this the final requirement? they may be given this grace of faith so that the vast and complex world can at last be acceptable to them and they can proceed with it as working entities.[53]

Bridgman's reply was gentle and appreciative, but nevertheless uncomprehending. He wrote that while he was flattered that Polly had "thought it worth the trouble to take so much pains with me," he did not know what to say, nor could he imagine what she might have expected him to say. Some of her remarks, he admitted, were "quite profound," but her final statements were less easy to understand.

I guess you and I use the word "faith" in different senses. I agree that there are many things that we do not understand, but it seems to me that it is possible to meet this situation by adopting tentative programs, which one admits have very little likelihood of being final, but which one follows because they are the best that one can do. I think there is no place for the conventional faith of my religious bringing up in the armory of a thinking man. I do not know how much your faith has in common with this conventional faith, but it seems to me that there are at least some elements in common. And of course your conclusion that there must be something back of it which you call "divine" seems to me a non-sequitur.[54]

At the time, neither Bridgman nor his family could know how comforting the openness of Polly's faith would be in times of crisis.

Standing in almost ludicrous contrast to the personal love of Polly Campbell and the learned humanism that informed the MIT symposium was the scientific chauvinism of the social physics project organized by Princeton astrophysicist John Q. Stewart. Bridgman donated many hours of his time to this project, an attempt, encouraged by support from the Rockefeller Foundation, to apply concepts from physics to social phenomena.[55] As indicated by some of Stewart's publications, it was expected that social analogues to physical quantities could be immediately defined. For example, Stewart was the author of an article entitled "Demographic Gravitation," which appeared in 1948 in the journal *Sociometry*, in which he defined as fundamental quantities potential, which was population/distance, and energy, which was persons squared/mile. Other articles appeared in such journals as *Science*, *American Journal of Physics*, and *Theory of Marketing*.

Stewart's social physics program was yet another manifestation of the common attitude that if a field of study is not a species of physics, it is not scientific. And again, just as it had for Boring's psychology, dimensional analysis exerted its lure. Despite criticism from Bridgman and historian of science Duane Roller, Sr., Stewart put considerable effort into developing a theory of social dimensions. It is difficult to know just how seriously Bridgman took Stewart and his social physics, but there is no doubt that he enjoyed the company and the diversion. At the very least, it provided him with recreational and social pleasure.

In fact, Bridgman was not really interested in developing a systematic social theory, whatever its formal foundations. More important to him were his own social relations and how to maintain intellectual integrity in the face of what he perceived as antagonistic social forces. However, he did allow himself the indulgence of some utopian speculation, included in his 1959 book, *The Way Things Are*.

The kind of society in which he himself would feel comfortable, he projected, would be one in which the code of conduct was not predicated upon the existence of extrarational principles. He rejected Christian ethics as having been created for a subjugated people who were "partners in misery." Instead, he would have a rationally founded ethic derived from an analysis of all conceivable consequences of various possible courses of action.

Considering the foremost ethical question that of deciding under what conditions one should join in exerting influence upon one's fellow man, Bridgman proposed what he called a minimum code, a principle of the least interference with individual activity. It was also a minimum code insofar as it makes the least assumptions and generates the least expectations. In this spirit, an individual must never take from society more than he gives, and just for good measure, he should give more than he takes. The principle is not to be pushed to the extreme of martyrdom, however, because such sacrifice is inconsistent with rational judgment. Galileo, Bridgman maintained, acted perfectly rationally in yielding to the threats of his inquisitors. To do otherwise might indicate some kind of integrity, but whatever it is, it is not intellectual integrity.

But in the end, Bridgman concluded, even intellectual integrity is an "affair of the individual." And since it is internally judged, its practice is even more difficult than any challenge presented by society.[56]

Bridgman's loyalty to the principles that he espoused was tested in the spring of 1961 when he developed what was at first thought to be muscular rheumatism. During June his condition worsened, and on June 30 he wrote to Dr. George Waring, his personal physician in Cambridge:

Since coming here [to Randolph] a month ago my muscular rheumatism has been getting steadily worse so that I have had to consult our local doctor. He has tried a couple of things with no success and now thinks that he should have an X-ray of the sacro-iliac region. I told him of the plates made in the Cambridge Hospital in May and he thought that these would be adequate.[57]

In July, the correct diagnosis was obtained. In Bridgman's own words, taken from a letter to the physicist William Paul, dated July 20:

It is now perfectly certain that I have a nonoperable cancer of the bones of the pelvic region. This has been developing rapidly in the last few weeks. How much longer it will have to run is not now apparent, but it is perfectly evident that I cannot live to complete the preparation of my papers* for the press.[58]

At this moment of disconsolation, at this moment of hopelessness, appeared a brief yet uplifting expression of understanding and love that provided Bridgman with an unmeasurable source of support, which his family, because it meant so much to Bridgman, has treasured and preserved. The words came again from Polly Campbell. "Dear Uncle Peter," she wrote:

Toby [her husband] came home from a week-end in Randolph with the news . . . that you have been ill. We have been feeling very badly that you have been laid up and in pain and have wanted to send you our love. I hope that you are home again and feeling much more comfortable and will soon be fit again.

---

*The papers of which he is speaking are his complete scientific papers, which were published by Harvard University Press in 1964.

I'd like to take this opportunity with what grace I can muster to tell you how much I have always admired and revered you, Uncle Peter.

There is more than a pedantic line between sentiment and sentimentality and I would wish that my thought could be plainly placed in the former category—which I expect is vain and only rationalizing.

Nonetheless, I'll risk the stigma of sentimentality because I want so much to thank you for being as you have been in your life. I've not the background nor intelligence to understand your scientific contributions but I can see a little of your splendid, spare attitude toward life; can guess at some of the sacrifices you've made, perhaps only seeing them yourself in retrospect. And I can guess, too, at Aunt Olive's part in it all.

I thank you for being such a man and such an example for people,—for your ranging, tensile, vital and astringent mind and your integrity.

Please give our love to Aunt Olive and we hope that this finds you feeling a great deal better.

> Respectfully and with love,
> Polly Campbell[59]

Polly's letter went to Bridgman's heart, and he responded as warmly as he knew how—with self-effacing humility, avoiding sentimentality, deflecting attention away from himself and toward her generosity of spirit. "Dear Polly," he wrote:

Your letter moved me more deeply than I like to admit, and I want you to know how much I appreciated it. Not many people are gifted with the imagination to write as you did and still fewer would have taken the trouble to actually write. All in all it was a rather brave thing that you did and I hope you will be encouraged to look for other similar opportunities among your other friends. Only I must protest that you have grossly exaggerated any slight merits which I may possess, and I am sure that with your own increasing experience you will come to take a more moderate view of them.

> With much love, from your
> Uncle Peter[60]

In August, Bridgman reported even further deterioration. On August 2 he wrote again to Waring.

It is a little difficult for me to try to visualize the future until I know more in detail what the prospects are. It is my personal feeling that this thing is progressing much more rapidly than I judge the physicians estimate or than is normal. Something is spreading to my feet and my whole leg below the knee is developing a numbness and a curious dead sensation when I scratch the skin. This is particularly true for the right leg. The pattern of the pain is changing. I used to have occasional stabs of almost unbearable pain after changing position. This has almost disappeared, and the pain is less intense and more persistent—the location is mostly in the back of the leg. Formerly I had much discomfort in sitting in an ordinary chair—I can do this more easily now. It is becoming increasingly difficult to walk about—I am getting so that I mostly just hobble—sometimes a cane helps. What is going to be the end of this? Shall I presently be using crutches, and then shall I have to stay in bed? Sometimes the pain in bed is just as bad as when I am moving about.

But no matter what, Bridgman did not want to leave his beloved New Hampshire home.

Whatever is done, I am anxious to stay in Randolph as long as possible. Life would be unbearably boresome in Cambridge, whereas here I have at least my garden to look at and there is some technical work I can do here as well as in Cambridge. I feel that if I should go to Cambridge now I would never get back here. I am willing to stay here until the only means of transport back to Cambridge is by ambulance or if it seems probable that this thing will be over in a couple of months, I would prefer to stay here until the end, provided the treatment of pain could be arranged. I do not know whether this will require a hospital, or whether it can be done in my own home.

From the pain, he could find no relief. In the same letter, he wrote,

I have had no contact with Dr. Appleton, the local physician who communicated with you, since the local hospital made their diagnosis. I am, however, continuing with the anti-pain tablets which he gave me. I do not know what these are, but I enclose one, on the chance that you may be able to identify it. The dosage is one every four hours and I do not believe it does much good. Before that, we found that aspirin is no good whatever.[61]

The pain and the hopelessness of the prognosis surely weighed heavily on his mind. Yet he carried on with his work in the dedicated manner that was his trademark. To be sure, near the end he reserved his energies for those projects he considered of highest priority, but he did not waste time in self-indulgence. He completed *The Sophisticate's Primer of Relativity*, he prepared the index for his collected experimental papers, he graciously answered his correspondence, and as late as July 28, he consented to write a book review for *Science*, which he submitted on August 6.

On August 20, 1961, Bridgman was found dead in the pumphouse, a small structure outside the home that he had built for himself, set among the trees of the White Mountains, where he had spent thousands of hours working in peace and solitude. He had shot himself in the head. The note he left behind read, "It isn't decent for Society to make a man do this thing himself. Probably this is the last day I will be able to do it myself. P.W.B."[62]

Among Bridgman's private papers there are no clues as to precisely when he made the decision to end his own life. The only thing that can be said is that until July 20, the heading on his correspondence included with the return address the parenthetical "address until Oct. 1." What is evident, however, is that this act was his final repudiation of the mores of society and, implicitly, of the authoritarian Puritan God. Faced with the prospect of lapsing into a state of helplessness and dependency, when death seemed to be the preferable alternative, Bridgman had asked for help from his friend and colleague Dr. Jack Fine. Fine, who was professionally unable to respond, recorded in his personal memoirs the agony he experienced upon being so importuned. To emphasize the difficulty of reconciling his personal instincts and professional obligations, he recalled further his appreciation of the fact that Bridgman

knew he had inoperable cancer in his bones. . . . There was no way out of his dilemma. . . . His daughter at a memorial later said that he believed it worth more to him than his Nobel Prize if this act of self-immolation should convince society that euthanasia has its place in the human at least as much as in the animal world.[63]

Any speculation that might tend to suggest that Bridgman's mental faculties were diminished or impaired even slightly must be sternly rejected. There can be no doubt that he acted with full knowledge of the rational meaning of his act. We are left to wonder, however, to what extent his final decision was a triumph of the human spirit. The ultimate significance of Bridgman's suicide is clouded by the tragic ambiguity of excessive pride in self-sufficiency and the paradox of a life whose meaning has been resolved by a defiant act of self-destruction.

Yet it is Polly Campbell to whom we may most justly reserve the concluding words. "Dear Aunt Olive," she wrote,

After recovering some from the hard shock of the news yesterday, both Toby and I have come increasingly to feel that it all is the way it should best have been under the sad circumstances. Though I am sure it must have been fiercely hard on you, so would the other way have been, too, and perhaps even more so, for what is harder than to helplessly watch some one you love suffer? This way Uncle Peter maintained control and choice, choosing for himself and for you this clean, quick, free way. He was considerate of you and of his own deep human dignity.

> With our loving sympathy,
> Polly

Reference Matter

Notes

The following abbreviations are used in the Notes and Bibliography:

Am. J. Phys.	American Journal of Physics
Am. J. Psychol.	American Journal of Psychology
Am. J. Sci.	American Journal of Science
Am. Phys. Teach.	American Physics Teacher
Am. Scient.	American Scientist
Bull. Atom. Scient.	Bulletin of the Atomic Scientists
Harv. Grad. M.	Harvard Graduates' Magazine
Hist. Stud. Phys. Sci.	Historical Studies in the Physical Sciences
J. Am. Chem. Soc.	Journal. American Chemical Society
J. Applied Phys.	Journal of Applied Physics
J. Chem. Phys.	Journal of Chemical Physics
J. Franklin Inst.	Journal. Franklin Institute
J. Phil.	Journal of Philosophy
J. Phil. Studies	Journal of Philosophical Studies
Phil. Mag.	Philosophical Magazine
Phil. Rev.	Philosophical Review
Phil. Sci.	Philosophy of Science
Phys. Rev.	Physical Review
Proc. Am. Acad. Arts Sci.	Proceedings. American Academy of Arts and Sciences
Proc. Nat. Acad. Sci.	Proceedings. National Academy of Sciences
Proc. Royal Soc. London Ser A	Proceedings. Royal Society of London. Series A
Psychol. Bull.	Psychological Bulletin
Psychol. Rec.	Psychological Record
Psychol. Rev.	Psychological Review
Sci. Amer.	Scientific American
Scient. Mon.	Scientific Monthly
Yale R.	Yale Review

CHAPTER 1

1. There is as yet no complete biography of P. W. Bridgman. For biographical facts, see Francis Birch, Roger Hickman, Gerald Holton, Edwin C. Kemble, "An Ingenious Invention, Percy Williams Bridgman," in *The Lives of Harvard Scholars* (Cambridge, Mass.: Harvard University Information Center, 1968); Edwin C. Kemble and Francis Birch, "Percy Williams Bridgman, 1882–1961," in *Biographical Memoirs*, vol. 41 (New York: Columbia University Press for the National Academy of Sciences of the United States, 1970); E. C. Kemble, Francis Birch, and Gerald Holton, "Percy Williams Bridgman," in Charles Coulston Gillispie, ed., *Dictionary of Scientific Biography* (New York: Charles Scribner's Sons, 1970); Albert Moyer, "Percy Williams Bridgman," in John A. Garrity, ed., *Dictionary of American Biography*, supplement 7 (1961–65) (New York: Charles Scribner's Sons, 1981); D. M. Newitt, "Percy Williams Bridgman," in *Biographical Memoirs of the Royal Society*, vol. 8 (London: Royal Society, 1962); John H. Van Vleck, "Percy Williams Bridgman," in *Year Book of the American Philosophical Society, 1962* (Philadelphia: American Philosophical Society, 1963). To augment the above, I have been permitted, by courtesy of Jane Bridgman Koopman, access to a great deal of unpublished material held by the family. This includes a family chronicle, written by P. W. Bridgman and his wife, Olive, letters written by Bridgman to his children, Jane and Robert, as well as many other documents. Mrs. Koopman has also spent many hours showing me around in Randolph, N.H., introducing me to neighbors and relatives, and sharing personal reminiscences.

2. For further discussion see, e.g., Richard Hofstadter, *The Age of Reform* (New York: Vintage Books, 1955); Daniel J. Kevles, *The Physicists* (New York: Vintage Books, 1979); Samuel Eliot Morison, ed., *The Development of Harvard University* (Cambridge, Mass.: Harvard University Press, 1930); Nathan Reingold, ed., *The Sciences in the American Context* (Washington, D.C.: Smithsonian Institution Press, 1979); Nathan Reingold and Ida Reingold, eds., *Science in America* (Chicago: University of Chicago Press, 1981); Katherine Russell Sopka, "Quantum Physics in America, 1920–1935" (Ph.D. diss., Harvard University, 1976), published with the same title: New York: Arno, 1980; Robert H. Wiebe, *The Search for Order, 1877–1920* (New York: Hill & Wang, 1967).

3. Quoted in Kevles, *The Physicists*, p. 20.

4. Jane Bridgman Koopman, "Personal Recollections of PWB for AIRAPT" (unpublished memoir, July 25, 1977).

5. Bridgman to Klyce, Jan. 5, 1930, Percy Williams Bridgman Papers, Harvard University Library, Cambridge, Mass. Unless otherwise indicated, all correspondence is located in the Bridgman papers at Harvard.

6. The diary is in the possession of Jane Bridgman Koopman. For the Eugenics Society questionnaire, see the Bridgman papers, Oct. 25, 1926.

7. Raymond Bridgman to Percy Bridgman, Oct. 13, 1924.

8. Bridgman to Defrees, Mar. 23, 1919.

9. J. C. Slater, "Presentation of Bingham Medal to P. W. Bridgman," presented at the Annual Meeting of the Society of Rheology, Chicago, Ill., Oct. 24–27, 1951; published in *Journal of Colloid Science* 7, no. 3 (June 1952): 199–202.

10. Bridgman to Owens, Feb. 3, 1914. Owens was secretary of the Franklin Institute.

11. Bridgman to Pegram, Dec. 31, 1916.

12. Bridgman to Jane Bridgman Koopman, Feb. 25, 1933 (private collection).

13. Family Chronicles, p. 176.

14. *Ibid.*, p. 38.

15. *Ibid.*, p. 160.

16. Jane Bridgman Koopman, "Personal Recollections."

17. Interview with Jack Stewart in Randolph, N.H., July 1987.

18. Bridgman to Jane Bridgman Koopman, Jan. 7, 1936 (private collection).

19. Family Chronicles, p. 204.

20. Bridgman to Risdon, Dec. 10, 1922.

21. P. W. Bridgman, *The Physics of High Pressure* (London: G. Bell & Sons, 1931), p. 78.

22. Taken from P. W. Bridgman, "Some Results in the Field of High Pressure Physics," *Endeavour* 10, no. 38 (Apr. 1951). See also the discussion in P. W. Bridgman, *The Physics of High Pressure*, pp. 32–33, 35.

23. Ice IV at $76.35°$ C and $20,670$ kg/cm^2, reported in "Water, in the Liquid and Five Solid Forms Under Pressure," *Proc. Am. Acad. Arts Sci.* 47 (1911): 518.

24. McPike to Bridgman, Aug. 24, 1912.

25. Martin to Bridgman, June 10, 1923.

26. Bridgman to Martin, July 4, 1923.

27. Bridgman to Day, May 7, 1909.

28. Richards to Lloyd, Nov. 20, 1915, T. W. Richards Papers, Harvard University Library, Cambridge, Mass. Richards won the Nobel Prize in chemistry in 1914.

29. Bridgman to the editors of the *Bulletin of the Atomic Scientists*—Messrs. Davies and Rabinovitch, Oct. 17, 1950 (private collection).

30. Bridgman to Kemble, Mar. 16, 1919.

31. Bridgman to Silberstein, May 20, 1922. Silberstein worked at the research laboratory of the Eastman-Kodak Co. in Rochester, New York.

32. P. W. Bridgman, "The Present State of Operationalism," in Philipp G. Frank, ed., *The Validation of Scientific Theories* (New York: Collier Books, 1961), p. 76.

CHAPTER 2

1. See note 2, Chapter 1.

2. Bridgman to R. M. Hunter, Oct. 23, 1919.

3. Bridgman to P. J. Risdon, Dec. 10, 1922.

4. P. W. Bridgman, "Science and Freedom," *Isis* 37 (1947): 129.

5. P. W. Bridgman, "Recent Work in the Field of High Pressures," *Am. Scient.* 31 (1943): 10.

6. *Ibid.*, p. 3.

7. P. W. Bridgman, "The Technique of High Pressure Experimenting," *Proc. Am. Acad. Arts Sci.* 49 (1914): 629.

8. Bethlehem Steel to Bridgman, June 11, 1906.

9. Baldwin Steel to Bridgman, Oct. 31, 1907.

10. Bethlehem Steel to Bridgman, Oct. 29, 1908.

11. Braeburn Steel to Bridgman, Nov. 3, 1902.

12. Bridgman to Cross, Apr. 19, 1909.

13. Bridgman to Cole, Jan. 16, 1909.

14. Bridgman, "The Technique of High Pressure Experimenting," p. 638.

15. P. W. Bridgman, *The Physics of High Pressure* (London: G. Bell & Sons, 1931), p. 33.

16. Interview with Charles Chase, Jan. 25, 1982.

17. P. W. Bridgman, "The Measurement of High Hydrostatic Pressure. I. A Simple Primary Gauge," *Proc. Am. Acad. Arts Sci.* 44 (1909): 201–17; "II. A Secondary Mercury Resistance Gauge," *ibid.*, pp. 221–51; "III. An Experimental Determination of Certain Compressibilities," *ibid.*, pp. 255–79.

18. Watson-Stillman to Bridgman, June 1 and 15, 1908.

19. Johnston to Bridgman, Feb. 21, 1912.

20. Johnston to Bridgman, n.d. (1912?)

21. Greene to Bridgman, May 11, 1912.

22. Bridgman to Greene, May 20, 1912.

23. Bridgman, "The Measurement of High Hydrostatic Pressure, I," p. 208.

24. Bridgman, *The Physics of High Pressure*, p. 61.

25. P. W. Bridgman, "The Action of Mercury on Steel at High Pressures," *Proc. Am. Acad. Arts Sci.* 46 (1911): 325–41.

26. For a representative example, see "Water, in the Liquid and Five Solid Forms Under Pressure," *Proc. Am. Acad. Arts Sci.* 47 (1911): 441–558. I thank Dr. Stephen Hawley for calling this to my attention.

27. Clarke to Bridgman, Oct. 30, Dec. 4, Dec. 22, 1908.

28. Lothrup to Hawkridge Bros. Co., Nov. 8, 1909. The letter was forwarded to Bridgman.

29. Tuttle to Bridgman, Dec. 7, 1909.

30. Tuttle to Bridgman, Jan. 6, 1910.

31. Bridgman to New England Electrical Works, Nov. 15, 1909.

32. New England Electrical Works to Bridgman, May 26, 1915.

33. Bridgman to Cross, April 19, 1909.

34. P. W. Bridgman, "The Measurement of Hydrostatic Pressures up to 20,000 kilograms per Square Centimeter," *Proc. Am. Acad. Arts Sci.* 47 (1911): 341.

35. Bridgman, "Water, in the Liquid and Five Solid Forms Under Pressure," p. 444.

36. *Ibid.*, p. 441.

37. Bridgman to S. C. Chandler, June 17, 1912.

38. Quoted in Bessie Jones and Lyle Boyd, *The Harvard College Observatory* (Cambridge, Mass.: Harvard University Press, 1964), p. 436.

39. *Ibid.*, p. 437.

40. *Ibid.*, p. 438.

41. Bridgman to Remson, Sept. 30, 1915.

42. Cross to Bridgman, Oct. 1, 1915.

43. The correspondence with these firms can be found in the Bridgman Papers.

44. Correspondence with Submarine Signal Company, 1911; Stevens Sound-Proofing Company, 1928; Korfund Company, 1929. The last includes a lengthy description of the device and the circumstances of its development.

45. Bridgman's attitude toward the Arsenal job is expressed in the Family Chronicles. The reports on the properties of armor plating can be found in the Bridgman papers. The article on gun construction, entitled "An Experiment in One-piece Gun Construction," was published in *Mining and Metallurgy*, Feb. 1920, pp. 1–16. It can also be found in the *Collected Experimental Papers*, 3: 1325–40.

46. Correspondence with Russ Manufacturing, 1927; Standard Development, 1927; Rheinische Metallwaaren und Maschinenfabrik, 1927; Schneider and Cie, 1922.

47. Bridgman to Greenewalt, May 23, 1928.

48. Greenewalt to Bridgman, May 25, 1928.

49. Bridgman to Lee, Jan. 9, 1916. There is no other correspondence in the Bridgman Papers about this proposition.

50. Family Chronicles, p. 21.

51. For the development of the artificial diamond, see George Wise, *The Science-Technology Spiral: Interaction of Science and Technology in General Electric's Research, 1909–1955*, Report no. 78CRD049 (Apr. 1978), internal publ., GE Research and Development Center, Schenectedy, N.Y.; and idem, "Research and Results: A History of the GE Research and Development Center 1945–1978" (Unpublished MS, 1982, on file at GE Research and Development Center, Schenectady, N.Y.).

52. Bridgman to Willis Whitney, Mar. 10, 1914.

53. See Wise, *The Science-Technology Spiral*, for a thoughtful analysis and discussion of pure versus applied research.

54. The 1917 nomination is recorded in Jones and Boyd, *The Harvard College Observatory*, p. 439. "In 1917 and 1918, asked to nominate a candidate for the Nobel Prize, [Edward C.] Pickering named Bridgman." (Pickering was the director of the Harvard Observatory.) Bridgman was informed of the 1933 nomination by E. H. Hall, Nov. 21, 1932. The recommendation was signed by Hall, Theodore Lyman, George W. Pierce, Frederick A. Saunders, E. Leon Chaffee, Edwin Kemble, and Otto Oldenberg, Nov. 9, 1931. It read in part, "We recommend the award to Professor Bridgman not because of any one discovery, but rather upon the broad ground of his great achievements in experimental research of a fundamental character and his outstanding contributions to scientific theory and philosophy." Listed as examples were Bridgman's *Dimensional*

Analysis, Condensed Collection of Thermodynamic Formulas, The Logic of Modern Physics, and *The Physics of High Pressure.* This letter is in the private collection held by the family.

55. *Nobel Lectures: Physics, 1942–1962* (Amsterdam: Elsevier Publishing Company for the Nobel Foundation, 1964), p. 50.

56. *Ibid.,* p. 52.

57. P. W. Bridgman, "A General Survey of Certain Results in the Field of High Pressure Physics," Nobel lecture delivered on Dec. 11, 1946, in *ibid.,* pp. 53–70.

CHAPTER 3

1. Bridgman to Korzybski, Nov. 13, 1927.

2. "The Beginnings of Solid State Physics," a symposium held April 30–May 2, 1979, organized by Sir Neville Mott, published in *Proc. Royal Soc. London Ser A* 371, no. 1744 (June 10, 1980): 1–177.

3. P. W. Bridgman, "Theoretical Considerations on the Nature of Metallic Resistance, with Especial Regard to the Pressure Effects," *Phys. Rev.* 9 (1917): 269.

4. *Ibid.*   5. *Ibid.,* p. 288.

6. *Ibid.,* pp. 288–89.   7. *Ibid.,* p. 270.

8. P. W. Bridgman, "The Electrical Resistance of Metals," *Phys. Rev.* 17 (1921): 161–94.

9. *Ibid.,* p. 183.

10. *Ibid., passim.*

11. Bridgman to Howard Trueblood, May 8, 1921.

12. Bridgman to Hall, July 24, 1921.

13. "The Effect of Tension on the Electrical Resistance of Certain Abnormal Metals"; "The Effect of Pressure on the Thermal Conductivity of Metals"; "The Failure of Ohm's Law in Gold and Silver at High Current Densities"; *Proc. Am. Acad. Arts Sci.* 57 (1922): 41–66, 77–127, 131–172.

14. Bridgman to Wilson, Aug. 7, 1921.

15. Wilson to Bridgman, Sept. 21, 1921.

16. "Measurements of the Deviation from Ohm's Law in Metals at High Current Densities," *Proc. Nat. Acad. Sci. U.S.,* 7 (1921): 299–303.

17. P. W. Bridgman, "The Electron Theory of Metals in the Light of New Experimental Data," *Phys. Rev.* 19 (1922): 114–34.

18. See Van Vleck's remarks in "Percy Williams Bridgman," *Year Book of the American Philosophical Society, 1962* (Philadelphia: American Philosophical Society, 1963).

19. *Phys. Rev.* 17 (1921): 169.

20. Buckingham to Bridgman, Nov. 6, 1925.

21. Hall to Bridgman, July 24, 1923.

22. Bridgman to Trueblood, Nov. 25, 1923.

23. A. H. Wilson, *The Theory of Metals,* 2nd ed. (Cambridge, Eng.: At the University Press, 1954), p. 12.

24. Banesh Hoffman, *The Strange Story of the Quantum*, 2nd ed. (New York: Dover Publications, 1959), p. 71.

25. P. W. Bridgman, "Permanent Elements in the Flux of Present-Day Physics," *Science* 71 (1930): 19; reprinted in P. W. Bridgman, *Reflections of a Physicist* (New York: Philosophical Library, 1950), p. 104 [quotation, p. 105].

26. P. W. Bridgman, *The Thermodynamics of Electrical Phenomena* and *A Condensed Collection of Thermodynamic Formulas* (New York: Dover Publications, 1961 [1924, 1925]), preface.

27. P. W. Bridgman, "The Physicist Today," *Harv. Grad. M.*, Mar. 1931, p. 297.

28. Bridgman, "Permanent Elements in the Flux of Present-Day Physics."

29. P. W. Bridgman, "Certain Aspects of High Pressure Research," *J. Franklin Inst.* 200 (1925): 157–58.

30. Lorentz–Bridgman correspondence, Feb.–Mar. 1924.

31. Oppenheimer to Bridgman, Feb. 12, 1926.

32. Bridgman to Oppenheimer, Apr. 3, 1927.

33. See "Beginnings of Solid State Physics"; and Wilson, *The Theory of Metals*, chap. 1; William Hume-Rothery, *Atomic Theory for Students of Metallurgy* (London: Institute of Metals, 1948), chap. 10.

34. See "Beginnings of Solid State Physics."

35. "Erinnern Sie sich unseres letzten Gespräches in München, als ich Sie zur Trambahn brachte? Es handelte sich um die Electronenleitung in Metallen; ich meinte damals, sie wäre hoffnungslos. Inzwischen bin ich zur Überzeugung gekommen, dass das Problem durch die neue Statistik der Wellenmechanik ohne besondere neue Annahmen behandelt werden kann. Meine Note die ich Ihnen zugeschickt habe (Sie haben wohl die Güte die überschüssigen Exemplare an Interessenten weiterzugeben) ist allerdings nur provisorisch. Eine ausführliche Arbeit wird bald folgen.

"Ich meine natürlich nicht, dass die neue Statistik alles leisten kann. Um die ganze Fülle Ihrer experimentallen Resultate zu beherrschen, bedarf es sicher spezieller Hypothesen über die Wechselwirkung zwischen Elektronen und Metallatomen in der Richtung, wie Sie sie ausgearbeitet haben. Aber es schien mir bemerkenswert, dass gewisse Hauptzüge des Problems schon schematische ohne besondere Hypothese erhalten werden können." Sommerfeld to Bridgman, Nov. 3, 1927.

36. Bridgman to Sommerfeld, Nov. 26, 1927.

37. Bridgman to Hume-Rothery, May 13, 1928.

38. Oppenheimer to Bridgman, May 16 (1928?).

39. See "Beginnings of Solid State Physics," p. 25.

40. *Ibid.*

41. Bridgman, *The Thermodynamics of Electrical Phenomena in Metals*, introduction, p. 1.

42. *Ibid.*

43. *Ibid.*

44. Bridgman to Hartley, Jan. 14, 1929.

45. Hartley to Bridgman, Jan. 29, 1929.

INTRODUCTION TO PART II

1. P. W. Bridgman, *The Logic of Modern Physics* (New York: Macmillan, 1927), p. vii.

2. In the 1905 paper Einstein wrote, "If we wish to describe the motion of a material point, we give the values of its co-ordinates as functions of the time. Now we must bear carefully in mind that a mathematical description of this kind has no physical meaning unless we are quite clear as to what we understand by 'time'. . . . We must have to take into account that all our judgments in which time plays a part are always judgments of simultaneous events. If, for instance, I say, 'That train arrives here at 7 o'clock,' I mean something like this: 'The pointing of the small hand of my watch to 7 and the arrival of the train are simultaneous events . . . .' [Or, for a rigid rod of length l, having been put into uniform translatory motion with respect to the *x*-axis of a stationary coordinate system,] imagine its length to be ascertained by the following two operations: (a) The observer moves together with the given measuring-rod and the rod to be measured, and measures the length of the rod directly by superposing the measuring rod, in just the same way as if all three were at rest. (b) By means of stationary clocks set up in the stationary system and synchronizing in accordance with [procedure described earlier], the observer ascertains at what points of the stationary system the two ends of the rod to be measured are located at a definite time. The distance between these two points, measured by the measuring-rod already employed, which in this case is at rest, is also a length which may be designated 'the length of the rod.'" Einstein went on to declare that the length, as determined by (a) will differ from the length as determined by (b). Albert Einstein, "On the Electrodynamics of Moving Bodies," trans. W. Perrett and G. B. Jeffery, in *The Principle of Relativity* (New York: Dover Publications, [1923]), pp. 37–65.

CHAPTER 4

1. Stanley Goldberg, "The Early Response to Einstein's Special Theory of Relativity, 1905–1911" (Ph.D. diss., Harvard University, 1969).

2. O. M. Stewart, *Phys. Rev.* 32 (1911): 418.

3. W. F. Magie, *Science* 25 (1912): 281–93.

4. G. N. Lewis and R. C. Tolman, *Phil. Mag.* 18 (1909): 510–33.

5. Dimensional analysis, or the dimensional method, is a technique of reasoning that permits the deduction of the mathematical form of a physical relationship by means of an analysis of the exponential powers of the fundamental measured units that enter into its functional expression. For example, to find the expression for the period of a simple pendulum (see Alfred W. Porter, *The Method of Dimensions* [London: Methuen & Co., 1933], p. 7), assume that the quantities entering into the relationship are t (time of period), m (mass of bob), w (weight of bob), and l (length of cord). Further assume that the function is of

the form (1) $t = Cl^a m^b w^c$, where a, b, and c are the *dimensions* of l, m, and w, respectively, and C is a constant. The variables l, m, and w are to be expressed in terms of the fundamental quantities mass (M), length (L), and time (T). Thus, since by definition (2) $w = mg = MLT^{-2}$, $l = L$, $m = M$, $t = T$, and C is some constant, by substituting (2) into (1), the relationship becomes (3) $T = C(L)^a (M)^b (MLT^{-2})^c$. Since it is further required that the expression must be physically homogeneous (that is, there are no additional terms), when we compare the left and right sides of the equation, the rules for combining exponents yield the following equations: with respect to length, $0 = a + c$; with respect to mass, $0 = b + c$; and with respect to time, $1 = -2c$. Solving this set of simultaneous equations yields $c = -\frac{1}{2}$, $b = \frac{1}{2}$, and $a = \frac{1}{2}$. Thus the expression for t (equation 1) becomes (4) $t = Cl^{\frac{1}{2}} m^{\frac{1}{2}} w^{-\frac{1}{2}}$. Or, gathering the factors of (4) together under one exponential sign and substituting (2) into (4), (4) becomes (5) $t = C(lm/w)^{\frac{1}{2}}, = C(l/g)^{\frac{1}{2}}$. The dimensional method is useful in situations where a detailed calculation is too complex to be carried out, particularly when the physical properties of systems are studied using scale models. The choice of quantities entering into the relationship is to some degree arbitrary and depends considerably on prior knowledge.

6. P. W. Bridgman, "The Present State of Operationalism," in Philipp G. Frank, ed., *The Validation of Scientific Theories* (New York: Collier Books, 1961), p. 76.

7. J. W. S. Rayleigh, *Nature* 95 (1915): 66.

8. See Planck, *Vorlesung über die Theorie der Wärmestrahlung* (Leipzig: Johann Ambrosius Barth, 1906). He aspired toward a system that possesses universal validity because its units do not depend on the properties of special substances or bodies.

9. G. N. Lewis, *Phys. Rev.* 3 (1914): 92–102.

10. R. C. Tolman, *Phys. Rev.* 3 (1914): 244–55.

11. In its application, Tolman used the principle of similitude to determine the form of functional relations between physical quantities. The procedure was to imagine a miniature universe that to a miniature observer O' there would appear the same as the actual universe does to an actual observer O in the actual universe. This requires that the form of physical laws be unchanged from one universe to the other. Once he had specified the transformation equations, it was necessary merely to substitute the transformed quantities into the general expression for the desired function and find the solution. A simple example might be helpful. If L is a length unit for O (the observer in the ordinary universe), and if L/x (x is arbitrary) is the length unit, L', for O' (the observer in the miniature universe), then if O' measures what is L for Ô, he will arrive at the magnitude xL'. Tolman extended this kind of reasoning to determine transformation relations for the variables relevant to the solution of a given problem. In his example of the energy density in a *hohlraum* in thermodynamic equilibrium (*ibid.*, p. 249), Tolman assumed that the form of the expression is (1) $u = F(T)$ in the ordinary universe, and by the principle of similitude, (2) u'

= F(T') in the miniature universe. T is the temperature, and u is the energy density. The transformation relations, as determined by Tolman, are (3) $u' = u/x^4$ and (4) $T' = T/x$. Since by similitude he is permitted to substitute (3) and (4) into (2), he gets (5) $u/x^4 = F(T/x)$. By rearranging and treating $F(T/x)$ as a constant a, he reaches the solution (6) $u = aT^4$. This is the correct result—Stephan's law (but Tolman knew that all along).

12. Ibid., pp. 253–55.

13. Edgar Buckingham, Phys. Rev. 4 (1914): 357.

14. Joseph Fourier, The Analytical Theory of Heat, trans. Alexander Freeman (New York: C. E. Stechert & Co., 1878), pp. 126–30.

15. W. Williams, Phil. Mag. 34 (1892): 234–71.

16. The dilemma was reviewed by O. J. Lodge in his "Note on the Dimensions and Meaning of J, Usually Called the Mechanical Equivalent of Heat" (Nature). While acknowledging that his title implies that J is the ratio of a quantity of mechanical energy (work) to an equivalent quantity of heat, Lodge goes on to state that the assumption that heat is energy implies that J should be unity (the number whose exponent is zero). On the other hand, he notes, J can be regarded as the ratio designating the number of work units per heat unit. Still, he acknowledges, J is spoken of as having the dimensions work/(mass × temp) rather than zero. The best way to resolve the problem, in his opinion, is to interpret J as the specific heat of water (the amount of heat required to raise the temperature of one gram of water one degree centigrade).

A. W. Rucker, in a well-known article (Phil. Mag. 27 [1889]: 104–14), remarked that "the satisfactory expression of the dimensions of the various thermal units is not possible so long as there is any doubt as to the mechanical definition of temperature" (p. 105). Therefore, he reasoned, as long as heat is measured in calories instead of ergs (units of work), "temperature must be regarded as a fundamental quantity," a unit independent of the units of mass, length, and time. He then proposed to "place temperature in a class of secondary fundamental units, which owing either to our ignorance or to our artificial methods of measurement, cannot be expressed in terms of length, mass and time" (p. 106).

An even more serious irregularity was the discrepancy between the units of the electromagnetic and electrostatic systems of measurement. Here the difficulty was the cause for W. Williams's complaint that "we get two different absolute dimensions for the same physical quantity [the electrical charge], each of which involves a different physical [mechanical] interpretation" (Phil. Mag. 34 [1892]: 234–71). Reasoning from another standpoint, two dimensional constants, u (the magnetic permeability) and k (the inductive capacity), appeared which were difficult to interpret mechanically. (In the preceding argument, Williams assumed that u and k were dimensionless.) It was easily shown that (1) $uk = l^{-2}t^2$ or that the product of the permeability and inductive capacity had the dimensions of inverse velocity squared. Professor G. F. FitzGerald (Phil. Mag. 27 [1889]: 323) called attention to the "naturalness" of the "assumption these inductive capacities are really of the nature of a slowness." Furthermore, he speculated, "It seems possible that they are related to the reciprocal of the square root of the mean energy of turbulence of the ether." On the other hand, W. Williams (*Phil. Mag.* 34 [1892]) observed that u and k need not be of the same dimensions, and thus one may possibly be a strain, and the other, a vortex motion. Lodge interpreted u as a density and 1/k as a rigidity.

17. R. C. Tolman, *Phys. Rev.* 3 (1914): 244–55.

18. T. Ehrenfest-Afanassjewa, *Phys. Rev.* 8 (1916): 1.

19. *Ibid.*, p. 6.

20. *Ibid.*, p. 7.

21. R. C. Tolman, *Phys. Rev.* 8 (1916): 9.

22. R. C. Tolman, *Phys. Rev.* 9 (1917): 237–53.

23. Bertrand Russell, *Principles of Mathematics* (New York: W. W. Norton & Co., 1903). Fundamental to Russell's argument was the distinction between magnitude and quantity, which he used to argue against both the traditional and the relative views of quantity and in favor of the absolute theory of magnitude. A magnitude was defined as anything that can enter into the relationships "greater than" and "less than" with something else of its kind. (It is not inappropriate to think of a magnitude as a "property.") The relationship "equal" was reserved for quantity. Quantity is particularized magnitude. Thus, "an actual footrule is a quantity; its length is a magnitude. Magnitudes are more abstract than quantities: when two quantities are equal, they have the *same* magnitude" (p. 159). Tolman introduced these definitions as basic to his rationalization of the foundations of physics.

While it is immediately obvious how Russell's axiomatics could provide the blueprint for Tolman's formalization of physics, it is not as easy to see why it was important for Tolman to emphasize the quantity-magnitude distinction. Only once in his entire article did he explicitly invoke the doctrine, and this was to justify including vector multiplication among his fundamental operations. However, the argument is revealing: because quantities are particularized magnitudes and spatial particularization has direction, vector operations are needed to complete the particularization. More important, Russell's theory of absolute quantity (or magnitude—Russell used the terms interchangeably) permitted Tolman to treat quantities as concrete entities. It provided the theoretical sanction for attributing physical reality to that which the symbols and formulas of physics represent. Magnitude, for Russell and Tolman, was a property possessed by a quantity. Thus, for Tolman's purpose, Russell's theory made it possible to grant quantities absolute existence.

24. *Phys. Rev.* 9 (1917): 245, 242.

25. Tolman to Bridgman, May 9, 1916.

26. Bridgman, *Phys. Rev.* 8 (1916): 423.

27. *Ibid.* Tolman's system of units, Bridgman stated, was the set, length, time, mass, temperature, and the charge of the electron, chosen so the universal constants h (Planck's constant), c (velocity of light), k (gas constant), and E

(the constant in Coulomb's law) would be unchanged in a transformation defined by a reduction of the unit of length by a factor of x.

28. Tolman to Bridgman, May 9, 1919.

29. Bridgman to Tolman, Nov. 9, 1919.

30. Bridgman to Tolman, Feb. 22, 1920.

31. Tolman to Bridgman, July 28, 1920.

32. Bridgman to Tolman, Aug. 8, 1920.

33. Bridgman to Tolman, Sept. 3, 1920.

34. Tolman to Bridgman, Sept. 8, 1920.

35. Tolman to Bridgman, Mar. 1921.

36. P. W. Bridgman, *Dimensional Analysis* (New Haven: Yale University Press, 1922), p. 46.

37. Bridgman to Bentley, Sept. 21, 1936.

38. Bridgman to Lamo, June 24, 1922.

39. Bridgman to Buckingham, June 10, 1923.

40. Bridgman, *Dimensional Analysis*, p. 17.

41. *Ibid.*, p. 18.

42. *Ibid., passim.*

43. *Ibid.*, p. 19.

44. Bridgman to Bentley, Dec. 18, 1937.

45. Silberstein to Bridgman, Jan. 1923.

46. Hersey to Bridgman, Nov. 15, 1922.

47. Day to Bridgman, Dec. 18, 1921.

48. Bridgman to Day, Dec. 23, 1921.

49. Buckingham to Bridgman, Nov. 13, 1922.

50. Bridgman, *Dimensional Analysis*, p. 105.

51. G. N. Lewis, *Phil. Mag.* 45 (1923): 266–75.

52. O. J. Lodge, *Phil. Mag.* 45 (1923): 275.

53. Bridgman to Lewis, June 21, 1924.

54. Lewis to Bridgman, July 16, 1924.

55. Norman Campbell, *Physics, the Elements* (Cambridge, Eng.: Cambridge University Press, 1920), pp. 395–96.

56. N. Campbell, *Phil. Mag.* 47 (1924): 168.

57. *Ibid.*, p. 494.

58. E. Buckingham, *Phil. Mag.* 48 (1924): 142.

59. *Ibid.*, pp. 143–44.

60. Bridgman to Lewis, May 27, 1924.

61. Lewis to Bridgman, June 4, 1924.

62. G. N. Lewis, *Phil. Mag.* 49 (1925): 739–52.

63. *Ibid.*, p. 746.

64. *Ibid.*, p. 751.

65. T. Ehrenfest-Afanassjewa, *Phil. Mag.*, series 7, vol. 1 (Jan. 1926): 257–72.

66. N. Campbell, *Phil. Mag.* 47 (1926): 1145–51.

67. *Ibid.*

68. P. W. Bridgman, *Phil. Mag.*, series 7, vol. 2 (1926): 1263–66.

69. *Ibid.*, p. 1266.

CHAPTER 5

1. Peter Dixon, "Popular Response to Relativity" (Senior thesis, Harvard University, 1982); Stanley Goldberg, "The Early Response to Einstein's Special Theory of Relativity, 1905–1911" (Ph.D. diss., Harvard University, 1969); Ronald C. Tobey, *The American Ideology of National Science, 1919–1930* (Pittsburgh: University of Pittsburgh Press, 1971).

2. P. W. Bridgman, *The Logic of Modern Physics* (New York: Macmillan, 1927).

3. Unpublished notes, P. W. Bridgman Papers.

4. Bridgman to Hoernle, Dec. 26, 1926.

5. Bridgman, *Logic*, p. 1.

6. *Ibid.*, pp. 1–2.

7. P. W. Bridgman, "*The Logic of Modern Physics* After Thirty Years," *Daedalus* 88 (1959): 518.

8. Bridgman, *Logic*, p. 26.

9. *Ibid.*, p. 6.

10. *Ibid.*, p. 10.

11. Unpublished notes, Bridgman Papers.

12. Bridgman, *Logic*, p. 24.

13. *Ibid.*, p. 171.

14. F.E.B., Review of *The Logic of Modern Physics*, *Am. J. Sci.* 14 (Oct. 1927): 326.

15. Tolman to Bridgman, Mar. 19, 1928.

16. Liljegren to Bridgman, Feb. 2, 1928.

17. Redman to Bridgman, June 16, 1927.

18. Harold Jeffreys, Review of *The Logic of Modern Physics*, *Nature* 121 (1928): 86.

19. Paul Weiss, Review of *The Logic of Modern Physics*, *Nation* 125 (Aug. 3, 1927): 115.

20. L. J. Russell, Review of *The Logic of Modern Physics*, *Mind* 47 (1928): 355.

21. In an unpublished paper, Albert Moyer cites an earlier enunciation of an operational philosophy by William Franklin, which evidently was unknown to Bridgman.

22. Unpublished paper, Bridgman Papers.

23. Bridgman, "*The Logic of Modern Physics* After Thirty Years," p. 518.

24. Unpublished manuscript, Bridgman Papers.

25. Paul Arthur Schilpp, ed., *Albert Einstein, Philosopher-Scientist*, 2 vols. (La Salle, Ill.: Open Court, 1949), p. 679.

26. Bridgman, *Logic*, p. 25.

27. P. W. Bridgman, *A Sophisticate's Primer of Relativity* (Middletown, Conn.: Wesleyan University Press, 1962).

28. Bridgman, *Logic*, p. 47.

29. *Ibid.*, p. 163.

30. *Ibid.*, p. 165.

31. Unpublished manuscript, Bridgman Papers.

32. Adolph Grünbaum, epilogue to Bridgman, *Sophisticate's Primer*, 1st ed., p. 191. See also Arthur Miller's introduction to the 2nd ed. for a detailed examination of Bridgman's theoretical investigation of relativity.

33. P. W. Bridgman, *The Nature of Physical Theory* (New York: Dover Publications, 1936), p. 92.

34. Unpublished manuscript, Bridgman Papers.

35. Bridgman to Korzybski, Oct. 17, 1928.

36. Mach expressed similar misgivings. See Gerald Holton, "Mach, Einstein, and the Search for Reality," in his *Thematic Origins of Scientific Thought, Kepler to Einstein* (Cambridge, Mass.: Harvard University Press, 1973), pp. 219–59.

37. Bridgman, *Logic*, pp. 167–68.

38. P. W. Bridgman, "Einstein's Theories and the Operational Point of View," in Schilpp, *Einstein, Philosopher-Scientist*, p. 344.

39. Bridgman, *Logic*, p. 169.

40. Bridgman, "Einstein's Theories and the Operational Point of View," p. 345.

41. *Ibid.*, p. 343.

42. *Ibid.*, p. 349.

43. Bridgman, *The Nature of Physical Theory*, p. 83.

44. *Ibid.*, p. 136.

45. These quotations are taken from Holton, "Mach, Einstein, and the Search for Reality."

46. Albert Einstein, "Religion and Science: Irreconcilable?" in Einstein, *Ideas and Opinions* (New York: Dell Publishing Co., 1954), p. 61.

47. Einstein, "Autobiographical Notes," in Schilpp, *Einstein, Philosopher-Scientist*, p. 5.

48. P. W. Bridgman, *The Way Things Are* (Cambridge, Mass.: Harvard University Press, 1959), p. 6.

INTRODUCTION TO PART III

1. See the long quotation from Bridgman at the beginning of the introduction to Part II.

2. See, e.g., Ernst Cassirer, *The Problem of Knowledge* (New Haven: Yale University Press, 1950); Mary Douglas and Steven M. Tipton, *Religion and America* (Boston: Beacon Press, 1982); Philipp Frank, *Modern Science and Its Philosophy* (Cambridge, Mass.: Harvard University Press, 1949); Morris Kline, *Mathematics: The Loss of Certainty* (Oxford: Oxford University Press, 1980); Joseph Wood Krutch, *A Krutch Omnibus* (New York: William Morrow & Co., 1970), esp. the essay "The Modern Temper"; and Edward Purcell, *The Crisis of Democratic Theory* (Lexington: University Press of Kentucky, 1973).

3. Max Black, *Language and Philosophy* (Ithaca, N.Y.: Cornell University Press, 1949), p. 112, quoting Bertrand Russell.

4. P. W. Bridgman, *The Logic of Modern Physics* (New York: Macmillan, 1927), p. 94.

5. Bridgman to Adams, Feb. 24, 1930.

6. Bridgman to Feigl, n.d. [1930].

7. Bridgman to Pegram, Oct. 31, 1927.

8. Bridgman to Korzybski, Aug. 5, 1928.

9. Bridgman to Bentley, Dec. 14, 1936.

10. Bridgman-Macmillan correspondence, 1926–27.

CHAPTER 6

1. Klyce to Bridgman, Apr. 11, 1927.

2. Burton E. Livingstone (secretary of the AAAS) to Bridgman, Feb. 16, 1927.

3. Bridgman to Livingstone, Mar. 6, 1927.

4. Scudder Klyce, *Universe* (Winchester, Mass.: the author, 1921).

5. Bridgman to Korzybski, Oct. 10, 1928.

6. Klyce to Bridgman, Apr. 18, 1927.

7. Klyce to Bridgman, Dec. 2, 1927.

8. Bridgman to Klyce, Apr. 10, 1927.

9. Klyce to Bridgman, Apr. 11, 1927.

10. Bridgman to Klyce, June 12, 1927.

11. Klyce to Bridgman, Dec. 2, 1927.

12. Bridgman to Klyce, Jan. 1929. For Klyce's book, see *Dewey's Suppressed Psychology* (Winchester, Mass.: the author, 1928).

13. Reprinted in P. W. Bridgman, *Reflections of a Physicist* (New York: Philosophical Library, 1950), pp. 81–103.

14. Bridgman to Korzybski, Mar. 18, 1928.

15. Bridgman to *Harper's* (Lee Foster, assoc. ed.), Nov. 27, 1928.

16. *Harper's* to Bridgman, Dec. 17, 1928.

17. Bridgman to *Harper's*, Jan. 21, 1929.

18. Bridgman, *Reflections*, pp. 99–100.

19. *Ibid.*, p. 94.

20. *Ibid.*, pp. 101–3.

21. Mentioned in correspondence between Bridgman and Hugo Dingler: Dingler to Bridgman, Apr. 25, 1930; Bridgman to Dingler, May 18, 1930.

22. Bridgman, *Reflections*, p. 103.

23. Klyce to Bridgman, Mar. 23, 1929.

24. Bridgman to Klyce, Mar. 31, 1929.

25. Klyce to Bridgman, Apr. 1, 1929.

26. Klyce to Bridgman, Jan. 3, 1930, p. 7.

27. Bridgman to Klyce, Jan. 5, 1930.

28. Klyce to Bridgman, Jan. 6, 1930.

29. Klyce to Bridgman, Jan. 6, 1930, afternoon.

30. Reprinted in Bridgman, *Reflections*, pp. 104–19.

31. *Ibid.*, p. 105.

32. *Ibid.*, pp. 108, 109, 111.

33. *Ibid.*, pp. 116, 119.

34. Klyce to Bridgman, Jan. 21, 1930.

35. Alfred Korzybski, *Science and Sanity* (New York: Science Press Printing Co. for the International Non-Aristotelian Library Publishing Co., 1933).

36. Korzybski to Bridgman, Oct. 14, 1928.

37. See critique by Max Black in Max Black, *Language and Philosophy* (Ithaca, N.Y.: Cornell University Press, 1949), pp. 223–46.

38. Korzybski, *Science and Sanity*, pp. 28, 50, 476.

39. *Ibid.*, p. viii.

40. Bridgman to Korzybski, Sept. 24, 1927.

41. Korzybski, *Science and Sanity*, pp. 34, 476–77.

42. *Ibid.*, p. 194.

43. *Ibid.*, p. 30.

44. *Ibid.*, p. 32.

45. Korzybski to Bridgman, Sept. 13, 1927.

46. Bridgman to Korzybski, Nov. 4, 1928.

47. Korzybski, *Science and Sanity*, p. 23.

48. Bridgman to Korzybski, Nov. 5, 1932.

49. P. W. Bridgman, "Einstein's Theories and the Operational Point of View," in Paul Arthur Schilpp, ed., *Albert Einstein, Philosopher-Scientist* (La Salle, Ill.: Open Court, 1949), pp. 338, 344.

50. P. W. Bridgman, "A Physicist's Second Reaction to *Mengenlehre*," *Scripta Mathematica* 2 (1934): 101–17, 224–34.

51. A sense of Brouwer's mathematical philosophy may be gained from the following description taken from Raymond Wilder, *Introduction to the Foundations of Mathematics* (New York: John Wiley & Sons, 1952), pp. 231–33. "The most striking aspect of it [intuitionism] is what we might call its *self-sufficiency*. It is *self-generating*, relying in no way on other philosophies or logic. Its basic ideas are to be found in the *intuition*, which seems to be similar to the time (not the spatial) intuition of Kant. Specifically, it recognizes the ability of the individual person to perform a series of mental acts consisting of a first act, then another, then another, and so on endlessly. In this way one attains 'fundamental series,' the best known of which is the series of natural numbers.

"This operation is not dependent upon the use of language. To quote Brouwer in this regard: 'neither the ordinary language nor any symbolic language can have any other role than that of serving as a nonmathematical auxiliary, to assist the mathematical memory or to enable different individuals to build up the same set.' As a consequence of this principle, mathematics is basically *independent of language*. For the *communication* of mathematics the usual symbolic devices, including ordinary language, are necessary, but this is their only function. This seems to make of mathematics virtually an *individual* affair rather than an organized or *cultural* phenomenon, and is perhaps the tenet of Intuitionism that it is most difficult to accept. For mathematics as it is known to most of us appears to be the result of the research of many different investigators who worked out and handed on their ideas by means of a symbolic faculty; without the latter not only would it be necessary to start a new mathematics with each generation, but also each individual would of necessity create his own mathematics. However, it is possible that the Brouwer thesis only requires that one set apart the basic ideas—the intuition of the natural numbers and the ideas regarding set-formation—from linguistic influences, while acknowledging the subsequent role of language in the development, on this basis, of mathematics as a cultural phenomenon. Whether this is a tenable doctrine or not is still debatable."

52. Bridgman, "Physicist's Second Reaction," pp. 226–27, 229.

53. Kreisel to Fraenkel, July 6, 1951; in the private possession of Mrs. Fraenkel, obtained by courtesy of Joseph Dauben. Professor Kreisel would like it known that he has recently gone back to the points of that letter, stating them in "more appropriate, less majestic language" (Georg Kreisel to author, Nov. 23, 1988).

54. Quine to Bridgman, Apr. 15, 1946.

55. I wish to express my thanks to Bill Aspray for helping me through this difficult subject.

56. Bridgman, "Physicist's Second Reaction," p. 224.

57. Bridgman to Schrödinger, Aug. 18, 1935.

58. Schrödinger to Bridgman, Sept. 3, 1935.

59. Korzybski to Bridgman, Mar. 11, 1934.

60. *Ibid.*

61. Bridgman to Korzybski, Mar. 18, 1934.

62. *Ibid.*

63. P. W. Bridgman, *The Nature of Physical Theory* (New York: Dover Publications, 1936).

64. Bridgman to Bentley, Aug. 2, 1936.

65. Korzybski to Bridgman, May 15, 1936.

66. Bridgman to Korzybski, May 31, 1936.

67. Korzybski, *Science and Sanity*, p. 754, item Z3.

68. See review by William Marras Malisoff, ed., *Phil. Sci.* 3 (July 1936): 360.

69. P. W. Bridgman, "How Much Rigor Is Possible in Physics?" in L. Henkin et al., eds., *The Axiomatic Method, with Special Reference to Geometry and Physics* (Amsterdam: North-Holland Publishing Co., 1959), p. 227.

70. Bridgman, *The Nature of Physical Theory*, pp. 13–14.

71. *Ibid.*, p. 15.

72. P. W. Bridgman, "Some Implications of Recent Points of View in Physics," *Revue Internationale de Philosophie* 3 (1949): 17.

73. Malisoff, review in *Phil. Sci.*, p. 361.

74. See entry on Arthur Bentley in Erwin R. A. Seligman, ed., *Encyclopedia of the Social Sciences* (New York: Macmillan, 1963).

75. John Dewey and Arthur Bentley, *Knowing and the Known* (Boston: Beacon Press, 1949).

76. Bridgman to Bentley, June 16, 1929.

77. Bridgman to Bentley, May 4, 1936.

78. Bridgman to Bentley, Aug. 12, 1936.

79. Bentley to Bridgman, Feb. 10, 1938.

80. Bentley to Bridgman, Feb. 10, 1938.

81. Bridgman to Bentley, Sept. 29, 1939.

82. Jane Bridgman to Bentley, Aug. 16, 1936.

CHAPTER 7

1. See S. S. Stevens, "Psychology and the Science of Science," *Psychol. Bull.* 36 (Apr. 1939): 236. This is an important review article and a primary document for anyone interested in the history of psychology in the United States. See also Edwin Boring, *A History of Experimental Psychology*, 2nd ed. (New York: Appleton-Century-Crofts, 1950), p. 656. B. F. Skinner appears to have arrived at his operationist ideas independently of the Feigl-Stevens effort (see his Ph.D. diss.).

2. Herbert Feigl, "The Wiener Kreis in America," in Donald Fleming and Bernard Bailyn, eds., *The Intellectual Migration: Europe and America, 1930–1960* (Cambridge, Mass.: Harvard University Press, Belknap Press, 1969), pp. 644–65. This is a lively and personal account of logical positivism and its career in America from Feigl's point of view.

3. *Ibid.*, p. 645.

4. Feigl to Bridgman, Mar. 2, 1930.

5. Bridgman to Feigl, Mar. 23, 1930.

6. Feigl, "The Wiener Kreis in America," p. 647; Albert Blumberg and Herbert Feigl, "Logical Positivism, a New Movement in European Philosophy," *J. Phil.* 28, no. 11 (May 21, 1931): 281.

7. *Ibid.*, pp. 287, 288.

8. P. A. Schilpp, *Phil. Sci.* 2 (1935): 129.

9. *Ibid.*, p. 129.

10. Blumberg and Feigl, "Logical Positivism," p. 286.

11. P. W. Bridgman, "Some General Principles of Operational Analysis," in Bridgman, *Reflections of a Physicist* (New York: Philosophical Library, 1950), p. 39.

12. Feigl, "The Weiner Kreis in America," p. 622.

13. Blumberg and Feigl, "Logical Positivism," pp. 285, 292.

14. Stevens, "Psychology and the Science of Science," p. 238.

15. Blumberg and Feigl, "Logical Positivism," pp. 292–93; W. M. Malisoff, *Phil. Sci.* 2 (1935): 339.

16. Bentley to Bridgman, Mar. 15, 1936. What Bentley said was true—Bridgman had added his name to the list of supporters of the Unity of Science movement and Bridgman's subjectivism was indeed antithetical to the positivist aims.

17. Bridgman to Bentley, Mar. 19, 1936.

18. P. W. Bridgman, "Science: Public or Private," in *Reflections*, pp. 43–61.

19. *Ibid.*, pp. 44, 48, 49.

20. *Ibid.*, p. 56.

21. *Ibid.*

22. P. W. Bridgman, "The Prospect for Intelligence," in *Reflections*, p. 351.

23. P. W. Bridgman, "New Vistas for Intelligence," in *Reflections*, p. 370.

24. Bridgman, "Science: Public or Private," in *Reflections*, pp. 57–58.

25. *Ibid.*, pp. 59–60.

26. P. W. Bridgman, "Freedom and the Individual," in *Reflections*, p. 62.

27. *Ibid.*, pp. 63, 68.   28. *Ibid.*, p. 72.

29. *Ibid.*, p. 75.   30. *Ibid.*, p. 69.

31. Bridgman to Hart, Feb. 25, 1953.

32. Bridgman, "Freedom and the Individual," in *Reflections*, p. 79.

33. *Ibid.*, p. 80.

34. P. W. Bridgman, "The Present State of Operationalism," in Philipp G. Frank, ed., *The Validation of Scientific Theories* (New York: Collier Books, 1961), p. 76.

35. All papers are published in Frank, *The Validation of Scientific Theories*, pp. 45–92.

36. *Ibid.*, p. 77.

37. See Boring, *A History of Experimental Psychology*; and especially Sigmund Koch, "The 'Operational Principle' in Psychology: A Case Study in Cognitive Pathology" (Paper presented at the symposium "Reflections on P. W. Bridgman: A Centenary Symposium," Boston University, Apr. 24, 1982).

38. Boring to Stevens, running comments on "The Operational Basis of Psychology," 1934; by courtesy of Mrs. Stevens.

39. S. S. Stevens, "Psychology, the Propaedeutic Science," *Phil. Sci.* 3 (Jan. 1936): 90–103.

40. E. Boring, *The Physical Dimensions of Consciousness* (New York: Century Co., 1933), pp. 22–23.

41. S. S. Stevens, "The Operational Basis of Psychology," *Am. J. Psychol.* 47 (Apr. 1935): 323.

42. S. S. Stevens, "The Operational Definition of Psychological Concepts," *Psychol. Rev.* 42 (Nov. 1935): 520.

43. Stevens, "Psychology, the Propaedeutic Science," p. 95.

44. Stevens, "The Operational Definition of Psychological Concepts," p. 518.

45. E. Boring, "Temporal Perception and Operationism," *Am. J. Psych.* 48 (1936): 519.

46. J. A. McGeoch, "A Critique of Operational Definition," *Psychol. Bull.* 34 (1937): 703–4. See also McGeoch, "Learning as an Operationally Defined Concept," *Psychol. Bull.* 32 (1935): 688: "Learning can be more adequately defined in terms of operations of measurement than in terms of phenomenal properties."

47. D. McGregor, "Scientific Measurement and Psychology," *Psychol. Rev.* 42 (1935): 246–66. In this article, which Stevens says was written in close collaboration with Boring, we may note the influence of dimensional analysis, as well as ideas from Norman Campbell's *Physics, the Elements,* and Whitehead and Russell's *Principia Mathematica.* For other examples, see J. R. Kantor, "The Operational Principle in the Physical and Psychological Sciences," *Psychol. Rec.* 2 (1938): 3–32; and R. H. Seashore and B. Katz, "An Operational Definition and Classification of Mental Mechanisms," *Psychol. Rec.* 1 (1937): 3–24.

48. E. C. Tolman, "An Operational Analysis of 'Demands,'" *Erkenntnis* 6 (1936): 383–90.

49. Stevens, "The Operational Basis of Psychology," p. 327.

50. Stevens, "The Operational Definition of Psychological Concepts," p. 522.

51. Stevens, "The Operational Basis of Psychology," pp. 328–29.

52. Bridgman to Bentley, May 4, 1936.

53. R. H. Waters and L. A. Pennington, "Operationism in Psychology," *Psychol. Rev.* 45 (1938): 414–23; quotation on p. 422.

54. S. S. Stevens, "Psychology and the Science of Science," *Psychol. Bull.* 36 (Apr. 1939): 221–62; quotation on p. 228.

55. H. Israel and B. Goldstein, "Operationism in Psychology," *Psychol. Rev.* 51 (1944): 185.

56. *Ibid.*

57. *Ibid.*, p. 187.

58. Boring, *A History of Experimental Psychology,* p. 663.

59. "Symposium on Operationism," *Psychol. Rev.* 52 (Sept. 1945): 241–94.

60. P. W. Bridgman, "Some General Principles of Operational Analysis," *Psychol. Rev.* 52 (Sept. 1945): 246.

61. Edwin Boring, "The Use of Operational Definitions in Science," *Psychol. Rev.* 52 (Sept. 1945): 244.

62. Harold E. Israel, "Two Difficulties in Operational Thinking," *Psychol. Rev.* 52 (Sept. 1945): 260–61.

63. Carroll C. Pratt, "Operationism in Psychology," *Psychol. Rev.* 52 (Sept. 1945): 263–64.

64. Herbert Feigl, "Operationism and Scientific Method," *Psychol. Rev.* 52 (Sept. 1945): 257–58.

65. B. F. Skinner, "The Operational Analysis of Psychological Terms," *Psychol. Rev.* 52 (Sept. 1945): 270–77.

66. *Ibid.*, p. 274.

67. P. W. Bridgman, "Rejoinders and Second Thoughts," *Psychol. Rev.* 52 (Sept. 1945): 281, 283.

68. *Ibid.*, pp. 282–83.

69. *Ibid.*, p. 283.

70. Bridgman to Ames, Jan. 15, 1951.

71. "Materials Relating to Bridgman's Writings," unpublished, dated Feb. 14, 1953, Bridgman Papers.

72. *Ibid.*, Nov. 19, 1952.   73. *Ibid.*, Apr. 20, 1953, p. 5.

74. *Ibid.*, Nov. 7, 1953.   75. *Ibid.*, Jan. 21, 1958.

76. Letter in Bridgman Papers.

INTRODUCTION TO PART IV

1. Interview at author's home, May 6, 1981. The interview supports the above characterization.

CHAPTER 8

1. Besides being evident from statements indicating his disillusionment with the cognitive restrictions imposed by the new physics, this position is confirmed in an essay written by Bridgman, probably when he was a senior in high school, and now in the possession of Mrs. Jane Koopman.

2. Curiously, references to Bridgman's operational statements appear to have been introduced more in the way of sanction or permission for the strangeness of quantum mechanics, rather than in the spirit of interpretation. See, e.g., Edward Condon and Philip Morse, *Quantum Mechanics* (New York: McGraw-Hill, 1929), pp. 17–18; and Edwin C. Kemble, *Fundamental Principles of Quantum Mechanics* (New York: McGraw-Hill, 1937), pp. 58, 76. In both, operationism seems to have little to do with the major part of the discussion. Moreover, there is no reference to Bridgman's operationism in the articles collected in the April 1929 issue of the *Journal of the Franklin Institute* (no. 207), which was devoted to quantum mechanics; among the authors were individuals who must certainly have been familiar with Bridgman's views—Slater, Van Vleck, Kemble, E. H. Kennard, Swann, and Condon. (Here apology would not have been necessary.) This would be consistent with the recollections of Edwin Kemble and Francis Birch, who wrote in their biographical memoir of Bridgman that Bridgman's operationism helped them to *accept* quantum mechanics. See also Silvan Schweber, "Empiricism Regnant," *Hist. Stud. Phys. Sci.* 17 (1986): 55–98; and the footnote on p. 201. Max Jammer, *The Conceptual Development of Quantum Mechanics* (New York: McGraw-Hill, 1966), is also permeated with operational language, but there is only one reference to Bridgman and that is in connection with the problem of the identifiability of particles.

3. Quoted in Jammer, *The Conceptual Development of Quantum Mechanics*, p. 206.

4. *Ibid.* The summary of developments in quantum mechanics is based primarily on this source.

5. P. W. Bridgman, "The New Vision of Science"; reprinted in *Reflections of a Physicist* (New York: Philosophical Library, 1950), p. 100.

6. P. W. Bridgman, "The Principles of Thermodynamics," in *Thermodynamics in Physical Metallurgy* (Novelty, Ohio: American Society for Metals, 1950), p. 14.

7. P. W. Bridgman, "The New Vision of Science," in *Reflections*, p. 103.

8. Bridgman to Carnap, July 7, 1942.

9. P. W. Bridgman, *The Nature of Physical Theory* (New York: Dover Publications, 1936), p. 118.

10. *Ibid.*, pp. 118–19.

11. *Ibid.*

12. *Ibid.*, p. 65.

13. Jammer, *The Conceptual Development of Quantum Mechanics*, p. 198.

14. *Ibid.*, chap. 5.

15. Bridgman, *The Nature of Physical Theory*, p. 65.

16. *Ibid.*, p. 118.		17. *Ibid.*, p. 112.

18. *Ibid.*, pp. 113, 122.		19. *Ibid.*, p. 130.

20. *Ibid.*, p. 129.

21. P. W. Bridgman, *The Nature of Thermodynamics* (Cambridge, Mass.: Harvard University Press, 1943), p. 180.

22. *Metaphysics*, 1053a 25–28. Trans. Hippocrates G. Apostle (Grinnell, Iowa: Peripatetic Press).

23. The presumption of theoretically unlimited accuracy of measurement in classical physics is a commonly repeated theme. For some examples, see Bridgman, *The Nature of Physical Theory*, pp. 94, 105, 111–12; C. G. Darwin, "Observation and Interpretation," p. 209, and G. Sussmann, "An Analysis of Measurement," p. 131, both in S. Korner, ed., *Observation and Interpretation in the Philosophy of Physics* (New York: Dover Publications, 1957); John McKnight, "The Quantum Theoretical Concept of Measurement," in C. West Churchman and Philburn Ratoosh, *Measurement: Definitions and Theories* (New York: John Wiley & Sons, 1959), p. 192; Fritz London and Edmond Bauer, "The Theory of Observation in Quantum Mechanics," in John Archibald Wheeler and Wojcieck Hubert Zurek, *Quantum Theory and Measurement* (Princeton, N.J.: Princeton University Press, 1983), p. 217. This idea warrants further research. For example, in what precise sense, philosophical, mathematical, or physical, was this principle understood? Did the idea appear in the process of contrasting classical and quantum mechanical concepts of measurement or was it assumed by physicists of the classical era? The reason for these questions is the fact that even in classical physics (disregarding human blunders), a "perfect" or exact measurement was already impossible in principle due to the infinity of possible values on the continuum. (This is an aspect of the problem to which Kreisel was referring in his commentary on Bridgman's article on *Mengenlehre*. See Chapter 6.) If every interval, no matter how small, can be further subdivided, what could be the meaning of an exact measure when we are talking

about comparing continuous magnitudes? Even if the standard is assumed to be exact by definition, at what point can we say, in principle, that we have perfect correspondence between the standard and its reference object? There are at least two difficulties involved, neither of which involves ineptitude in any practical sense. First, an actual physical standard has been defined as a unitary rational entity (a unit), when in fact it is not and cannot be such. (This is one reason that one might be attracted to consider the possibility of a system of ultimate rational units. See Chapter 4.) Second, while there is a formal definition (after Cantor) for the equality of infinite sets—a one-to-one correspondence between the elements of the sets—it is of no use for making a physical measurement of a continuous magnitude, as Kreisel recognized. ("Perfection" or exactness of measurement is possible only for integers. There is no error in counting, say, a dozen oranges or eggs.) We can hardly fault Bridgman if he felt uncomfortable with Cantor's theory of infinite sets. It had no relevance to measurement as he practiced and understood it.

24. Euclid, *Elements*, trans. Thomas L. Heath (Chicago: Encyclopedia Britannica, 1952), p. 2.

25. Bertrand Russell, *Principles of Mathematics*, 2nd ed. (New York: W. W. Norton & Co., 1938), p. 405.

26. *Metaphysics*, 1075a 4–6 (see note 22 to this chapter).

27. Morris Kline, *Mathematics: The Loss of Certainty* (Oxford: Oxford University Press, 1980), chaps. 4–5, and p. 20.

28. For a detailed discussion, see Stanley Goldberg, *Understanding Relativity* (Boston: Birkhauser, 1984).

29. See P. W. Bridgman, *The Logic of Modern Physics* (New York: Macmillan, 1927), pp. 97–101.

30. P. W. Bridgman, *The Nature of Physical Theory*, chap. IX.

31. *Ibid.*, pp. 128–29.   32. *Ibid.*, p. 101.

33. *Ibid.*, p. 102.   34. *Ibid.*, p. 122.

CHAPTER 9

1. P. W. Bridgman, "Reflections on Thermodynamics," *Am. Scient.*, Nov. 1953, p. 554 (also published in *Proc. Am. Acad. Arts Sci.* 82 [Dec. 1953]: 301–9).

2. P. W. Bridgman, *The Nature of Thermodynamics* (Cambridge, Mass.: Harvard University Press, 1941), pp. 3–4.

3. *Ibid.*

4. *Ibid.*, Introduction.

5. Bridgman, "Reflections on Thermodynamics," pp. 552–53.

6. *Ibid.* Bridgman did not elaborate this idea in any of his writings.

7. *Ibid.*, p. 554.

8. Bridgman, *The Nature of Thermodynamics*, chap. 1.

9. *Ibid.*, pp. 111, 113, 86.

10. Bridgman, "Reflections on Thermodynamics," p. 552.

11. P. W. Bridgman, "The Principles of Thermodynamics," in *Thermodynamics in Physical Metallurgy* (Novelty, Ohio: American Society for Metals, 1950), p. 6.

12. Bridgman, *The Nature of Thermodynamics*, p. 115.

13. Carnap to Bridgman, June 17, 1942.

14. Bridgman to Carnap, July 7, 1942.

15. Bridgman, *The Nature of Thermodynamics*, pp. 133–34.

16. *Ibid.*, pp. 122–23.

17. *Ibid.*

18. P. W. Bridgman, "The Scientist's Commitment," *Bull. Atom. Scient.*, July 1949, p. 196.

19. P. W. Bridgman, "The Principles of Thermodynamics," pp. 6–7.

20. *Ibid.*

21. *Ibid.*, p. 8.

22. P. W. Bridgman, *The Nature of Physical Theory* (New York: Dover Publications, 1936), p. 165.

23. *Ibid.*, p. 125.

24. Bridgman, *The Nature of Thermodynamics*, p. 170.

25. Bridgman, "The Principles of Thermodynamics," pp. 12–13. This viewpoint is also recorded in notes preserved by John W. Stewart from Physics 261 (fall 1949), Bridgman's course on thermodynamics (supplied to the author by courtesy of John W. Stewart).

26. *Ibid.*, "The Principles of Thermodynamics."

27. Bridgman, *The Nature of Physical Theory*, pp. 62, 131.

28. Boring to Bridgman, June 6, 1945. The letter is in the possession of Mrs. Jane Koopman.

29. May 15, 1957. The document is in the possession of Mrs. Jane Koopman.

30. Leon Brillouin, *Science and Information Theory* (New York: Academic Press, 1956), preface.

31. P. W. Bridgman, "Probability, Logic, and ESP," *Science* 123 (Jan. 6, 1956): 15–17. See also Sidney Hook, ed., *Determinism and Freedom in the Age of Modern Science* (New York: New York University Press, 1958).

INTRODUCTION TO PART V

1. Max Weber, "Science as a Vocation," in Max Weber, *Essays in Sociology*, ed. and trans. H. H. Gerth and C. Wright Mills (Oxford: Oxford University Press, 1946), chap. 5.

2. See, e.g., George C. Bedell, Leo Sandon, Jr., and Charles T. Wellborn, *Religion in America*, 2nd ed. (New York: Macmillan, 1982); Mary Douglas and Steven M. Tipton, eds., *Religion and America* (Boston: Beacon Press, 1982); *Eerdman's Handbook to Christianity in America* (Grand Rapids, Mich.: William B. Eerdman's Publishing Co., 1983); Edward Purcell, *The Crisis of Democratic Theory* (Lexington: University Press of Kentucky, 1973); and Ronald C. Tobey, *The*

American Ideology of National Science, 1919–1930 (Pittsburgh: University of Pittsburgh Press, 1971).

3. For short summaries, see *Eerdman's Handbook to Christianity in America*, pp. 368–87; and Bedell et al., *Religion in America*, pp. 251–60. A highly respected work on this subject is William R. Hutchison, *The Modernist Impulse in American Protestantism* (Cambridge, Mass.: Harvard University Press, 1976).

4. Purcell, *The Crisis of Democratic Theory*, p. 11.

5. *Eerdman's Handbook to Christianity in America*, p. 417.

6. Bridgman to Whitney, Nov. 4, 1957.

7. P. W. Bridgman, *The Intelligent Individual and Society* (New York: Macmillan, 1938), pp. 10–11.

8. P. W. Bridgman, "The Scientist's Commitment," *Bull. Atom. Scient.*, July 1949, p. 193.

9. P. W. Bridgman, "The Struggle for Intellectual Integrity," *Harper's*, Dec. 1933; reprinted in P. W. Bridgman, *Reflections of a Physicist* (New York: Philosophical Library, 1950), p. 224.

10. Bridgman, "Intellectual Integrity," in *Reflections*, p. 228.

11. P. W. Bridgman, "The Prospect for Intelligence," *Yale Review* 54 (1945); reprinted in Bridgman, *Reflections*, p. 347.

12. Michael Walzer, *The Revolution of the Saints: A Study in the Origins of Radical Politics* (Cambridge, Mass.: Harvard University Press, 1965).

13. P. W. Bridgman, "Society and the Intelligent Physicist," *Am. Phys. Teach.* 6 (1939); reprinted in P. W. Bridgman, *Reflections*, p. 243.

CHAPTER 10

1. Michael Walzer, *The Revolution of the Saints: A Study in the Origins of Radical Politics* (Cambridge, Mass.: Harvard University Press, 1965), p. 35.

2. This discussion relies heavily on Walzer's interpretation of the sociology of Puritanism.

3. Reprinted in P. W. Bridgman, *Reflections of a Physicist* (New York: Philosophical Library, 1950), pp. 314–16.

4. Born to Bridgman, May 1, 1939.

5. Shapley to Bridgman, Feb. 24, 1939.

6. Shapley to Ellis Freeman, Apr. 4, 1939, Harlow Shapley Papers, Harvard University Library, Cambridge, Mass.

7. P. W. Bridgman, "Science, and Its Changing Social Environment," *Science* 97 (1943); reprinted in Bridgman, *Reflections*, p. 271.

8. Bridgman to Kilgore, June 1, 1943.

9. P. W. Bridgman, "Scientific Freedom and National Planning," paper presented on Dec. 7, 1947; printed in Bridgman, *Reflections*, p. 326.

10. P. W. Bridgman, "Society and the Individual," *Bull. Atom. Scient.* 14 (Dec. 1958): 413.

11. Paul Sabine to Bridgman, undated.

12. Bridgman to Sabine, Nov. 16, 1948.

13. P. W. Bridgman, "The Physicist Today," *Harv. Grad. M.*, Mar. 1931, p. 289.

14. P. W. Bridgman, "Science and Freedom," *Isis* 37 (1947): 128; reprinted in Bridgman, *Reflections*, p. 299.

15. P. W. Bridgman, "Scientific Freedom and National Planning," in Bridgman, *Reflections*, p. 328.

16. P. W. Bridgman, "Scientists and Social Responsibility," *Scient. Mon.* 45 (1947); reprinted in Bridgman, *Reflections*, pp. 283, 280.

17. *Ibid.*, pp. 285, 286.　　　　　　18. *Ibid.*, pp. 283, 291, 292.

19. *Ibid.*, p. 287.　　　　　　　　　20. Bridgman, *Reflections*, p. 334.

21. *Ibid.*, pp. 336, 338.　　　　　　22. *Ibid.*, p. 340.

23. Bridgman, "Society and the Individual," in Bridgman, *Reflections*, p. 415.

CHAPTER 11

1. Max Weber, "Science as a Vocation," in Max Weber, *Essays in Sociology*, trans. and ed. H. H. Gerth and C. Wright Mills (Oxford: Oxford University Press, 1946), p. 151.

2. P. W. Bridgman, "The Struggle for Intellectual Integrity," *Harper's*, Dec. 1935; reprinted in P. W. Bridgman, *Reflections of a Physicist* (New York: Philosophical Library, 1950), p. 236.

3. *Ibid.*, p. 238.

4. *Ibid.*, pp. 230–31.

5. *Ibid.*, pp. 233–34.

6. P. W. Bridgman, *The Intelligent Individual and Society* (New York: Macmillan, 1938), p. 258.

7. Bridgman to Seeger, Dec. 19, 1960.

8. See note 6 to this chapter.

9. See especially E.N., *J. Phil.* 35 (May 26, 1938): 304; and George Sabine, *Phil. Rev.* 48 (1939): 221–23. L. J. Russell, in *Philosophy* 13 (1938): 496, called Bridgman's view "operational liberalism."

10. Bridgman to Whitney, Dec. 27, 1951.

11. Bridgman, *The Intelligent Individual and Society*, p. 1.

12. *Ibid.*, *passim*.　　　　　　　　13. *Ibid.*, p. 111.

14. *Ibid.*, p. 115.　　　　　　　　　15. *Ibid.*, p. 142.

16. *Ibid.*, p. 143.　　　　　　　　　17. *Ibid.*, pp. 143, 230.

18. *Ibid.*, p. 151.　　　　　　　　　19. *Ibid.*, p. 153.

20. *Ibid.*, p. 156.　　　　　　　　　21. *Ibid.*, p. 164.

22. *Ibid.*, p. 241.　　　　　　　　　23. *Ibid.*, pp. 200, 224–28.

24. *Ibid.*, pp. 199, 273, 263.　　　　25. *Ibid.*, pp. 179, 197.

26. *Ibid.*, p. 189.　　　　　　　　　27. *Ibid.*, p. 214.

28. *Ibid.*, pp. 268–9.　　　　　　　29. *Ibid.*, p. 270.

30. Bridgman to Ashley-Montagu, May 17, 1945.

31. Bridgman, *The Intelligent Individual and Society*, pp. 294, 284.

32. P. W. Bridgman, *The Way Things Are* (Cambridge, Mass.: Harvard University Press, 1959). The letter is preserved in the Bridgman Papers, filed under Harvard University Press correspondence re/The Way Things Are, 1958.

33. P. W. Bridgman, "The Task Before Us," *Proc. Am. Acad. Arts Sci.* 83 (1954): 95–112.

34. Conclusion to untitled paper delivered at the symposium. The paper and the program are preserved in the Bridgman Papers.

35. Bridgman to Parsons, Feb. 11, 1950. The class notes and student papers are preserved in the Bridgman Papers.

36. The papers presented were published in *Bulletin of the Atomic Scientists*, July 1949. For a more complete account, see John Ely Burchard, ed., *Mid-Century: The Social Implications of Scientific Progress* (Cambridge, Mass.: Technology Press of MIT; New York: John Wiley & Sons, 1950), chap. 5.

37. *Bull. Atom. Scient.*, p. 193.   38. *Ibid.*, p. 196.

39. *Ibid.*, p. 201.   40. Burchard, *Mid-Century*, p. 244.

41. *Ibid.*, p. 245.   42. *Ibid.*

43. *Ibid.*, p. 247.

44. This simple statement belies the multiplicity of ways in which orders of truth are proportionally interconnected in Maritain's philosophy. The categories of knowledge are related in a much more complex manner than mere ranking. For a short discussion of Maritain's thought in the context of the philosophy of physics, see Yves R. Simon, "Maritain's Philosophy of the Sciences," in *The Philosophy of Physics* (Jamaica, N.Y.: St. John's University Press, 1961). For a more general exposition of these relationships, see James F. Anderson, "The Role of Analogy in Maritain's Thought," in Joseph W. Evans, ed., *Jacques Maritain: The Man and His Achievement* (New York: Sheed & Ward, 1963).

45. Burchard, *Mid-Century*, p. 248*n*.   46. *Ibid.*

47. *Ibid.*, p. 249*n*.   48. *Ibid.*

49. *Ibid.*, p. 250*n*.   50. *Ibid.*, p. 250*n*.

51. See note 13 to the introduction to this section.

52. Polly (Chen Sun) Campbell to Bridgman, Nov. 25, 1957. This letter and the subsequently quoted correspondence between the Bridgman family and Polly Campbell are in the possession of Mrs. Jane Koopman.

53. *Ibid.*

54. Bridgman to Polly (Chen Sun) Campbell, Dec. 15, 1957.

55. See Stewart's letter to Bridgman, Jan. 12, 1951. The Bridgman Papers contain a good amount of material on this subject, including lists of participants in conferences and many of Stewart's articles. We can sense the direction of Stewart's ambitions from the following selections taken from an abstract of a lecture he planned to give in Philadelphia on April 29, 1953, sponsored by the Foundation for Integrated Education. He wrote, "With aid from Princeton University, from the Institution for Advanced Study, Research Corporation, and, currently, from the Rockefeller Foundation, the Princeton social physics

project has been engaged for the past half-dozen years in studies of how the methods and some of the principles of physical science can be transferred to the social field. At first our work was strictly inductive, directed toward the recognition of mathematical regularities among social phenomena, and particularly to the discovery of empirical regularities in demographic statistics. A number of such regularities have been published over many years by several investigators. The statement that human relations are not susceptible of physical treatment is no longer made by informed people. . . .

"First, the concepts and relations of physics and chemistry must be described at a high level of generality and abstraction. Second, we seek to establish point-to-point correspondences, called isomorphisms, with the concepts and relations of social science at an equivalent level. Here and there the description of physics, in turn, may be improved as the result of suggestions arising in social science: thus the mutual relation of the two fields is bilateral, by no means implying unilateral dictatorship by the logic of existing standard physics. . . .

"For the uses of social physics, physical science to begin with can be described and interpreted in terms of its 'dimensions' (a list of six has been published), and of the corresponding sorts of physical energy. It is found that the descriptions can then be transferred to social science, and in a manner so broad and flexible as to touch most or all of the phenomena there encountered but as yet imperfectly catalogued. The 'social energies' may be considered to be the social values—meaning the forms of wealth, material and psychic, which are the stuff of living. . . .

"To each sort of value corresponds its special type of motivation, or generalized force. In such wise, guided by the amazingly intricate and integrated physical model (where velocity, force, gravitational potential, temperature, electromotive force, and chemical potential are the generalized forces, or 'intensive quantities'), the social physicist can fashion his network of fundamental social concepts and relations."

56. *Ibid.*, p. 325.
57. Bridgman to Waring, June 30, 1961.
58. Bridgman to Paul, July 20, 1961.
59. Polly (Chen Sun) Campbell to Bridgman, July 18, 1961.
60. Bridgman to Polly (Chen Sun) Campbell, July 23, 1961.
61. Bridgman to Waring, Aug. 2, 1961.
62. This note appears in Edwin C. Kemble and Francis Birch, "Percy Williams Bridgman, 1882–1961," in *Biographical Memoirs*, vol. 41 (New York: Columbia University Press for the National Academy of Sciences, 1970).
63. Unpublished memoirs. Jane Bridgman Koopman does not remember making this statement.

Bibliography

The bibliography has been divided into sections. The first section covers general sources. Following this are five sections corresponding to the parts of the book. See p. 313 for a list of the abbreviations used here.

GENERAL SOURCES

Unpublished Documents

Cambridge, Mass. Harvard University. Percy Williams Bridgman Papers; Harlow Shapley Papers; T. W. Richards Papers.

Interviews and Personal Communications

Interview with Bernard and Jane Bridgman Koopman, May 6, 1981. (Mrs. Koopman, P. W. Bridgman's daughter, has also been very helpful in providing photographs and other information throughout the entire time of writing.)
Interview with B. F. Skinner, June 29, 1982.
Interview with Charles Chase, Jan. 25, 1982.
Interview with Edwin C. Kemble, Nov. 11, 1980.
Interview with Jack Stewart, July 1987.
Interviews with residents of Randolph, New Hampshire, July 1987.

Books by P. W. Bridgman

Dimensional Analysis. New Haven: Yale University Press, 1922.
The Thermodynamics of Electrical Phenomena and *A Condensed Collection of Thermodynamic Formulas.* New York: Dover Publications, 1925, 1934, [1961].
The Logic of Modern Physics. New York: Macmillan, 1927.
The Physics of High Pressure. London: G. Bell and Sons, 1931.
The Nature of Physical Theory. New York: Dover Publications, 1936.
The Intelligent Individual and Society. New York: Macmillan, 1938.
The Nature of Thermodynamics. Cambridge, Mass.: Harvard University Press, 1941.

Reflections of a Physicist. New York: Philosophical Library, 1950.
The Nature of Some of Our Physical Concepts. New York: Philosophical Library, 1952.
The Way Things Are. Cambridge, Mass.: Harvard University Press, 1959.
A Sophisticate's Primer of Relativity. Middletown, Conn.: Wesleyan University Press, 1st ed., 1962; 2nd, rev. ed., 1983. (The major differences between these two editions are in the commentary rather than in the text. The first edition contains an epilogue by Adolph Grünbaum that is highly critical of Bridgman's interpretation of relativity. In the second edition, Grünbaum's remarks have been replaced by an introduction by Arthur Miller drawing on unpublished notes written by Bridgman in the early 1920's when he first began his examination of relativity, as well as drafts of *A Sophisticate's Primer* that had only shortly before been added to the archives.)
Collected Experimental Papers. 7 vols. Cambridge, Mass.: Harvard University Press, 1964.

PART I: A CAREER IN PHYSICS

Unpublished Documents

Cambridge, Mass. Harvard University. T. W. Richards Papers.
Koopman, Jane Bridgman, "Personal Recollections of PWB for AIRAPT, July 25, 1977."

Theses

Bridgman, Percy Williams. "Mercury Resistance as a Pressure Gauge." Ph.D. dissertation, Harvard University, 1908.
Sopka, Katherine Russell. "Quantum Physics in America, 1920–1935." Ph.D. dissertation, Harvard University, 1976. Published under the same title: New York: Arno, 1980.

Memoirs

Birch, Francis, Roger Hickman, Gerald Holton, and Edwin C. Kemble. "An Ingenious Invention, Percy Williams Bridgman." In *The Lives of Harvard Scholars.* Cambridge, Mass.: Harvard University Information Center, 1968.
"Expressions of Appreciation as Arranged in the Order Given at the Memorial Meeting for Professor Percy Williams Bridgman." Oct. 24, 1961.
Kemble, Edwin C., and Francis Birch. "Percy Williams Bridgman, 1882–1961." In *Biographical Memoirs*, vol. 41. New York: Columbia University Press for the National Academy of Sciences of the United States, 1970.
Kemble, Edwin C., Francis Birch, and Gerald Holton. "Percy Williams Bridgman." In Charles Coulston Gillispie, ed., *Dictionary of Scientific Biography.* New York: Charles Scribner's Sons, 1970.
Moyer, Albert. "Percy Williams Bridgman." In John A. Garraty, ed., *Dictionary of American Biography*, supplement 7 (1961–1965). New York: Charles Scribner's Sons, 1981.

Newitt, D. M. "Percy Williams Bridgman." In *Biographical Memoirs of the Royal Society*, vol. 8. London: Royal Society, 1962.

Van Vleck, John H. "Percy Williams Bridgman." In *Year Book of the American Philosophical Society, 1962*. Philadelphia: American Philosophical Society, 1963.

Articles by P. W. Bridgman

The numbers in brackets indicate the page numbers in the *Collected Experimental Papers.*

"The Measurement of High Hydrostatic Pressure. I. A Simple Primary Gauge; II. A Secondary Mercury Resistance Gauge: III. An Experimental Determination of Certain Compressibilities." *Proc. Am. Acad. Arts Sci.* 44 (1909): 201–79 [1–75].

"The Action of Mercury on Steel at High Pressures." *Proc. Am. Acad. Arts Sci.* 46 (1911): 325–41 [77–93].

"The Measurement of Hydrostatic Pressures up to 20,000 Kilograms per Square Centimeter." *Proc. Am. Acad. Arts Sci.* 47 (1911): 321–43 [94–117].

"Water, in the Liquid and Five Solid Forms Under Pressure." *Proc. Am. Acad. Arts Sci.* 47 (1911): 441–558 [213–370].

"The Technique of High Pressure Experimenting." *Proc. Am. Acad. Arts Sci.* 49 (1914): 627–43 [593–609].

"Theoretical Considerations on the Nature of Metallic Resistance, with Especial Regard to the Pressure Effects." *Phys. Rev.* 9 (1917): 269–289 [1137–57].

"The Electrical Resistance of Metals." *Phys. Rev.* 17 (1921): 161–94.

"The Electron Theory of Metals in the Light of New Experimental Data." *Phys. Rev.* 19 (1922): 114–34.

"Certain Aspects of High Pressure Research." *J. Franklin Inst.* 200 (1925): 147–60.

"Permanent Elements in the Flux of Present-Day Physics." *Science* 71 (1930): 19–23.

"The Physicist Today." *Harv. Grad. M.*, Mar. 1931, pp. 289–97.

"Recent Work in the Field of High Pressures." *Am. Scient.* 31 (1943): 1–35 [3505–39].

"A General Survey of Certain Results in the Field of High Pressure Physics" [1946]. In *Nobel Lectures: Physics, 1942–1962*. Amsterdam: Elsevier Publishing Company for the Nobel Foundation, 1964, pp. 53–70. Also published in *Journal of the Washington Academy of Sciences* 38 (1948): 145–56 [3873–90].

"Science and Freedom." *Isis* 37 (1947): 128–31.

"An Experimental Contribution to the Problem of Diamond Synthesis." *J. Chem. Phys.* 15 (1947): 92–98 [3784–90].

"Some Results in the Field of High Pressure Physics." *Endeavour* 10, no. 38 (Apr. 1951): 63–69 [4105–11].

"Synthetic Diamonds." *Sci. Amer.* 193 (1955): 42–46 [4519–29].

"General Outlook on the Field of High-Pressure Research." In W. Paul and

D. M. Warschauer, eds., *Solids Under Pressure*. New York: McGraw-Hill, 1963, pp. 1–13 [4625–37].

Other Books

Conductibilité electrique des métaux et problèmes connexes: Rapports et discussions, avril 24–29, 1924. Paris: Gauthier-Villars, 1927. Contains Bridgman's contribution to the 1924 Solvay Conference.

Frank, Philipp G. *The Validation of Scientific Theories*. Boston: Beacon Press, 1961. Includes Bridgman's essay "The Present State of Operationalism," in which he expresses his dismay at being misunderstood.

Hoffman, Banesh. *The Strange Story of the Quantum*. 2nd ed. New York: Dover Publications, 1959.

Hofstadter, Richard. *The Age of Reform*. New York: Vintage Books, 1955.

Hume-Rothery, William. *Atomic Theory for Students of Metallurgy*. London: Institute of Metals, 1948.

Jones, Bessie, and Lyle Boyd. *The Harvard College Observatory*. Cambridge, Mass.: Harvard University Press, 1964.

Kevles, Daniel J. *The Physicists*. New York: Vintage Books, 1979.

Morison, Samuel Eliot, ed. *The Development of Harvard University*. Cambridge, Mass.: Harvard University Press, 1930.

Nobel Lectures: Physics, 1942–1962. Amsterdam: Elsevier Publishing Company for the Nobel Foundation, 1964.

Rosenberg, Charles E. *No Other Gods: On Science and American Social Thought*. Baltimore: Johns Hopkins University Press, 1976.

Slater, John C. *Solid-State and Molecular Theory: A Scientific Biography*. New York: John Wiley and Sons, 1975.

Wiebe, Robert H. *The Search for Order, 1877–1920*. New York: Hill and Wang, 1967.

Wilson, A. H. *A Theory of Metals*. 2nd ed., Cambridge, Eng.: At the University Press, 1954.

Other Articles

"The Beginnings of Solid State Physics: A Symposium Held April 30–May 2, 1979." *Proc. Royal Soc. London Ser A* 371, no. 1744 (June 10, 1980): 1–177. This set of papers is important for anyone interested in the development of solid-state physics.

Slater, J. C. "Presentation of Bingham Medal to P. W. Bridgman." *Journal of Colloid Science* 7, no. 3 (1952): 199–202.

PART II: THE MEANING OF MEASUREMENT

Theses

Dixon, Peter. "Popular Response to Relativity." Senior thesis, Harvard University, 1982.

Goldberg, Stanley. "The Early Response to Einstein's Special Theory of Relativity, 1905–1911." Ph.D. dissertation, Harvard University, 1969.

Articles by P. W. Bridgman

"Tolman's Principle of Similitude." *Phys. Rev.* 8 (1916): 423–31.
"Dimensional Analysis Again." *Phil. Mag.* series 7, vol. 2 (1926): 1263–66.
"*The Logic of Modern Physics* After Thirty Years." *Daedalus* 88 (1959): 518–26.
"The Significance of the Mach Principle." *Am. J. Phys.* 29 (1961): 32–36.

Book Reviews of 'The Logic of Modern Physics'

Benjamin, A. Cornelius. *J. Phil.* 24 (Nov. 1927): 663.
Boston Evening Transcript, Aug. 6, 1927, p. 2.
Constantinides, P. A. *School Science and Mathematics*, June 1929, p. 608.
Dingler, H. *Physikalische Zeitschrift* 24 (1928): 710.
F.E.B. *Am. J. Sci.* 14 (Oct. 1927): 326.
Jeffreys, Harold. *Nature* 121 (1928): 86.
Lindsay, R. B. *Phil. Sci.* 4 (Oct. 1937): 456.
New York Times Book Review, Mar. 4, 1928, p. 38.
Oppenheimer, J. R. *Phys. Rev.* 32 (1928): 145.
Quarterly Review of Biology, Dec. 1927, p. 578.
Russell, L. J. *Mind* 47 (1928): 355.
Sachs, Albert Parsons. *Saturday Review of Literature*, Aug. 6, 1927, p. 24.
Stebbing, L. S. *J. Phil. Studies*, June 1928, p. 96.
Weiss, Paul. *The Nation* 125 (Aug. 3, 1927): 115.
Whyte, L. L. *London Observer*, Oct. 2, 1927.
Wilmer, C. P. *Union Seminary Review* 41 (Oct. 1929): 77.

Other Books

Benjamin, A. Cornelius. *Operationism*. Springfield, Ill.: Charles C. Thomas, 1955.
Campbell, Norman. *Physics, the Elements*. Cambridge, Eng.: Cambridge University Press, 1920.
Danto, Arthur, ed. *Philosophy of Science*. Cleveland: World Publishing Co., 1968.
Eddington, Arthur. *Space, Time, and Gravitation*. Cambridge, Eng.: At the University Press, 1923.
———. *The Nature of the Physical World*. Cambridge, Eng.: At the University Press, 1953.
Einstein, Albert. *Ideas and Opinions*. New York: Dell Publishing Co., 1954.
Ellis, Brian. *Basic Concepts of Measurement*. Cambridge, Eng.: Cambridge University Press, 1966.
Fourier, Joseph. *The Analytical Theory of Heat*. Trans. Alexander Freeman. New York: C. E. Stechert and Co., 1878.
Frank, Philipp. *Einstein: His Life and Times*. New York: Alfred A. Knopf, 1947.
———. *The Validation of Scientific Theories*. Boston: Beacon Press, 1954.
Holton, Gerald. *Thematic Origins of Scientific Thought, Kepler to Einstein*. Cambridge, Mass.: Harvard University Press, 1973.
Huntley, H. E. *Dimensional Analysis*. New York: Rinehart and Co., 1955.
Lewis, G. N. *The Anatomy of Science*. New Haven: Yale University Press, 1926.

Newton, Isaac. *Principia*, vol. 1. Trans. Andrew Notte. 1729. Rev. and supplemented Florian Cajori. Berkeley: University of California Press, 1934.

Perrett, W., and G. B. Jeffery, trans. and eds. *The Principle of Relativity*. N.p.: Dover Publications, [1923].

Planck, Max. *Vorlesung über die Theorie der Wärmestrahlung*. Leipzig: Johann Ambrosius Barth, 1906.

Porter, Alfred W. *The Method of Dimensions*. London: Methuen and Co., 1933.

Russell, Bertrand. *Principles of Mathematics*. New York: W. W. Norton and Co., 1903.

Schilpp, Paul Arthur, ed. *Albert Einstein, Philosopher-Scientist*, 2 vols. La Salle, Ill.: Open Court, 1949. Includes Bridgman's article "Einstein's Theories and the Operational Point of View" and Einstein's response to Bridgman's epistemology.

Tobey, Ronald C. *The American Ideology of National Science, 1919–1930*. Pittsburgh: University of Pittsburgh Press, 1971. Contains informative discussion of the Einstein controversy in the context of American culture.

Tolman, Richard C. *The Theory of the Relativity of Motion*. Berkeley: University of California Press, 1917.

Woolf, Harry, ed. *Quantification*. Indianapolis and New York: Bobbs Merrill Co., 1961.

Articles on Dimensional Analysis and URU

Blakesley, T. H. "On Some Facts Connected with the Systems of Scientific Units of Measurement." *Phil. Mag.* 27 (1889): 178–86.

Buckingham, Edgar. "On Physically Similar Systems: Illustrations of the Use of Dimensional Equations." *Phys. Rev.* 4 (1914): 345–70.

———. "The Principle of Similitude." *Nature* 96 (1915): 396–97.

———. "Notes on the Method of Dimensions." *Phil. Mag.* 42 (1921): 696–719.

———. "Dimensional Analysis." *Phil. Mag.* 48 (1924): 141–45.

Campbell, Norman. "Ultimate Rational Units." *Phil. Mag.* 47 (1924): 159–72.

———. "Dimensional Analysis." *Phil. Mag.* 47 (1924): 481–94.

———. "Dimensional Analysis." *Phil. Mag.* 47 (1926): 1145–51.

Ehrenfest-Afanassjewa, T. "On Mr. R. C. Tolman's Principle of Similitude." *Phys. Rev.* 8 (1916): 1–11.

———. "Dimensional Analysis Viewed from the Standpoint of the Theory of Similitudes." *Phil. Mag.*, series 7, vol. 1 (1926): 257–72.

Fitzgerald, G. F. "On the Dimensions of Electromagnetic Units." *Phil. Mag.* 27 (1889): 323.

Larmor, J. "The Principle of Similitude." *Nature* 95 (1915): 644.

Lewis, G. N. "Physical Constants and Ultimate Rational Units." *Phil. Mag.* 45 (1923): 266–275.

———. "Ultimate Rational Units and Dimensional Theory." *Phil. Mag.* 49 (1925): 739–52.

Lewis, G. N., and E. Q. Adams. "Notes on Quantum Theory: A Theory of Ultimate Rational Units." *Phys. Rev.* 3 (1914): 92–102.

Lewis, G. N., and R. C. Tolman. "The Principle of Relativity and Non-Newtonian Mechanics." *Phil. Mag.* 18 (1909): 510–33.

Lodge, O. J. "Note on the Dimensions and Meaning of *J*, Usually Called the Mechanical Equivalent of Heat." *Nature* (1888): 320.

Magie, W. F. "The Primary Concepts of Physics." *Science* 25 (1912): 281–93.

Rayleigh, J. W. S. "The Principle of Similitude." *Nature* 95, no. 2368 (1915): 66.

———. "The Principle of Similitude." *Nature* 95, no. 2389 (1915): 644.

Riabouchinsky, D. "The Principle of Similitude." *Nature* 95 (1915): 591.

Rucker, A. W. "On the Suppressed Dimensions of Physical Quantities." *Phil. Mag.* 27 (1889): 104–14.

Stewart, O. M. "The Second Postulate of Relativity and the Electromagnetic Emission Theory of Light." *Phys. Rev.* 32 (1911): 418.

Thompson, D'Arcy. "Galileo and the Principle of Similitude." *Nature* 95 (1915): 644.

Tolman, R. C. "The Principle of Similitude." *Phys. Rev.* 3 (1914): 244–55.

———. "The Principle of Similitude and the Principle of Dimensional Homogeneity." *Phys. Rev.* 6 (1915): 219–33.

———. "The Measureable Quantities of Physics." *Phys. Rev.* 9 (1917): 237–53.

———. "The Entropy of Gases." *J. Am. Chem. Soc.* 42 (1920): 1185–91.

———. "The Entropy of Gases and the Principle of Similitude." *Phys. Rev.* 15 (1920): 521–22.

———. "Relativity Theories in Physics." *General Electric Review* 23 (June 1920): 486–92.

Williams, W. "On the Relation of the Dimensions of Physical Quantities to Directions in Space." *Phil. Mag.* 34 (1892): 234–71.

Miscellaneous Articles

Brock, W. H., and D. M. Knight. "The Atomic Debates: Memorable and Interesting Evenings in the Life of the Chemical Society." *Isis* 56 (1965): 5–25.

Kennedy, Hubert C. "Peano's Concept of Number." *Historia Mathematica* 1 (1974): 387–408.

Moyer, Albert E. "P. W. Bridgman's Operational Perspective on Physics: Origins, Development, and Reception." Paper read at the Bridgman Centenary Symposium, Apr. 1982.

PART III:
OPERATIONAL REASONING AND THE TYRANNY OF LANGUAGE

Unpublished Sources

Private Papers of Georg Kreisel, in the possession of Mrs. Kreisel.

Private Papers of S. S. Stevens, in the possession of Mrs. Stevens.

Thesis

Skinner, B. F. "The Concept of the Reflex in the Description of Behavior." Ph.D. dissertation, Harvard University, 1930.

Reviews of Bridgman's Books and Articles

E.N. Review of *The Intelligent Individual and Society*. *J. Phil*. 35 (May 26, 1938): 304.

Malisoff, William Marras. Review of *The Nature of Physical Theory*. *Phil. Sci*. 3 (July 1936): 360.

Russell, L. J. Review of *The Intelligent Individual and Society*. *Philosophy* 13 (1938): 496.

Sabine, George. Review of *The Intelligent Individual and Society*. *Phil. Rev*. 48 (1939): 221.

Weiss, A. P. Review of "A New Vision of Science." *Scient. Mon*., Dec. 1929, pp. 506–14.

Articles by Bridgman

"A New Vision of Science." *Harper's*, Mar. 1929.

"Permanent Elements in the Flux of Present-Day Physics." *Science* 71 (1930): 19–23.

"A Physicist's Second Reaction to *Mengenlehre*." *Scripta Mathematica* 2 (1934): 101–17, 224–34.

"Freedom and the Individual." In Ruth Nanda Anshen, ed., *Freedom: Its Meaning*. New York: Harcourt Brace and Co., 1940.

"Science: Public or Private?" *Phil. Sci*. 7 (1940): 36–48.

"The Prospect for Intelligence." *Yale R*. 34 (1945): 444–61.

"Some General Principles of Operational Analysis." *Psychol. Rev*. 52 (Sept. 1945): 246–49.

"New Vistas for Intelligence." In *Physical Science and Human Values*. Princeton, N.J.: Princeton University Press, 1947.

"Some Implications of Recent Points of View in Physics." *Revue Internationale de Philosophie* 3 (1949): 17.

"The Operational Aspect of Meaning." *Synthèse* 8 (1950–51): 251–59.

"Remarks on the Present State of Operationalism." *Sci. Mon*. 70 (Oct. 1954): 224–26.

"Science and Broad Points of View." *Proc. Nat. Acad. Sci*. 42 (June 1956): 315–25.

"How Much Rigor Is Possible in Physics?" In L. Henkin et al., eds., *The Axiomatic Method, with Special Reference to Geometry and Physics*. Amsterdam: North-Holland Publishing Co., 1959, pp. 225–37.

"The Nature of Physical 'Knowledge.'" In L. W. Friedrich, ed., *The Nature of Physical Knowledge*. Bloomington: Indiana University Press; Milwaukee: Marquette University Press, 1960, pp. 13–24.

"The Present State of Operationalism." In Philipp G. Frank, ed., *The Validation of Scientific Theories*. New York: Collier Books, 1961, pp. 74–80.

Other Books

Bentley, Arthur. *Inquiry into Inquiries*. Ed. Sidney Ratner. Gainesville: University of Florida Press, 1960.

Black, Max. *Language and Philosophy*. Ithaca, N.Y.: Cornell University Press, 1949.

Boring, Edwin G. *The Physical Dimensions of Consciousness*. New York: Century Co., 1933.

———. *A History of Experimental Psychology*. 2nd ed. New York: Appleton-Century-Crofts, 1950.

———. *History, Psychology, and Science*. New York: John Wiley and Sons, 1963.

Cassirer, Ernst. *The Problem of Knowledge*. New Haven: Yale University Press, 1950.

Chase, Stuart. *The Tyranny of Words*. New York: Harcourt Brace Jovanovich, 1938.

Douglas, Mary, and Steven M. Tipton. *Religion and America*. Boston: Beacon Press, 1982.

Frank, Philipp G. *Modern Science and Its Philosophy*. Cambridge, Mass.: Harvard University Press, 1949.

———, ed. *The Validation of Scientific Theories*. New York: Collier Books, 1961.

Friedrich, L. W., ed. *The Nature of Physical Knowledge*. Bloomington: Indiana University Press; Milwaukee: Marquette University Press, 1960.

Hayakawa, S. I. *Language in Action*. New York: Harcourt Brace and Co., 1941.

Kline, Morris. *Mathematics: The Loss of Certainty*. Oxford: Oxford University Press, 1980.

Klyce, Scudder. *Universe*. Winchester, Mass.: the author, 1921.

———. *Sins of Science*. Boston: Marshall Jones Co., 1925.

Korzybski, Alfred. *Science and Sanity*. New York: Science Press Printing Co. for the International Non-Aristotelian Library Publishing Co., 1933.

Krutch, Joseph Wood. *A Krutch Omnibus*. New York: William Morrow and Company, 1970.

Kuklick, Bruce. *The Rise of American Philosophy*. New Haven: Yale University Press, 1977.

Ogden, C. K., and I. A. Richards. *The Meaning of Meaning*. New York: Harcourt Brace and Co., 1936.

Perrett, Geoffrey. *America in the Twenties*. New York: Simon and Schuster, 1982.

Planck, Max. *Where Is Science Going?* New York: W. W. Norton and Co., 1932.

Pratt, Carroll C. *The Logic of Modern Psychology*. New York: Macmillan, 1939.

Purcell, Edward. *The Crisis of Democratic Theory*. Lexington: University Press of Kentucky, 1973. Contains the fullest discussion of the scientific ideology and its relationship to the intellectual and political crisis in America between the two world wars.

Taylor, Richard W., ed. *Life, Language, Law: Essays in Honor of Arthur F. Bentley*. Yellow Springs, Ohio: Antioch Press, 1957.

Wilder, Raymond. *Introduction to the Foundations of Mathematics*. New York: John Wiley and Sons, 1952.

Other Articles

Blumberg, Albert, and Herbert Feigl. "Logical Positivism, a New Movement in European Philosophy." *J. Phil.* 28, no. 11 (May 21, 1931): 281–96.

Boring, Edwin. "Temporal Perception and Operationism." *Am. J. Psychol.* 48 (1936): 519–22.

Feigl, Herbert. "The Logical Character of the Principle of Induction." *Phil. Sci.* 1 (1934): 20–29.

———. "Logical Analysis of the Psycho-physical Problem." *Phil. Sci.* 1 (1934): 420–45.

———. "The Wiener Kreis in America." In Donald Fleming and Bernard Bailyn, eds. *The Intellectual Migration: Europe and America, 1930–1960.* Cambridge, Mass.: Harvard University Press, Belknap Press, 1969, pp. 630–73.

Israel, H., and B. Goldstein. "Operationism in Psychology." *Psychol. Rev.* 51 (1944): 177–88.

Kantor, J. R. "The Operational Principle in the Physical and Psychological Sciences." *Psychol. Rec.* 2 (1938): 3–32.

Koch, Sigmund. "The Cognitive Pathology." Paper delivered at the symposium "Reflections on P. W. Bridgman: A Centenary Symposium," Boston University Center for the Philosophy and History of Science, Apr. 24, 1982.

Malisoff, W. M. "Quantum Theories." *Phil. Sci.* 2 (1935): 334–43.

McGeoch, J. A. "Learning as an Operationally Defined Concept." *Psychol. Bull.* 32 (1935): 688.

———. "A Critique of Operational Definition." *Psychol. Bull.* 34 (1937): 703–4.

McGregor, D. "Scientific Measurement and Psychology." *Psychol. Rev.* 42 (1935): 246–66.

Schilpp, P. A. "The Nature of the Given." *Phil. Sci.* 2 (1935): 128–38.

Seashore, R. H., and B. Katz. "An Operational Definition and Classification of Mental Mechanisms." *Psychol. Rec.* 1 (1937): 3–24.

Stevens, S. S. "The Operational Basis of Psychology." *Am. J. Psychol.* 47 (Apr. 1935): 323–30.

———. "The Operational Definition of Psychological Concepts." *Psychol. Rev.* 42 (Nov. 1935): 517–27.

———. "Psychology, the Propaedeutic Science." *Phil. Sci.* 3 (Jan. 1936): 90–103.

———. "Psychology and the Science of Science." *Psychol. Bull.* 36 (Apr. 1939): 221–62. An important review article.

"Symposium on Operationism." *Psychol. Rev.* 52 (Sept. 1945): 241–94. Organized by Edwin Boring. Papers were presented by E. Boring, P. W. Bridgman, H. Israel, C. Pratt, H. Feigl, and B. F. Skinner. Discussion and rejoinders are included.

Tolman, E. C. "An Operational Analysis of 'Demands.'" *Erkenntnis* 6 (1936): 383–90.

Waters, R. H., and L. A. Pennington. "Operationism in Psychology." *Psychol. Rev.* 45 (1938): 414–23.

PART IV: MEASUREMENT AND COMMUNICATION

Books by P. W. Bridgman

The Logic of Modern Physics. New York: Macmillan, 1927.
The Nature of Physical Theory. New York: Dover Publications, 1936.
The Nature of Thermodynamics. Cambridge, Mass.: Harvard University Press, 1941.

Reflections of a Physicist. New York: Philosophical Library, 1950.
The Thermodynamics of Electrical Phenomena in Metals. New York: Dover Publications, 1961.

Articles by P. W. Bridgman

"The Scientist's Commitment." *Bull. Atom. Scient.*, July 1949, pp. 192–200.
"The Principles of Thermodynamics." In *Thermodynamics in Physical Metallurgy*. Novelty, Ohio: American Society for Metals, 1950, pp. 1–15.
"Reflections on Thermodynamics." *Proc. Am. Acad. Arts Sci.* 82 (Dec. 1953): 301–9; *Am. Scient.* 41, no. 4 (Oct. 1953): 549–55.
"Probability, Logic, and ESP." *Science* 123 (Jan. 6, 1956): 15–17.

Other Books and Articles

Atkins, P. W. *The Second Law*. New York: W. H. Freeman and Co., 1984.
Brillouin, Leon. *Science and Information Theory*. 2nd ed. New York: Academic Press, 1962.
Brush, Stephen G. *The Kind of Motion We Call Heat*. Amsterdam: North-Holland Publishing, 1976.
———. *Statistical Physics and the Atomic Theory of Matter*. Princeton, N.J.: Princeton University Press, 1983.
Buchdahl, H. A. *The Concepts of Thermodynamics*. Cambridge, Eng.: Cambridge University Press, 1966.
Churchman, C. West, and Philburn Ratoosh, eds. *Measurement: Definitions and Theories*. New York: John Wiley and Sons, 1959.
Condon, Edward, and Philip Morse. *Quantum Mechanics*. New York: McGraw-Hill, 1929.
Euclid. *Elements*. Trans. Sir Thomas L. Heath. Chicago: Encyclopedia Britannica, 1952.
Feynman, Richard P., Robert Leighton, and Matthew Sands. *The Feynman Lectures on Physics*, vol. 3, *Quantum Mechanics*. Reading, Mass.: Addison-Wesley, 1965.
Goldberg, Stanley. *Understanding Relativity*. Boston: Birkhauser, 1984.
Jammer, Max. *The Conceptual Development of Quantum Mechanics*. New York: McGraw-Hill, 1966.
Journal of the Franklin Institute, Apr. 1929.
Kemble, Edwin C. *The Fundamental Principles of Quantum Mechanics*. New York: McGraw-Hill, 1937.
Kennard, E. H. "Quantum Mechanics of an Electron." *J. Franklin Inst.*, Jan. 1929.
Kline, Morris. *Mathematics: The Loss of Certainty*. Oxford: Oxford University Press, 1980.
Klotz, Irving. *Chemical Thermodynamics*. New York: W. A. Benjamin, 1964.
Korner, S., ed. *Observation and Interpretation in the Philosophy of Physics*. New York: Dover Publications, 1957.
Krantz, David, R. Duncan Luce, Patrick Suppes, and Amos Tversky. *Foundations of Measurement*. New York: Academic Press, 1971.

Prigogine, Ilya. *From Being to Becoming.* New York: W. H. Freeman and Co., 1980.
Rae, Alastair. *Quantum Physics: Illusion or Reality?* Cambridge, Eng.: Cambridge University Press, 1986.
Rothstein, Jerome. *Communication, Organization, and Science.* Indian Hills, Colo.: Falcon's Wing Press, 1958.
Russell, Bertrand. *Principles of Mathematics.* 2nd ed. New York: W. W. Norton and Co., 1938.
Schweber, Silvan. "Empiricism Regnant." *Hist. Stud. Phys. Sci.* 17 (1986): 55–98.
Susman, Warren I. *Culture as History: The Transformation of American Society in the Twentieth Century.* New York: Pantheon Books, 1973; reissued 1984.
Tolman, Richard C. *Relativity, Thermodynamics, and Cosmology.* Oxford: Clarendon Press, 1934.
Tribus, Myron. *Thermostatics and Thermodynamics.* Princeton, N.J.: D. Van Nostrand, 1961.
Wheeler, John Archibald, and Wojcieck Hubert Zurek. *Quantum Theory and Measurement.* Princeton, N.J.: Princeton University Press, 1983.

PART V: SCIENCE AS A VOCATION

Unpublished Sources

Correspondence of P. W. Bridgman, held by Mrs. Jane Bridgman Koopman.
Memoirs of Dr. Jack Fine, in the possession of Mrs. Fine.

Articles by P. W. Bridgman

"The Physicist Today." *Harv. Grad. M.*, Mar. 1931, pp. 289–97.
"The Struggle for Intellectual Integrity." *Harper's*, Dec. 1935.
"'Manifesto' by a Physicist." *Science* 89 (1939): 179.
"Society and the Intelligent Physicist." *Am. Phys. Teach.* 6 (1939): 109–16.
"A Challenge to Physicists." *J. Applied Phys.* 4 (Apr. 1942): 209.
"Science, and Its Changing Social Environment." *Science* 97 (1943): 147–50.
"The Prospect for Intelligence." *Yale R.* 54 (1945): 444–60.
"Science and Freedom." *Isis* 37 (1947): 128–31.
"Scientific Freedom and National Planning." Presented on Dec. 7, 1947. Printed in P. W. Bridgman, *Reflections of a Physicist.* New York: Philosophical Library, 1950, pp. 320–31.
"Scientists and Social Responsibility." *Scient. Mon.* 45 (1947): 148–54.
"The Scientist's Commitment." *Bull. Atom. Scient.*, July 1949, pp. 192–193, 196.
"Sentimental Democracy and the Forgotten Physicist." A 1949 speech to the American Physical Society. Reprinted in P. W. Bridgman, *Reflections of a Physicist.* New York: Philosophical Library, 1950, pp. 332–41.
"Impertinent Reflections on the History of Science." *Phil. Sci.* 17 (Jan. 1950): 63–73.
"The Discovery of Science." *Harvard Alumni Bulletin*, Nov. 8, 1952, pp. 580–81.

"Science and Common Sense." *Scient. Mon.* 79 (July 1954): 32–39.
"The Task Before Us." *Proc. Am. Acad. Arts Sci.* 83 (1954): 95–112.
"Some of the Broader Implications of Science." *Physics Today* 10 (Oct. 1957): 17–24.
"Quo Vadis?" *Daedalus* 87 (1958): 85–93.
"Society and the Individual." *Bull. Atom. Scient.* 14 (Dec. 1958): 413–16.

Other Books

Burchard, John Ely, ed. *Mid-Century: The Social Implications of Scientific Progress.* Cambridge, Mass.: Technology Press of MIT; New York: John Wiley and Sons, 1950.
Douglas, Mary, and Steven M. Tipton, eds. *Religion and America.* Boston: Beacon Press, 1982.
Eerdman's Handbook to Christianity in America. Grand Rapids, Mich.: William B. Eerdman's Publishing Co., 1983.
Evans, Joseph W., ed. *Jacques Maritain: The Man and His Achievement.* New York: Sheed and Ward, 1963.
Hutchison, William R. *The Modernist Impulse in American Protestantism.* Cambridge, Mass.: Harvard University Press, 1976.
Purcell, Edward. *The Crisis of Democratic Theory.* Lexington: University Press of Kentucky, 1973.
Susman, Warren I. *Culture as History: The Transformation of American Society in the Twentieth Century.* New York: Pantheon Books, 1973; reissued 1984.
Tobey, Ronald C. *The American Ideology of National Science, 1919–1930.* Pittsburgh: University of Pittsburgh Press, 1971.
Walzer, Michael. *The Revolution of the Saints: A Study in the Origins of Radical Politics.* Cambridge, Mass.: Harvard University Press, 1965.
Wellborn, Charles T. *Religion in America.* 2nd ed. New York: Macmillan, 1982.

Other Articles

Simon, Yves. "Maritain's Philosophy of the Sciences." In *The Philosophy of Physics.* Jamaica, N.Y.: St. John's University Press, 1961.
Weber, Max. "Science as a Vocation." In Max Weber, *Essays in Sociology.* Trans. and ed. H. H. Gerth and C. Wright Mills. Oxford: Oxford University Press, 1946, chap. 5.

Index

In this index an "f" after a number indicates a separate reference on the next page, and an "ff" indicates separate references on the next two pages. A continuous discussion over two or more pages is indicated by a span of page numbers, e.g., "57–59." *Passim* is used for a cluster of references in close but not consecutive sequence.

Library of Congress Cataloging-in-Publication Data

Walter, Maila L.
  Science and cultural crisis : an intellectual biography of Percy Williams
  Bridgman (1882–1961) / Maila L. Walter.
      p.   cm.
  Includes bibliographical references.
  ISBN 0-8047-1796-6 (alk. paper) :
    1. Bridgman, P. W. (Percy Williams),
  1882–1961.   2. Physics—History—20th
  century.   3. Physicists—United States—
  Biography.
  I. Title.
  QC16.B73W35   1990
  530′.092—dc20
  [B]   90-30511
                                                  CIP

⊗ This book is printed on acid-free paper

Printed in the USA
CPSIA information can be obtained
at www.ICGtesting.com
JSHW021319221024
72173JS00001B/4